Encyclopaedia of Mathematical Sciences

Volume 64

Editor-in-Chief: R.V. Gamkrelidze

M. A. Shubin (Ed.)

Partial Differential Equations VII

Spectral Theory of Differential Operators

Springer-Verlag
Berlin Heidelberg New York
London Paris Tokyo
Hong Kong Barcelona
Budapest

Title of the Russian edition:
Itogi nauki i tekhniki, Sovremennye problemy matematiki,
Fundamental'nye napravleniya, Vol. 64, Differentsial'nye
uravneniya s chastnymi proizvodnymi 7
Publisher VINITI, Moscow 1989

Mathematics Subject Classification (1991):
34L05, 35-02, 35P05, 35P10, 35P15, 35P20, 47F05, 58G03, 58G11, 58G18,
58G25

ISBN 3-540-54677-4 Springer-Verlag Berlin Heidelberg New York
ISBN 0-387-54677-4 Springer-Verlag New York Berlin Heidelberg

CIP data applied for

Printed in Germany
Typesetting: Camera-ready copy from the translator using a Springer TEX macro package
SPIN: 10015162 41/3140 - 5 4 3 2 1 0 - Printed on acid-free paper

List of Editors, Authors and Translators

Editor-in-Chief

R. V. Gamkrelidze, Russian Academy of Sciences, Steklov Mathematical Institute, ul. Vavilova 42, 117966 Moscow, Institute for Scientific Information (VINITI), ul. Usievicha 20a, 125219 Moscow, Russia, e-mail: gam@ips.ac.msk.su

Consulting Editor

M. A. Shubin, Department of Mathematics, Northeastern University, Boston, MA 02115, USA, e-mail: shubin@neu.edu

Authors

G. V. Rozenblum, University of Telecommunications, Department of Mathematics, Moika 61, 191065 St. Petersburg, Russia
M. A. Shubin, Department of Mathematics, Northeastern University, Boston, MA 02115, USA, e-mail:shubin@neu.edu
M. Z. Solomyak, Department of Theoretical Mechanics, The Weizmann Institute of Sciences, 76100 Rehovot, Israel, e-mail: solom@wisdom.weizmann.ac.il

Translator

T. Zastawniak, Department of Mathematics and Computing, University of Glamorgan, Pontypridd, Mid Glamorgan CF37 1DL, United Kingdom, e-mail: tzastawn§genvax.glam.ac.uk

Spectral Theory of Differential Operators

G.V. Rozenblum, M.A. Shubin, M.Z. Solomyak

Translated from the Russian
by T. Zastawniak

Contents

Preface

The spectral theory of operators in a finite-dimensional space first appeared in connection with the description of the frequencies of small vibrations of mechanical systems (see Arnol'd et al. 1985). When the vibrations of a string are considered, there arises a simple eigenvalue problem for a differential operator. In the case of a homogeneous string it suffices to use the classical theory of Fourier series. For an inhomogeneous string it becomes necessary to consider the general Sturm-Liouville problem, which is the eigenvalue problem for a simple one-dimensional differential operator with variable coefficients. Failing to be explicitly soluble, the problem calls for a qualitative and asymptotic study (see Egorov and Shubin 1988a, §9). When considering the vibrations of a membrane or a three-dimensional elastic body, we arrive at the eigenvalue problems for many-dimensional differential operators. Such problems also arise in the theory of shells, hydrodynamics, and other areas of mechanics. One of the richest sources of problems in spectral theory, mostly for Schrödinger operators, is quantum mechanics, in which the eigenvalues of the quantum Hamiltonian, and, more generally, the points of the spectrum of the Hamiltonian, are the possible energy values of the system.

The present article contains a survey of various aspects of the spectral theory of many-dimensional linear differential operators (mostly self-adjoint ones). Some relatively elementary problems of this theory have been briefly presented in the earlier volumes of the present series (Egorov and Shubin 1988a, 1988b). By no means do we aspire to give a complete presentation or bibliography. In particular, we restrict ourselves only to presenting the L_2-theory, laying aside everything concerned with the spectral theory of differential operators in non-Hilbert functional spaces. Nor do we touch upon such questions as scattering theory, inverse problems of spectral theory, or eigenfunction expansions, which are quite important for applications. Each of these topics deserves a separate article.

Different parts of the article are written with a varying degree of thoroughness and completeness. In particular, §§1–4, §7, and §8 have largely an introductory character. We have striven to achieve a greater degree of completeness in sections dealing with more up-to-date questions. In this respect the individual scientific interests of the authors have played a certain role.

G. V. Rozenblum and M. Z. Solomyak have written §§1–14. §13 and §14 have been written in collaboration with Yu. G. Safarov, to whom the authors wish to express their deep gratitude. §16 and §20 have been written by G. V. Rozenblum, and §15 and §§17–19 by M. A. Shubin.

The authors wish to thank M. S. Agranovich, V. Ya. Ivrii, S. Z. Levendorskij, L. A. Malozemov, S. A. Molchanov, Ya. G. Sinai, and D. R. Yafaev, who read the manuscript or separate parts of it and made a number of useful remarks.

The standard multi-index notation is used throughout the article. As usual, $D_{x_j} = -i\partial/\partial x_j$, H^r denotes the Sobolev space, and c (frequently without an index) designates various constants.

§1 Some Information on the Theory of Operators in a Hilbert Space

The language of the general theory of operators (mainly unbounded ones) in a Hilbert space is systematically used in the spectral theory of differential operators. Here we shall give a list of the principal terms and notions as well as the formulations of some theorems in operator theory to be used later on. For a systematic presentation, see, for example, Maurin (1959), Dunford and Schwartz (1963), Akhiezer and Glazman (1966), Kato (1966), and Birman and Solomyak (1980).

1.1. Linear Operators. Closed Operators

Let $\mathfrak{H}$ be a complex Hilbert space, let $D \subset \mathfrak{H}$ be a linear subset, and let $A : D \to \mathfrak{H}$ be a linear (not necessarily continuous) map. For brevity, A is said to be a *linear operator* in $\mathfrak{H}$. The set D is denoted by $D(A)$ and called the *domain* of the operator. If $D(A) = \mathfrak{H}$ and A is bounded, then we write $A \in B(\mathfrak{H})$. If D_0 is a linear subset of D, then $A_0 = A|_{D_0}$ (the notation $A \restriction D_0$ is also used) is said to be a *restriction* of A. The operator A is then called an *extension* of A_0. We shall write $A_0 \subset A$.

On $D(A)$ one can define the *graph norm* or *A-norm* $\|\cdot\|_A$ by

$$\|x\|_A^2 = \|Ax\|^2 + \|x\|^2, \quad x \in D(A). \tag{1.1}$$

A is said to be a *closed operator* if $D(A)$ is complete in the A-norm. An equivalent definition is this: A is closed if its *graph* $\mathcal{G}(A) = \{\{x, y\} \in \mathfrak{H} \oplus \mathfrak{H} : x \in D(A),\ y = Ax\}$ is closed in $\mathfrak{H} \oplus \mathfrak{H}$. We say that A is a *closable* operator if the closure of the graph of A in $\mathfrak{H} \oplus \mathfrak{H}$ is also the graph of an operator. An equivalent condition is that if $\{x_n\}$, where $x_n \in D(A)$, is a Cauchy sequence in the A-norm and $\|x_n\| \to 0$, then $\|x_n\|_A \to 0$. The latter property means that the topologies generated by the norm of $\mathfrak{H}$ and by the A-norm on $D(A)$ are compatible.

If A is closable, then the operator $\bar{A}$ defined by $\mathcal{G}(\bar{A}) = \overline{\mathcal{G}(A)}$ is called the *closure* of A. If A is bounded, then $\bar{A}$ coincides with the extension of A by continuity.

1.2. The Adjoint Operator

Let A be a *densely defined* operator, i.e., $\overline{D(A)} = \mathfrak{H}$. Then the *adjoint operator* A^* can be constructed as follows. The domain of A^* is

$$D(A^*) \stackrel{\text{def}}{=} \{y \in \mathfrak{H} : \exists\, h \in \mathfrak{H} \;\; (Ax, y) = (x, h) \;\; \forall\, x \in D(A)\}\,.$$

The vector h is uniquely determined by y, and we set $h = A^*y$. Thus

$$(Ax, y) = (x, A^*y) \quad \forall\, x \in D(A) \quad \forall\, y \in D(A^*).$$

As opposed to the case of $A \in B(\mathfrak{H})$, this equality is used not only to describe the 'action' of A^*, but, as we can see, also to describe the domain of A^*.

A^* is always a closed operator. $\overline{D(A^*)} = \mathfrak{H}$ if and only if A is closable. Under this assumption, $(A^*)^* = \bar{A}$. If $A_0 \subset A$ and $\overline{D(A_0)} = \mathfrak{H}$, then $A_0^* \supset A^*$.

1.3. Self-Adjoint Operators

An operator A such that $A^* = A$ is said to be *self-adjoint*. An operator A such that $\overline{D(A)} = \mathfrak{H}$ and

$$(Ax, y) = (x, Ay) \quad \forall\, x, y \in D(A)$$

is called *symmetric*. These two notions are equivalent for $A \in B(\mathfrak{H})$. If $A^* = \bar{A}$, then A is said to be *essentially self-adjoint*. If A is symmetric and $\bar{A} \neq A^*$, then A^* is seen not to be symmetric.

The self-adjointness of an operator can often be established by means of perturbation theory, i.e., from the fact that the operator is close to another operator known in advance to be self-adjoint. The following theorem is a typical result in this direction.

Theorem 1.1 (Kato-Rellich; see Kato 1966; Reed and Simon 1975, Vol. 2). *Let A be a self-adjoint operator and let B be a symmetric operator in a Hilbert space $\mathfrak{H}$ such that $D(B) \supset D(A)$ and*

$$\|Bx\| \le a\|Ax\| + b\|x\| \quad \forall\, x \in D(A) \tag{1.2}$$

for some $0 < a < 1$ and $b \ge 0$. Then $A + B$ is self-adjoint on $D(A)$ and essentially self-adjoint on any domain of essential self-adjointness of A.

1.4. The Spectrum of an Operator

Let A be a closed operator. By definition, the *resolvent set* $\rho(A)$ consists of points $\lambda \in \mathbb{C}$ such that there exists $(A - \lambda I)^{-1} \in B(\mathfrak{H})$ (I being the identity operator in $\mathfrak{H}$). The complement $\sigma(A) = \mathbb{C} \setminus \rho(A)$ of the resolvent set is called the *spectrum* of A. The set $\rho(A)$ is open and $\sigma(A)$ is closed. It is possible that $\sigma(A) = \mathbb{C}$ or $\sigma(A) = \emptyset$. (For $A \in B(\mathfrak{H})$ neither of these possibilities can be realized.)

If $A = A^*$, then the spectrum of A is non-empty and lies on the real axis. The spectrum $\sigma(A)$ of a self-adjoint operator can be represented as the union of the *point spectrum* $\sigma_{\mathrm{p}}(A)$ (i.e., the set of all eigenvalues) and the *continuous spectrum*

$$\sigma_{\mathrm{c}}(A) = \{\lambda \in \mathbb{R} : \mathrm{Im}\,(A - \lambda I) \text{ is a non-closed set}\}.$$

The spectra $\sigma_{\mathrm{p}}(A)$ and $\sigma_{\mathrm{c}}(A)$ can have a non-empty intersection. If $\sigma_{\mathrm{p}}(A) = \emptyset$, then A has a *purely continuous spectrum*. If the linear hull of the eigenspaces $\mathrm{Ker}\,(A - \lambda I)$, where $\lambda \in \sigma_{\mathrm{p}}(A)$, is dense in $\mathfrak{H}$, then A has a *pure point spectrum*. In this case the continuous spectrum coincides with the set of limit points of the point spectrum and, generally speaking, is non-empty.

The union of the continuous spectrum and the set of eigenvalues of infinite multiplicity is called the *essential spectrum* of a self-adjoint operator A ($\sigma_{\mathrm{ess}}(A)$). If $\sigma_{\mathrm{ess}}(A) = \emptyset$, then A is an operator with *discrete spectrum*. An equivalent condition for A to have discrete spectrum is that $(A - \lambda I)^{-1}$ be a compact operator for some $\lambda \in \rho(A)$ (and then for all such λ).

1.5. Spectral Measure. The Spectral Theorem for Self-Adjoint Operators

Suppose that associated with every Borel set $\delta \subset \mathbb{R}$ is an orthogonal projection $E(\delta)$ in $\mathfrak{H}$. Let $E(\mathbb{R}) = I$ and let the following condition of countable additivity be satisfied: if $\{\delta_n\}$, $n = 1, 2, \ldots$ are pairwise disjoint Borel sets, then $\sum_n E(\delta_n) = E(\bigcup_n \delta_n)$. (The series on the left-hand side converges in the strong operator topology.) Any such map $E : \delta \mapsto E(\delta)$ is called a *spectral measure* in $\mathfrak{H}$ (defined on the Borel subsets of the real axis).

If E is a spectral measure, then, for any $x \in \mathfrak{H}$, $E(\,\cdot\,)x$ is a vector-valued measure and $\mu_x(\,\cdot\,) = (E(\,\cdot\,)x, x)$ is a scalar-valued Borel measure normalized by $\mu_x(\mathbb{R}) = \|x\|^2$. For any $x, y \in \mathfrak{H}$, $\mu_{x,y}(\,\cdot\,) = (E(\,\cdot\,)x, y)$ is a complex-valued Borel measure.

As in the case of scalar measures, the *support of a spectral measure* $(\mathrm{supp}\, E)$ can be defined as the smallest closed subset $F \subset \mathbb{R}$ such that $E(F) = I$. The expression 'almost everywhere with respect to E' (E-a.e.) has the standard meaning.

Let E be a spectral measure and let φ be a Borel measurable scalar function defined E-a.e. on $\mathbb{R}$. Then one can define the integral

$$J_\varphi = \int \varphi \, dE \quad \left(= \int \varphi(s)\, dE(s) \right), \tag{1.3}$$

which is a closed operator in $\mathfrak{H}$ with dense domain

$$D(J_\varphi) = D_\varphi = \left\{ x \in \mathfrak{H} : \int |\varphi|^2 \, d\mu_x < \infty \right\}. \tag{1.4}$$

The integral (1.3) can be understood, for example, in the 'weak sense,' that is, $(J_\varphi x, y) = \int \varphi \, d\mu_{x,y}$ for $x \in D_\varphi$ and $y \in \mathfrak{H}$. The operator J_φ is self-adjoint if and only if φ is an E-a.e. real-valued function. J_φ is bounded if and only if φ is E-a.e. bounded.

The following *spectral theorem* plays a central role in the spectral analysis of self-adjoint operators.

Theorem 1.2. *To every self-adjoint operator A there corresponds a unique spectral measure E^A such that*

$$A = \int s \, dE^A(s).$$

It turns out that $\operatorname{supp} E^A = \sigma(A)$.

The relations

$$x = \int dE^A(s)x, \qquad \|x\|^2 = \int d\mu_x^A(s) \qquad \forall\, x \in \mathfrak{H},$$
$$D(A) = \left\{ x \in \mathfrak{H} : \int s^2 \, d\mu_x^A(s) < \infty \right\}, \tag{1.5}$$

where $\mu_x^A(\,\cdot\,) = (E^A(\,\cdot\,)x, x)$, follow from Theorem 1.2 together with (1.4). They express the *decomposition theorem*.

The operators $J_\varphi^A = \int \varphi \, dE^A$ can be regarded as functions $\varphi(A)$ of a self-adjoint operator A, which is consistent with the direct definition of the powers A^n (for $\varphi(s) = s^n$) and the resolvent $(A - \lambda I)^{-1}$ for $\lambda \notin \sigma_{\mathrm{p}}(A)$ (in which case $\varphi(s) = (s - \lambda)^{-1}$).

Example 1.1. Let A be a self-adjoint operator with pure point spectrum and let $P(\lambda)$, where $\lambda \in \sigma_{\mathrm{p}}(A)$, be the orthogonal projections onto the eigenspaces $\operatorname{Ker}(A - \lambda I)$. Then the spectral measure E^A can be defined by

$$E^A(\delta) = \sum_{\lambda \in \delta} P(\lambda)$$

and the decomposition theorem (1.5) takes the simpler form

$$x = \sum_{\lambda \in \sigma_{\mathrm{p}}(A)} P(\lambda)x, \qquad \|x\|^2 = \sum_{\lambda \in \sigma_{\mathrm{p}}(A)} \|P(\lambda)x\|^2. \tag{1.5'}$$

In this case

$$\varphi(A) = \sum_{\lambda \in \sigma_p(A)} \varphi(\lambda) P(\lambda). \tag{1.6}$$

If $\{e_n\}_1^\infty$ is a complete orthonormal system of eigenvectors of A, then (1.5′) means that the Fourier expansion and the Parseval formula

$$x = \sum_n (x, e_n) e_n, \qquad \|x\|^2 = \sum_n |(x, e_n)|^2$$

are both valid for any $x \in \mathfrak{H}$.

Example 1.2. Multiplication operators. Let $\mathcal{X}$ be a space with a positive measure μ and let $a(x)$ be a real-valued measurable function on $\mathcal{X}$ that is finite a.e. In $L_2 = L_2(\mathcal{X}, \mu)$ we consider the operator $Q_a : u(x) \mapsto a(x)u(x)$ with the natural domain $D(Q_a) = \{u \in L_2 : au \in L_2\}$. The operator Q_a is self-adjoint. If $a = \chi_e$ (= the indicator function of a measurable set $e \subset \mathcal{X}$), then Q_a is the orthogonal projection onto the subspace of functions equal to zero a.e. outside e. The spectral measure of any operator Q_a can be defined by

$$E^{Q_a}(\delta) = Q_{\chi_{e(\delta)}}, \qquad e(\delta) = a^{-1}(\delta)$$

($\delta \subset \mathbb{R}$ is a Borel set and $a^{-1}(\delta)$ is the inverse image of δ in $\mathcal{X}$.) A point $\lambda \in \mathbb{R}$ is an eigenvalue of Q_a if and only if

$$\mu\{x \in \mathcal{X} : a(x) = \lambda\} > 0.$$

$\lambda \in \sigma_c(Q_a)$ if and only if

$$\mu\{x \in \mathcal{X} : 0 < |a(x) - \lambda| < \varepsilon\} > 0 \quad \forall\, \varepsilon > 0.$$

The functions $\varphi(Q_a)$ are given by $Q_{\varphi \circ a}$.

1.6. The Pure Point, Absolutely Continuous, and Continuous Singular Components of a Self-Adjoint Operator

Let A be a self-adjoint operator in $\mathfrak{H}$ and let E^A be the spectral measure of A. One can distinguish the following subspaces: $\mathfrak{H}_p$ – the closure of the linear hull of all eigenspaces of A, $\mathfrak{H}_{ac}$ – the set of all $x \in \mathfrak{H}$ such that the measure $\mu_x^A(\cdot) = (E^A(\cdot)x, x)$ is absolutely continuous with respect to the Lebesgue measure, and $\mathfrak{H}_{cs}$ – the orthogonal complement of $\mathfrak{H}_p + \mathfrak{H}_{ac}$ in $\mathfrak{H}$. If $x \in \mathfrak{H}_{cs}$, then μ_x^A is a continuous measure singular relative to the Lebesgue measure.

The subspaces $\mathfrak{H}_p$, $\mathfrak{H}_{ac}$, and $\mathfrak{H}_{cs}$ are orthogonal to each other and invariant with respect to A. The parts A_p, A_{ac}, and A_{cs} of A in these subspaces (for example, $A_{ac} = A \restriction D(A) \cap \mathfrak{H}_{ac}$) are self-adjoint as operators in $\mathfrak{H}_p$, $\mathfrak{H}_{ac}$, and

$\mathfrak{H}_{cs}$, respectively. They are called the *pure point*, *absolutely continuous*, and *continuous singular* components of A. One can also talk about the *absolutely continuous component of the spectrum*, etc. If $A = A_{ac}$, then the spectrum of A is said to be *absolutely continuous*.

1.7. Other Formulations of the Spectral Theorem

The meaning of Theorem 1.2 is that any self-adjoint operator becomes a multiplication operator under a suitable transformation of $\mathfrak{H}$. There are formulations of the spectral theorem that express this fact explicitly.

Let $\mathfrak{H}_k$, where $k = 1, 2$, be Hilbert spaces with scalar products $(\cdot\, ,\cdot)_k$ and let $V : \mathfrak{H}_1 \to \mathfrak{H}_2$ be a *unitary operator* from $\mathfrak{H}_1$ onto $\mathfrak{H}_2$, that is, $\operatorname{Im} V = \mathfrak{H}_2$ and $(Vx, Vy)_2 = (x, y)_1\ \forall\, x, y \in \mathfrak{H}_1$. Now let A_k be an operator in $\mathfrak{H}_k$. It is said that A_1 is *unitarily equivalent* to A_2 if there is a unitary transformation $V : \mathfrak{H}_1 \to \mathfrak{H}_2$ such that

$$VD(A_1) = D(A_2), \qquad A_2Vx = VA_1x \quad \forall\, x \in D(A_1).$$

The relations can be written just as $A_2 = VA_1V^{-1}$. Any invariants of an operator under unitary equivalence are called *unitary invariants*.

Theorem 1.3. *For any self-adjoint operator A in a separable Hilbert space $\mathfrak{H}$ there exist a Borel measure μ concentrated on $\sigma(A)$ and a (finite or countable) sequence of nested sets*

$$\{\mathcal{X}_k\}_{k\in\mathcal{K}}, \qquad \mathcal{K} = \{1, \ldots, n\} \quad \text{or} \quad \mathcal{K} = \mathbb{N}, \qquad \mathcal{X}_1 = \sigma(A)$$

such that A is unitarily equivalent to the orthogonal sum of the operators of multiplication by s in the spaces $L_2(\mathcal{X}_k, \mu)$.

Let us explain the contents of this theorem. We denote by $\mathfrak{H}_1$ the Hilbert space of vector-valued functions $f = \{f_k\}_{k\in\mathcal{K}}$ such that, for each $k \in \mathcal{K}$, f_k is a measurable scalar function on $\sigma(A)$ equal to zero a.e. outside $\mathcal{X}_k$ with norm

$$\|f\|_1^2 = \sum_{k\in\mathcal{K}} \int_{\sigma(A)} |f_k|^2 d\mu \ (< \infty).$$

The multiplication operator $A_1 : f(s) \to sf(s)$ with the natural domain $D(A_1) = \{f \in \mathfrak{H}_1 : sf(s) \in \mathfrak{H}_1\}$ is self-adjoint in $\mathfrak{H}_1$. According to the theorem, A and A_1 are unitarily equivalent operators.

The new formulation of the spectral theorem enables one, in particular, to specify a complete system of unitary invariants of a self-adjoint operator A. The first of these invariants is the *type* of μ, i.e., the equivalence class with respect to the following relation: Borel measures μ_1 and μ_2 are equivalent if and only if they have the same sets of measure zero, i.e., $\mu_1(e) = 0 \iff$

$\mu_2(e) = 0$. The only other invariant of this system is the *multiplicity function*[1] n_A defined as follows: $n_A(s) = 0$ outside $\sigma(A)$ and $n_A(s) = j$ for $s \in \mathcal{X}_j \setminus \mathcal{X}_{j+1}$, where $j = 1, 2, \ldots$. If $\mathcal{K} = \{1, \ldots, n\}$, then $n_A(s) = n$ on $\mathcal{X}_n$; but if $\mathcal{K} = \mathbb{N}$, then $n_A(s) = \infty$ on $\bigcap_j \mathcal{X}_j$. The completeness of the system of invariants means that if the measures μ_1 and μ_2 that correspond, in accordance with Theorem 1.3, to two self-adjoint operators A_1 and A_2 are of the same type and $n_{A_1}(s) = n_{A_2}(s)$ a.e. relative to either of these measures, then A_1 and A_2 are unitarily equivalent.

If the sequence $\{\mathcal{X}_k\}$ reduces to the single set $\mathcal{X}_1 = \sigma(A)$, then A is a *self-adjoint operator with simple spectrum.*

Finally, the following formulation of the spectral theorem can sometimes be useful.

Theorem 1.4. *For any self-adjoint operator A in a separable Hilbert space there exist a measure space $(\mathcal{X}, \mu)$ and a real-valued measurable function a on $\mathcal{X}$ such that A is unitarily equivalent to the operator of multiplication by $a(x)$ in $L_2(\mathcal{X}, \mu)$ denoted by Q_a.*

A weakness of this version of the spectral theorem is that neither $(\mathcal{X}, \mu)$ nor a is an invariant object. Nevertheless, the weakness is often offset by the fact that the spectral characteristics of Q_a (and so those of A) can be easily computed; see Example 1.2.

1.8. Semi-Bounded Operators and Forms

The 'method of forms,' which is applicable in the semi-bounded case, is one of the principal methods for studying self-adjoint operators. A symmetric operator A is said to be *(lower) semi-bounded* if

$$\gamma_A \stackrel{\text{def}}{=} \inf_{x \in D(A),\ \|x\|=1} (Ax, x) > -\infty.$$

A is said to be *non-negative* ($A \geq 0$) if $\gamma_A \geq 0$, *positive* ($A > 0$) if $A \geq 0$ and $\operatorname{Ker} A = \{0\}$, and *positive definite* if $\gamma_A > 0$.

If $A \geq 0$ is a self-adjoint operator (and not merely symmetric), then its spectrum lies on the half-axis $s \geq 0$, which implies that the values of $\sqrt{s}$ are real E^A-a.e. As a consequence, the self-adjoint operator

$$A^{1/2} = \int s^{1/2} dE^A(s)$$

can be defined. If $x \in D(A)$, then $(Ax, x) = \|A^{1/2}x\|^2$.

Now let $d \subset \mathfrak{H}$ be a dense linear subset and let a be a complex-valued function defined on $d \times d$ that is linear in the first argument and Hermitian,

[1] Or, more precisely, its equivalence class with respect to μ.

i.e., $a[x,y] = \overline{a[y,x]}\ \forall\, x,y \in d$. Hence it follows that the function is conjugate-linear in the second argument. It also follows that the *quadratic form* $a[x] = a[x,x]$, where $x \in d[a] = d$, has real values. The function $a[x,y]$ itself is called a *Hermitian sesquilinear form*. It can be uniquely reconstructed from $a[x]$.

A form a is said to be *(lower) semi-bounded* if

$$\gamma_a \stackrel{\text{def}}{=} \inf_{x\in d[a],\ \|x\|=1} a[x] > -\infty.$$

One can also talk of *non-negative* ($\gamma_a \geq 0$), *positive* ($\gamma_a \geq 0$ and $a[x] > 0$ for $x \neq 0$), and *positive definite* ($\gamma_a > 0$) forms.

Any positive form defines a norm in $d[a]$ given by $\|x\|_a^2 = a[x]$ and called the *a-norm*. If the form is positive definite and $d[a]$ is complete in the a-norm, then the form is said to be *closed* (on $d[a]$). A positive definite form is *closable* if the topologies on $d[a]$ generated by the norm of $\mathfrak{H}$ and by the a-norm are compatible: if $\{x_n\}$ is a Cauchy sequence in $d[a]$ with respect to the a-norm and $x_n \to 0$ in $\mathfrak{H}$, then $a[x_n] \to 0$.

The method of forms rests on the *Friedrichs theorem*, which establishes a one-to-one correspondence between closed positive definite forms and positive definite self-adjoint operators.

Theorem 1.5. *1) To every positive definite self-adjoint operator A there corresponds a unique closed positive definite form a such that*

$$D(A) \subset d[a], \qquad (Ax,y) = a[x,y] \quad \forall\, x \in D(A) \quad \forall\, y \in d[a]. \tag{1.7}$$

2) Conversely, to every closed positive definite form a there corresponds a unique positive definite self-adjoint operator A such that (1.7) is satisfied.

If an operator A and a form a correspond to one another, then

$$\gamma_A = \gamma_a. \tag{1.8}$$

Let us explain how to determine a in terms of A:

$$d[a] = D(A^{1/2}), \qquad a[x,y] = (A^{1/2}x, A^{1/2}y).$$

On the other hand, given a, we first construct a bounded self-adjoint operator T in $\mathfrak{H}$. Namely, for any $x \in \mathfrak{H}$, we take $h = Tx$ to be an element in $d[a]$ such that $(x,y) = a[h,y]\ \forall\, y \in d[a]$. The existence and uniqueness of such h follows from the Riesz theorem on the general form of a linear functional in a Hilbert space. The desired operator A is given by $A = T^{-1}$.

Now let a be an arbitrary (lower) semi-bounded form. We say that a is *closed (closable)* if the positive definite form $a_c[x] = c\|x\|^2 + a[x]$ has the corresponding property for some (and then for all) $c > -\gamma_a$. If A_c is the operator corresponding to a_c, then the self-adjoint operator $A = A_c - cI$ is independent of the choice of c and semi-bounded. A is the operator to be associated with a. It satisfies the above relations (1.7) and (1.8).

For a self-adjoint operator A and a closed quadratic form a that correspond to one another we shall write

$$a = \mathrm{QF}\,(A), \qquad A = \mathrm{Op}\,(a).$$

The latter symbol is used in the theory of pseudodifferential operators, where it has a different meaning. There will be no risk of misunderstanding in our case.

We introduce a partial order in the set of closed semi-bounded forms. By definition, $a \le b$ if $d[a] \supset d[b]$ and $a[x] \le b[x]$ for all $x \in d[b]$. If, in addition, $a[x] = b[x]$ only for $x = 0$, then $a < b$. The order can be carried over in a natural way to semi-bounded self-adjoint operators: $A \le (<)\ B$ means that $\mathrm{QF}\,(A) \le (<)\ \mathrm{QF}\,(B)$.

1.9. The Friedrichs Extension

Let A_0 be a symmetric operator in $\mathfrak{H}$. The theory of extensions (see Akhiezer and Glazman 1966; Kato 1966; Birman and Solomyak 1980) answers the question concerning the conditions under which A_0 has a self-adjoint extension. The theory also provides an abstract description of all such extensions. If A_0 is semi-bounded, then self-adjoint extensions are certain to exist and we can select one that plays a special role. Namely, if A_0 is a positive definite symmetric operator, then the form $a_0[x] = (A_0 x, x)$ with $d[a_0] = D(A)$ is closable. On completing $D(A_0)$ in the a_0-norm and extending a_0 by continuity, we obtain a set $d \subset \mathfrak{H}$ and a closed positive definite form a defined on d. The operator $A = \mathrm{Op}\,(a)$ is an extension of A_0. It is called the *the Friedrichs extension*. If A_0 is merely semi-bounded, then the above procedure can be applied to $A_0 + cI$, where $c > -\gamma_A$.

1.10. Variational Triples

A version of the method of forms based on the notion of a variational triple proves useful in a number of cases, in particular in the study of spectral problems of the form

$$Bu = \lambda A u. \tag{1.9}$$

A *variational triple* $\{d; a, b\}$ consists of a Hilbert space d with a metric form $a[x]$ and a bounded sesquilinear Hermitian form $b[x,y]$ in d. (It suffices to specify the corresponding quadratic form $b[x]$.) The relation

$$b[x,y] = a[Tx, y] \quad \forall\, x, y \in d \tag{1.10}$$

assigns a unique operator $T = T(d; a, b)$ to $\{d; a, b\}$. The operator T is bounded and self-adjoint in d.

In particular, let a and b be the quadratic forms of operators A and B acting in a Hilbert space $\mathfrak{H}$. More precisely, let A be a positive definite self-adjoint operator and let $a = \mathrm{QF}\,(A)$. We assume that B is a symmetric operator defined on a dense set $\mathcal{V} \subset d[a]$ (in the simplest case $\mathcal{V} = D(A)$) and the quadratic form (Bx, x) is bounded in $d[a]$. Extending the form by continuity, we can obtain a bounded form $b[x]$ on $d[a]$. Hence we have constructed a variational triple $\{d[a]; a, b\}$. The operator determined by this triple coincides with $A^{-1}B : d[a] \to d[a]$ on $\mathcal{V}$. It is therefore natural to associate the spectrum of this operator with (1.9).

If $b[x] = \|x\|^2$, we have $T = A^{-1} \restriction d[a]$. Since $A^{1/2}$ is a unitary operator from $d = D(A^{1/2})$ onto $\mathfrak{H}$, it follows that T is unitarily equivalent to the operator $A^{-1} = A^{1/2}TA^{-1/2}$ in $\mathfrak{H}$, which implies that all the spectral characteristics of these operators are the same. In applications it is often expedient to go over from the unbounded operator A to the bounded operator T.

Let us note that there is no need to use an 'embracing' Hilbert space to construct a variational triple in the general case.

A simplified terminology is often used in the study of the spectra of operators determined by variational triples. One can talk of the 'spectrum of a variational triple,' etc.

1.11. The Distribution Function of the Spectrum. The Spectral Function

The *distribution function*

$$N(\Sigma; A) = \dim E^A(\Sigma)\mathfrak{H} \tag{1.11}$$

($\Sigma \subset \mathbb{R}$ being an arbitrary interval) serves as an important characteristic of the location of the spectrum of a self-adjoint operator A. The case when $N(\Sigma; A) < \infty$ is the most interesting one. In this case the spectrum of A on Σ consists of finitely many eigenvalues of finite multiplicity and (1.11) is equal to the sum of the multiplicities. If $N(\Sigma; A) = \infty$ and Σ is a bounded interval, then the closure of Σ contains at least one point belonging to $\sigma_{\mathrm{ess}}(A)$. If $\Sigma \cap \sigma_{\mathrm{ess}}(A) = \emptyset$, then the spectrum of A is said to be *discrete on* Σ.

For a lower semi-bounded self-adjoint operator, we set

$$N(\lambda; A) = N((-\infty, \lambda); A) \quad \forall \lambda \in \mathbb{R}. \tag{1.12}$$

For the study of the spectrum it is essential that (1.12) can be expressed explicitly in terms of $a = \mathrm{QF}\,(A)$. The equality (*Glazman's lemma;* cf. Glazman 1963)

$$N(\lambda; A) = \max\left\{\dim F : F \subset d[a],\ a[x] < \lambda\|x\|^2,\ 0 \neq x \in F\right\} \tag{1.13}$$

serves as one of the most widely used versions of such an expression.

In (1.13) the inclusion $F \subset d[a]$ can be replaced by $F \subset \mathcal{F}$, where $\mathcal{F} \subset d[a]$ is an arbitrary linear subset that is dense in $d[a]$ relative to the a-norm (the norm $c\|x\|^2 + a[x]$ with $c > -\gamma_a$ if $\gamma_a \leq 0$).

Formula (1.13) as well as other analogous formulae are said to provide a *variational description of the spectrum*. The following important result is a consequence of (1.13).

Theorem 1.6. *Let A and B be semi-bounded self-adjoint operators such that $A \leq B$. Then*

$$N(\lambda; A) \geq N(\lambda; B)$$

for any $\lambda \in \mathbb{R}$.

Let us now assume that A is a semi-bounded self-adjoint operator in a space $L_2(\mathcal{X}, \mu)$. In cases that are of interest in applications, it often turns out that the *spectral projection* $E_\lambda^A = E^A(-\infty, \lambda)$ is an integral operator. The kernel $e^A(\lambda; x, y)$ of this operator is called the *spectral function of A*. For example, let the spectrum of A be discrete and let $\{\lambda_j\}_1^\infty$ be the eigenvalues of A with a complete orthonormal sequence $\{\varphi_j\}_1^\infty$ of corresponding eigenfunctions. Then

$$e^A(\lambda; x, y) = \sum_{\lambda_j < \lambda} \varphi_j(x)\overline{\varphi_j(y)} \tag{1.14}$$

is a degenerate kernel (of rank $N(\lambda; A)$). For operators acting on vector-valued functions of dimension k,

$$e^A(\lambda; x, y) = \sum_{\lambda_j < \lambda} \langle\ \cdot\ , \varphi_j(y)\rangle \varphi_j(x), \tag{1.15}$$

$\langle \cdot\, , \cdot \rangle$ being the scalar product in $\mathbb{C}^k$.

In the case of a non-discrete spectrum the spectral function can be expressed in terms of the *generalized eigenfunctions* of A, that is, the solutions of the equation $Au = \lambda u$ that do not belong to L_2. This subject is presented in more detail in Maurin (1959), Dunford and Schwartz (1963), Berezanskij (1966), and Berezin and Shubin (1983).

As opposed to $N(\lambda; A)$, the spectral function is not a unitarily invariant object, since it requires that the Hilbert space $\mathfrak{H}$ be realized as an L_2 space.

The two *distribution functions*

$$N_\pm(\lambda; A) = N((0, \lambda); \pm A), \qquad \lambda > 0 \tag{1.16}$$

are usually considered for self-adjoint operators that are not semi-bounded. The corresponding spectral functions $e_\pm^A(\lambda; x, y)$ are the kernels of the integral operators representing the projections $E^{\pm A}(0, \lambda)$.

1.12. Compact Operators

Here we shall briefly present some information about compact (completely continuous) operators in a Hilbert space $\mathfrak{H}$. With every operator T of this kind one can associate the sequence $s_n(T) = (\lambda_n(T^*T))^{1/2}$ of its *s-numbers*, $\lambda_n(T^*T)$ being the eigenvalues of the self-adjoint operator $T^*T \geq 0$ arranged into a decreasing sequence with their multiplicity taken into account. If $\sum_n s_n(T) < \infty$, then T is called a *trace class* operator. For trace class operators, the series

$$\operatorname{Tr} T = \sum_n \lambda_n(T),$$

in which the eigenvalues of T are numbered so that their algebraic multiplicity is taken into account, is absolutely convergent.

In particular, let $\mathfrak{H} = L_2(X)$, where $X \subset \mathbb{R}^n$ is a bounded domain, and let T be an *integral operator* such that

$$(Tu)(x) = \int_X T(x,y)u(y)\,dy.$$

If T is a trace class operator and $T(x,y)$ is continuous in $\bar{X} \times \bar{X}$, then

$$\operatorname{Tr} T = \int_X T(x,x)\,dx. \tag{1.17}$$

If one is dealing with integral operators in a space of vector-valued functions on X, then $T(x,y)$ is matrix-valued, and, under the above assumptions,

$$\operatorname{Tr} T = \int_X \operatorname{tr} T(x,x)\,dx, \tag{1.18}$$

where tr denotes the trace of a matrix.

It is customary to express the rate at which the s-numbers decrease in terms of the distribution function

$$n(s;T) = \operatorname{card}\{n \in \mathbb{N} : s_n(T) > s\} = N\left((s^2,\infty); T^*T\right), \quad s > 0. \tag{1.19}$$

For compact self-adjoint operators one can set

$$n_\pm(\lambda;T) = N((\lambda,\infty); \pm T), \quad \lambda > 0. \tag{1.20}$$

There holds an analogue of (1.13) for $n_\pm(\lambda, T)$:

$$n_\pm(\lambda;T) = \max\left\{\dim F : F \subset \mathfrak{H},\ \pm(Tx,x) > \lambda\|x\|^2,\ 0 \neq x \in F\right\}. \tag{1.21}$$

If $A > 0$ is a self-adjoint operator with discrete spectrum, then, obviously,

$$N(\lambda; A) = n_+(\lambda^{-1}; A^{-1}).$$

§2
Defining Differential Operators. Essential Self-Adjointness

In accordance with §1, by defining a *differential operator* we mean specifying its domain and the 'rule of action.' We shall usually be interested in 'self-adjoint realizations of differential operators,' or, in other words, in differential operators on a domain on which they are self-adjoint. It is not always easy to verify whether a differential operator is self-adjoint or not. In a number of cases the domain of a self-adjoint differential operator admits an indirect description only. In addition, the action of the operator on such a domain may not always be given explicitly. This feature appears to be particularly distinct in the 'method of quadratic forms,' to which we devote §3.

2.1. Differential Expressions and Their Symbols

Let $X \subseteq \mathbb{R}^n$ be a domain. In what follows we take $L_2(X)$ as our basic Hilbert space. Let a *differential expression*

$$(\mathcal{L}u)(x) = \mathcal{L}(x, D)u = \sum_{|\alpha|\le m} a_\alpha(x) D^\alpha u(x) \tag{2.1}$$

of order $m \ge 1$ be given in X. To avoid unnecessary complications, we first assume that

$$a_\alpha \in C^\infty(X) \quad (\alpha \ne 0), \qquad a_0 \in L_{2,\,\mathrm{loc}}(X). \tag{2.2}$$

Under these assumptions, the differential expression (2.1) is certainly defined for all $u \in L_2(X)$ (in the distribution sense). The *formally adjoint expression*

$$(\mathcal{L}^+ u)(x) = \sum_{|\alpha|\le m} D^\alpha \left[\overline{a_\alpha(x)} u(x)\right] \tag{2.3}$$

can also be defined. It is obvious that

$$(\mathcal{L}u, v) = (u, \mathcal{L}^+ v) \tag{2.4}$$

for $u, v \in C_0^\infty(X)$. If $\mathcal{L}^+ = \mathcal{L}$, then $\mathcal{L}$ is called a *formally self-adjoint* differential expression. The condition $\mathcal{L}^+ = \mathcal{L}$ means that the operator $\mathcal{L} \restriction C_0^\infty(X)$ is symmetric. It is a necessary condition for $\mathcal{L}$ to have self-adjoint realizations.

The polynomials (in $\xi \in \mathbb{R}^n$)

$$\mathcal{L}(x, \xi) = \sum_{|\alpha|\le m} a_\alpha(x) \xi^\alpha, \tag{2.5}$$

$$\mathcal{L}^0(x, \xi) = \mathcal{L}_m(x, \xi) = \sum_{|\alpha|=m} a_\alpha(x) \xi^\alpha \tag{2.6}$$

are referred to as the *symbol* (or the *complete symbol*) and the *principal symbol* of the differential expression (2.1). It is obvious that $(\mathcal{L}^+)^0(x,\xi) = \overline{\mathcal{L}^0(x,\xi)}$ is the principal symbol of $\mathcal{L}^+$. Thus, if $\mathcal{L}$ is formally self-adjoint, then its principal symbol is real.

Let X be a smooth n-dimensional manifold (with or without boundary). A differential expression $\mathcal{L}$ on X (more precisely, on $\operatorname{Int} X$) can be defined by a formula of the form (2.1) in any local coordinate system on X. The principal symbol $\mathcal{L}^0(x,\xi)$ transforms as a function on the cotangent bundle T^*X when the coordinate system is changed (see Egorov and Shubin 1988b).

If $d\mu$ is a fixed smooth positive density on X (Tréves 1982, Vol. 2), then one can define the space $L_2(X;d\mu)$ and the corresponding operation of taking the adjoint expression $\mathcal{L} \mapsto \mathcal{L}^+$. (Formula (2.3) defines such an adjoint expression in the case when $X \subset \mathbb{R}^n$ and $d\mu = dx$ is the n-dimensional Lebesgue density.) Equality (2.4) is satisfied, as before. The most important case is that of a Riemannian manifold X. In this case the 'Riemannian volume element' is taken as the standard choice for $d\mu$, and $L_2(X)$ is written in place of $L_2(X;d\mu)$.

One can also consider operators of the form (2.1) acting in the space of vector-valued functions of dimension k defined on a domain $X \subset \mathbb{R}^n$. Then the coefficients $a_\alpha(x)$ are $(k \times k)$-matrices. In (2.3) $\overline{a_\alpha(x)}$ must be replaced by the adjoint matrix $a_\alpha^*(x)$. The complete and principal symbols are defined by (2.5) and (2.6), as before. If $\mathcal{L}$ is formally self-adjoint, then $\mathcal{L}^0(x,\xi)$ has values in the set of Hermitian matrices.

We can obtain an even more general situation by considering differential expressions on the sections of a Hermitian vector bundle $\mathcal{E}$ over a manifold X. For any local trivialization the coefficients $a_\alpha(x)$ are homomorphisms of the fibres $\mathcal{E}_x$. The principal symbol can be defined in an invariant way. It is a homomorphism of the induced bundle $\tilde{\mathcal{E}}$ over $T'X = T^*X \setminus \{0\}$ obtained from $\mathcal{E}$ by the natural projection $\pi : T'X \to X$.

2.2. Elliptic Differential Expressions

A scalar differential expression $\mathcal{L}$ of order m is said to be *elliptic in a domain* $X \subset \mathbb{R}^n$ if its principal symbol satisfies the inequality

$$|\mathcal{L}^0(x,\xi)| \geq \gamma(x)|\xi|^m, \qquad \gamma(x) > 0$$

in $X \times \mathbb{R}^n$, and it is said to be *uniformly elliptic* if

$$|\mathcal{L}^0(x,\xi)| \geq \gamma|\xi|^m, \qquad \gamma > 0. \tag{2.7}$$

On many occasions it is also said that $\mathcal{L}(x,\xi)$ is an *elliptic symbol*. The greatest possible value of γ in (2.7) is called the *ellipticity constant* of $\mathcal{L}$ and denoted by $\gamma_{\mathcal{L}}$. A formally self-adjoint scalar elliptic differential expression must obviously have an even order.

For an operator on vector-valued functions (in other words, for systems of equations) to be elliptic, $\det \mathcal{L}^0(x,\xi)$ must, by definition, be a scalar elliptic symbol. The property of being elliptic is preserved under smooth changes of variables, which enables one to extend the notion of ellipticity to the case of operators on manifolds (including operators on vector bundles).

Example 2.1. Formally self-adjoint second-order elliptic operators. Let

$$(\mathcal{L}u)(x) = \sum_{1\le i,j\le n} D_{x_i}\left(a_{ij}(x) D_{x_j} u(x)\right) + a_0(x)u(x) \tag{2.8}$$

with $a_{ij} = \bar{a}_{ji}$ and $\operatorname{Im} a_0 = 0$ in a domain $X \subset \mathbb{R}^n$, and let

$$\sum_{1\le i,j\le n} a_{ij}(x)\xi_i\xi_j \ge \gamma|\xi|^2, \qquad \gamma > 0 \tag{2.9}$$

in $X \times \mathbb{R}^n$. As opposed to the general uniform ellipticity condition (2.7), condition (2.9) determines the sign of $\mathcal{L}^0$.

The *Laplace operator*

$$\mathcal{L} = -\Delta = \sum_{1\le i\le n} D^2_{x_i},$$

for which $\mathcal{L}(\xi) = \mathcal{L}^0(\xi) = |\xi|^2$, serves as the leading special case.

Example 2.2. The Laplace-Beltrami operator on a Riemannian manifold X. Let $\{g_{ij}\}$, where $1 \le i, j \le n$, be the matrix of the metric tensor in local coordinates, let $\{g^{ij}\}$ be the inverse matrix, and let $g = \det\{g_{ij}\}$. By definition,

$$-\Delta = \frac{1}{\sqrt{g}} \sum_i D_{x_i}\left(\sum_j g^{ij}\sqrt{g} D_{x_j}\right) \tag{2.10}$$

in these coordinates. If $X \subset \mathbb{R}^n$ with the standard Euclidean metric, then (2.10) reduces to the ordinary Laplace operator.

The Laplace-Beltrami operator Δ_p on p-forms over X (Trèves 1982) represents an important example of an elliptic operator on a bundle.

Example 2.3. The polyharmonic operator $\mathcal{L} = (-\Delta)^r$. For this operator $\operatorname{ord}\mathcal{L} = 2r$ and $\mathcal{L}(\xi) = \mathcal{L}^0(\xi) = |\xi|^{2r}$.

2.3. The Maximal and Minimal Operators

It has already been mentioned that, under the conditions (2.2), the differential expression $\mathcal{L}$ is defined in the distribution sense for any $u \in L_2(X)$. To talk of $\mathcal{L}$ as an operator in $L_2 = L_2(X)$, one has to fix a domain of $\mathcal{L}$, that is, a linear set $D \subset L_2$ such that $\mathcal{L}D \subset L_2$. The operator $A = \mathcal{L} \restriction D$ is often called a *realization* of $\mathcal{L}$. By definition, we take the symbol (the principal symbol) of $\mathcal{L}$ as the symbol (the principal symbol) of $A = \mathcal{L} \restriction D$, that is, we set $A(x,\xi) = \mathcal{L}(x,\xi)$ and $A^0(x,\xi) = \mathcal{L}^0(x,\xi)$.

The *maximal operator* $\mathcal{L}_{\max} = \mathcal{L} \restriction D_{\max}$, where

$$D_{\max} = D_{\max}(\mathcal{L}) = \{u \in L_2 : \mathcal{L}u \in L_2\},$$

has the largest possible domain.

On the other hand, as a rule, only domains $D \supset C_0^\infty(X)$ are considered. The closure of $\mathcal{L} \restriction C_0^\infty(X)$ is called the *minimal operator* generated by $\mathcal{L}$. The completion $D_{\min} = D_{\min}(\mathcal{L})$ of $C_0^\infty(X)$ in the $\mathcal{L}$-norm of the form (1.1) serves as the domain of the minimal operator.

Example 2.4. Let X be a domain in $\mathbb{R}^n$ and let $\mathcal{L}$ be an elliptic differential expression of order m with constant coefficients. Then $D_{\min}(\mathcal{L}) = \mathring{H}^m(X)$, which is easily proved with the aid of the Fourier transform.

Example 2.4 can also be extended to uniformly elliptic differential expressions with variable coefficients (provided they are sufficiently regular).

For $u \in D_{\min}(\mathcal{L})$ some boundary conditions on ∂X can also be retained in the general (non-elliptic) case. On the other hand, the functions $u \in D_{\max}$ do not, in general, satisfy any boundary conditions. However, if X is an unbounded domain or the coefficients of $\mathcal{L}$ are rapidly increasing near ∂X, then the relation $\mathcal{L}u \in L_2(X)$ can involve implicit conditions at infinity or on ∂X that must be satisfied by the functions from $D_{\max}$.

Example 2.5. Let $X = \mathbb{R}^1$, $\mathcal{L} = D$, and $u \in D_{\max}(\mathcal{L}) = H^1(\mathbb{R})$. Then $u(x) \to 0$ as $|x| \to \infty$.

If $\mathcal{L}$ is a formally self-adjoint expression, then $\mathcal{L}_{\max} = (\mathcal{L}_{\min})^*$, which follows immediately from the definitions. It may turn out that

$$\mathcal{L}_{\max} = \mathcal{L}_{\min}, \tag{2.11}$$

or, equivalently, $D_{\max}(\mathcal{L}) = D_{\min}(\mathcal{L})$. Then $\mathcal{L}_{\min}$ is a self-adjoint operator. It is also said that $\mathcal{L}$ is an essentially self-adjoint differential expression on $C_0^\infty(X)$. In particular, this is the case in Example 2.5.

If $A = \mathcal{L} \restriction D$ is a realization of a differential expression $\mathcal{L} = \mathcal{L}^+$, then

$$D_{\min}(\mathcal{L}) \subset D \subset D_{\max}(\mathcal{L}). \tag{2.12}$$

It follows that essential self-adjointness on C_0^∞ means that $\mathcal{L}$ has a unique self-adjoint realization.

Self-adjoint realizations do not always exist. It follows from the general theory of extensions (see Akhiezer and Glazman 1966; Kato 1966; Birman and Solomyak 1980) that a formally self-adjoint expression $\mathcal{L}$ has a self-adjoint realization if and only if the *deficiency indices*

$$n_\pm(\mathcal{L}) = \dim \operatorname{Ker}(\mathcal{L}_{\max} \pm iI)$$

are equal to one another. This condition is necessarily satisfied if $\mathcal{L}_{\min}$ is lower semi-bounded, since then $n_\pm(\mathcal{L}) = \dim \operatorname{Ker}(\mathcal{L}_{\max} - cI)$ for $c < \gamma_{\mathcal{L}_{\min}}$.

Everything that has been said above can be carried over in a natural way to the more general settings mentioned in Sect. 2.1.

2.4. Essential Self-Adjointness of Elliptic Operators

One of the simplest cases in which a differential expression has a unique self-adjoint realization is the case of an elliptic operator on a manifold without boundary.

Theorem 2.1. *Let X be a smooth compact manifold with fixed smooth positive density and let $\mathcal{L} = \mathcal{L}^+$ be an elliptic differential expression on X. Then the operator $\mathcal{L}$ is essentially self-adjoint on $C^\infty(X)$. The equality*

$$D_{\max}(\mathcal{L}) = H^m(X), \quad m = \operatorname{ord} \mathcal{L}$$

is satisfied.

The theorem can be extended to the case of elliptic differential expressions on bundles.

We shall now discuss the case when $\mathcal{L}_{\max} \neq \mathcal{L}_{\min}$, confining ourselves to elliptic differential expressions. If D is the domain of a self-adjoint realization $\mathcal{L} \restriction D$, then it satisfies (2.12) and is closed in the $\mathcal{L}$-norm (of course, these conditions are only necessary). It follows from (2.12) that, for $u \in D_{\max}$, the condition $u \in D$ depends only on the behaviour of u in the vicinity of ∂X (or at infinity if X is unbounded). From this point of view, any self-adjoint realization of $\mathcal{L}$ is determined by the boundary conditions.

The above discussion admits localization. Let Γ be a closed subset of ∂X and let

$$D_{\min}^\Gamma = D_{\min}^\Gamma(\mathcal{L}) = \{u \in D_{\max}(\mathcal{L}) : u = 0 \text{ in a neighbourhood of } \Gamma\}.$$

If the closure of $\mathcal{L} \restriction D_{\min}^\Gamma$ differs from $\mathcal{L}_{\max}$, then we say that $\mathcal{L}$ requires boundary conditions on Γ. Similarly, we can talk of boundary conditions at infinity, etc.

The problem of boundary conditions at infinity was fist studied by Weyl (see Dunford and Schwartz 1963; Glazman 1963; Levitan and Sargsyan 1970) in the case of second-order differential expressions on the half-axis $\mathbb{R}_+$. Weyl considered this problem as a limit of problems on the expanding system of intervals $(0, l)$. Assigned to every l was a certain disc $\mathbf{D}_l \subset \mathbb{C}^1$ such that $\mathbf{D}_{l_2} \subset \mathbf{D}_{l_1}$ if $l_2 > l_1$. It follows that the discs $\mathbf{D}_l$ contract either to a 'limit-point' or a 'limit-circle' as $l \to \infty$. In the former case the operator does not require any boundary conditions at infinity, but it does in the latter. Because of this, one often refers to the *limit-point (limit-circle)* case at $+\infty$ for a second-order operator on a half-axis, which means that no boundary conditions are required (that boundary conditions are required) at $+\infty$. For an operator on the real axis one can talk of the limit-point and limit-circle cases at $+\infty$ as well as $-\infty$.

We shall present the classic examples of self-adjoint operators defined by a formally self-adjoint differential expression with boundary conditions.

Example 2.6. The basic boundary value problems for a second-order elliptic differential expression. Let a uniformly elliptic formally self-adjoint differential expression (2.8) with coefficients in $C^\infty(\bar{X})$ be given in a bounded domain $X \subset \mathbb{R}^n$ with smooth boundary ∂X. We set

$$\mathcal{L}_{\mathrm{I}} = \mathcal{L} \restriction D_{\mathrm{I}}, \quad D_I = \left\{u \in H^2(X) : u|_{\partial X} = 0\right\}, \tag{2.13}$$

$$\mathcal{L}_{\mathrm{II}} = \mathcal{L} \restriction D_{\mathrm{II}}, \quad D_{\mathrm{II}} = \left\{u \in H^2(X) : \left.\frac{\partial u}{\partial \nu}\right|_{\partial X} = 0\right\}, \tag{2.14}$$

$$\mathcal{L}_{\mathrm{III}} = \mathcal{L} \restriction D_{\mathrm{III}}, \quad D_{\mathrm{III}} = \left\{u \in H^2(X) : \left.\left(\frac{\partial u}{\partial \nu} + \sigma u\right)\right|_{\partial X} = 0\right\}. \tag{2.15}$$

In (2.14) and (2.15) $\partial/\partial\nu$ is the *co-normal derivative* at $x \in \partial X$:

$$\frac{\partial u}{\partial \nu} = \sum_{i,j} a_{ij}(x) \frac{\partial u}{\partial x_j} \cos(n, x_i),$$

where $n = n(x)$ is the exterior normal vector. In (2.15) $\sigma = \sigma(x)$ is a smooth real-valued function on ∂X. The boundary conditions in (2.13) and (2.15) are meaningful by virtue of the embedding theorems for H^2. The differential operators $\mathcal{L}_{\mathrm{I}}$, $\mathcal{L}_{\mathrm{II}}$, and $\mathcal{L}_{\mathrm{III}}$ are called the *operators of the first, second, and third boundary value problems*, respectively.

Example 2.7. The operator of the Dirichlet problem for an equation of order $2r$. Let $X \subset \mathbb{R}^n$ be a bounded domain with smooth boundary and let $\mathcal{L} = \mathcal{L}^+$ be a differential expression of even order $m = 2r$ that is uniformly elliptic in $\bar{X}$ with real leading coefficients a_α, $|\alpha| = m$. The operator

$$\mathcal{L}_D = \mathcal{L} \restriction H^{2r}(X) \cap \mathring{H}^r(X) \tag{2.16}$$

is called the *operator of the Dirichlet problem for $\mathcal{L}$*. If $m = 2$, then $\mathcal{L}_D$ coincides with the operator (2.13).

The operators (2.13)–(2.16) are the most important special cases of operators of regular elliptic boundary value problems. Let $X \subset \mathbb{R}^n$ be a bounded domain with smooth boundary and let $\mathcal{L}$ be a uniformly elliptic differential expression of order $m = 2r$ with smooth coefficients in $\bar{X}$ that acts on scalar functions in X. Let $\mathcal{B} = \{\mathcal{B}_1, \dots, \mathcal{B}_r\}$ be a system of differential expressions defined in a neighbourhood of ∂X and satisfying the following conditions.

1° $0 \le m_j = \operatorname{ord} \mathcal{B}_j < m$ for $j = 1, \dots, r$, and $m_i \ne m_j$ for $i \ne j$.

2° The principal symbols $\mathcal{B}_j^0(x, \xi)$ are such that $\mathcal{B}_j^0(x, n(x)) \ne 0 \ \forall\, x \in \partial X$.

3° For any vector $0 \neq \xi \in \mathbb{R}^n$ tangent to ∂X at $x \in \partial X$, the polynomials $B_j^0(x, \xi + \tau n(x))$, $j = 1, \ldots, r$ of a complex variable τ are linearly independent modulo the polynomial

$$\prod_{1 \leq i \leq r} \left(\tau - \tau_i^+(x, \xi)\right),$$

where τ_i^+ are the roots of the polynomial $\mathcal{L}^0(x, \xi + \tau n(x))$ with positive imaginary parts.

Under these conditions, $\{\mathcal{L}, \mathcal{B}\}$ is said to be the *operator of a regular elliptic boundary value problem.*

Theorem 2.2. *Let $\{\mathcal{L}, \mathcal{B}\}$ be the operator of a regular elliptic boundary value problem. In this case, if the operator*

$$A = \mathcal{L} \restriction \left\{u \in H^m(X) : \mathcal{B}_j u\big|_{\partial X} = 0,\ j = 1, \ldots, r\right\}$$

is symmetric, then it is self-adjoint. In particular, the operators (2.13)–(2.16) are self-adjoint.

The theorem follows from the theory of solubility of elliptic boundary value problems (Lions and Magenes 1968; Hörmander 1983–1985). The symmetry condition assumed in the theorem implies formal self-adjointness of $\mathcal{L}$ and also stipulates certain consistency between $\mathcal{L}$ and the boundary conditions (the operators (2.14) and (2.15) serve as examples).

In general, Theorem 2.2 is no longer valid for problems in domains with non-smooth boundary. For instance, if $X \subset \mathbb{R}^2$ is a domain whose boundary contains a corner point, the interior angle being greater than π, then Δ_D is not a self-adjoint operator (in this connection, see Example 3.2).

Regular elliptic boundary value problems can also be stated for operators acting in spaces of sections of vector bundles. Theorem 2.2 can be extended to this case. We shall collect all such operators as well as elliptic differential operators on compact manifolds without boundary under the name of *regular elliptic differential operators*.

2.5. Singular Differential Operators

Let us agree to refer to differential operators other than regular elliptic ones as *singular* operators. An operator may be singular because of various reasons such as the non-compactness of the manifold, violation of the uniform ellipticity condition, non-regularity of the coefficients or the boundary, and the like. From this list it is clear that it is a matter of convention to collect all such operators under one name. Many singular differential operators have properties similar to those of regular elliptic operators.

In the singular case the domain of a self-adjoint realization of a differential expression $\mathcal{L} = \mathcal{L}^+$ rarely admits an explicit description. The following result, which can be easily proved by going over to the Fourier transforms, is one of the exceptions.

Theorem 2.3. *Let $X = \mathbb{R}^n$ and let $\mathcal{L} = \mathcal{L}^+$ be a differential expression with constant coefficients. Then $\mathcal{L}$ is essentially self-adjoint on $C_0^\infty(\mathbb{R}^n)$ and self-adjoint on the domain*

$$D(\mathcal{L}) = \left\{ u \in L_2(\mathbb{R}^n) : \int \left(1 + |\mathcal{L}(\xi)|^2\right) |\tilde{u}(\xi)|^2 \, d\xi < \infty \right\}. \tag{2.17}$$

In particular, if $\mathcal{L}$ is an elliptic operator, then it is self-adjoint on $H^m(\mathbb{R}^n)$, where $m = \operatorname{ord} \mathcal{L}$.

2.6. The Schrödinger Operator

In applications the most important example of a singular differential operator is the *Schrödinger operator*

$$\mathcal{L}u = -\Delta u + Vu \tag{2.18}$$

in $L_2(\mathbb{R}^n)$. The real-valued (within the scope of the self-adjoint theory) function V is usually called the *potential*. In accordance with (2.2), in this section we assume that

$$V \in L_{2,\mathrm{loc}}(\mathbb{R}^n). \tag{2.19}$$

The operator (2.18) in $L_2(X)$ is also singular whenever $X \subset \mathbb{R}^n$ is an unbounded domain. We shall only touch upon the case when $\mathbb{R}^n \setminus X$ is compact, and also, for $n = 1$, the case of the half-axis $X = \mathbb{R}_+ = (0, \infty)$. We can set any self-adjoint boundary condition on ∂X (for example, $u|_{\partial X} = 0$ or $\partial u / \partial n|_{\partial X} = 0$).

The equation $D_t u = \mathcal{L}u$ with $\mathcal{L}$ given by (2.18) describes the evolution of the wave function of a quantum system with the Hamiltonian $\mathcal{L}(x, \xi) = |\xi|^2 + V(x)$. In order that the evolution operator of this equation be unitary, the realization of $\mathcal{L}$ must be a self-adjoint operator. Roughly speaking, the essential self-adjointness of $\mathcal{L}$ on $C_0^\infty(\mathbb{R}^n)$ means that a wave at first localized in space cannot escape to infinity in finite time. The *completeness* of a Hamiltonian system with Hamiltonian $\mathcal{L}(x, \xi)$, that is, the existence of a solution of the system for all $t \in \mathbb{R}$ and any initial data (x_0, ξ_0) such that $\xi_0 \neq 0$, serves as the classical analogue of essential self-adjointness. See (Reed and Simon 1975, Vol. 2) for a discussion of the limits of applicability of this analogy.

Only in rare cases can one succeed in giving an exact description of the domain on which (2.18) is a self-adjoint differential operator. For the most part the domain can be studied on the basis of Theorem 1.1. In application to the Schrödinger operator, one usually sets $A = -\Delta \restriction H^2(\mathbb{R}^n)$ (which, by Theorem 2.3, is a self-adjoint operator) and $Bu = Vu$. Estimates of the form

(1.2) can be deduced from embedding theorems. Below we present a typical result based on Theorem 1.1.

Theorem 2.4 (Stummel; see Reed and Simon 1975, Vol. 2). *Let V be a real-valued potential that satisfies the following conditions:*

a) if $n \geq 4$, then

$$\sup_{x \in \mathbb{R}^n} \int_{|x-y|<1} |x-y|^{-(n-4+\alpha)} V^2(y)\, dy < \infty$$

for any α such that $0 < \alpha < 4$;

b) if $n \leq 3$, then

$$\sup_{x \in \mathbb{R}^n} \int_{|x-y|<1} V^2(y)\, dy < \infty .$$

Then the operator (2.18) is self-adjoint on $H^2(\mathbb{R}^n)$ and essentially self-adjoint on $C_0^\infty(\mathbb{R}^n)$.

In other words, under the assumptions of the theorem, the operator (2.18) requires no boundary conditions at infinity. In this formulation the theorem can be extended to the case of the Schrödinger operator in a domain $X \neq \mathbb{R}^n$ (when verifying the assumptions of the theorem, one should set $V \equiv 0$ outside X).

Example 2.8. The *three-dimensional Schrödinger operator with Coulomb potential* $V(x) = -c|x|^{-1}$ is self-adjoint on $H^2(\mathbb{R}^3)$.

Example 2.9. The *N-particle Schrödinger operator* in $L_2(\mathbb{R}^{nN})$:

$$\begin{aligned} H = &- \sum_{1 \leq j \leq N} (2\mu_j)^{-1} \Delta_j + \sum_{1 \leq j \leq N} V_j(x_j) \\ &+ \sum_{1 \leq k < l \leq N} W_{kl}(x_k - x_l), \quad x_j \in \mathbb{R}^n. \end{aligned} \tag{2.20}$$

Here μ_j is the mass of the j-th particle and Δ_j is the Laplace operator with respect to the j-th group of variables. The operator (2.20) is self-adjoint on $H^2(\mathbb{R}^{nN})$ if each of the potentials V_j and W_{kl} satisfies the assumptions of Theorem 2.4. In particular, this is the case for the Schrödinger operator of an atom with Coulomb interaction between the particles.

The assumptions of Theorem 2.4 are certainly not satisfied if $|V(x)| \to \infty$ as $|x| \to \infty$. For such potentials the domain on which the operator (2.18) is self-adjoint must depend on V. The set

$$\hat{D} = \{u \in H^2(\mathbb{R}^n) : Vu \in L_2(\mathbb{R}^n)\}$$

is the natural domain on which (2.18) is defined as an operator in $L_2(\mathbb{R}^n)$. If (2.18) is a self-adjoint operator on $\hat{D}$ (in which case it must be essentially self-adjoint on $C_0^\infty(\mathbb{R}^n)$), then it is said to be *separated*. Separation is always connected with a specific regular behaviour of the potential.

Theorem 2.5 (Rozenblum 1974). *Let $V(x) = \bar{V}(x) \geq 1$ and let the condition*

$$V(y) \leq cV(x) \quad \forall\, x, y \in \mathbb{R}^n : |x-y| < 1 \tag{2.21}$$

be satisfied for some $c > 0$. Then the operator (2.18) is separated.

Firstly, (2.21) means that the potential does not grow too fast as $|x| \to \infty$ (not faster then exponentially). Secondly, it rules out rapid oscillations and other irregularities in the behaviour of $V(x)$. Another condition, which also ensures separation (see Bojmatov 1984) and allows $V(x)$ to grow rapidly as $|x| \to \infty$, reads

$$V(x) \geq 0, \quad |\nabla V(x)| \leq c[V(x)]^{3/2}, \quad c < 2.$$

There are also separation conditions that admit local irregularities of the potential. They can be stated in terms of an auxiliary 'averaged' potential of the type

$$V^*(x) = \inf\left\{ d : d^{n-2} \geq c_n \int_{|y-x|<d} V(y)\, dy \right\}.$$

(The precise definition uses capacity terms.) Regarding this point, see (Mynbaev and Otelbaev 1988), which also contains further references.

In a more complex situation one fails to give an exact description of the domain of a self-adjoint realization of the Schrödinger operator. Then the investigation of conditions for V under which (2.11) is satisfied, that is, no boundary condition is required at infinity, comes to the fore. The theorem below is one of the basic results in this direction.

Theorem 2.6 (Kato; see Reed and Simon 1975, Vol. 2). *Let $V \in L_{2,\mathrm{loc}}(X)$ be a real-valued potential bounded from below, that is, $V(x) \geq c > -\infty$. Then the operator (2.18) requires no boundary condition at infinity.*

The result remains valid if the potential tends to $-\infty$ not too rapidly as $|x| \to \infty$.

Theorem 2.7 (Sears; see Reed and Simon 1975, Vol. 2; Berezin and Shubin 1983). *Let $V(x)$ be a real-valued potential that satisfies the condition*

$$V(x) \geq -Q(|x|),$$

where $Q(r)$ is a non-decreasing positive continuous function on $\mathbb{R}_+$ such that

$$\int^{\infty} [Q(r)]^{-1/2}\, dr = \infty.$$

Then (2.18) is an essentially self-adjoint operator on $C_0^\infty(\mathbb{R}^n)$.

The assumption of the theorem is satisfied, in particular, for $Q(r) = r^\alpha$, where $\alpha \le 2$. For more general results of this kind (in particular, for $n = 1$), see (Dunford and Schwartz 1963) and (Reed and Simon 1975, Vol. 2). As an illustration of the theorem, we consider the following example.

Example 2.10. The operator $\mathcal{L}u = -u'' - x^2u$ *on an axis.* The equation $\mathcal{L}u = iu$ has solutions

$$u = \exp\left(-ix^2/2\right)\left(c_1 + c_2 \int_0^x \exp(it^2)\,dt\right).$$

Since $\int_{\mathbb{R}} \exp(it^2)\,dt = \sqrt{\pi/2} \ne 0$, it follows that $u \equiv 0$ is the only solution in L_2. The same is true for the equation $\mathcal{L}u = -iu$. Therefore $\mathcal{L}$ is essentially self-adjoint on C_0^∞.

We remark that the operator (2.18) with $V(x) = -|x|^a$, where $a > 2$, is no longer self-adjoint on C_0^∞.

2.7. The Schrödinger Operator: Local Singularities of the Potential

We shall discuss the case when the potential V in (2.18) has *singularities* such that (2.19) is not satisfied. We assume that the singularities of the potential are *localized*, that is, there exists a closed set $F \subset X$ such that $\operatorname{mes}_n F = 0$ and $V \in L_{2,\text{loc}}(X \setminus F)$. Then $\mathcal{L}$ can be considered as a differential expression in $X \setminus F$. In accordance with Sect. 2.3, $\mathcal{L}_{\min}$ and $\mathcal{L}_{\max}$ can be defined as operators in $L_2(X \setminus F)$, which can be identified with $L_2(X)$ in the natural way. There arises the question of whether or not $\mathcal{L}$ requires any boundary conditions on F.

We shall restrict ourselves to examples in which F is the one-element set $F = \{0\}$. For $n = 1$ we shall consider an operator on the half-axis $\mathbb{R}_+$.

Theorem 2.8 (Reed and Simon 1975, Vol. 2). *Let* $V = \bar{V} \in C(\bar{\mathbb{R}}_+)$ *be positive near* $x = 0$ *and let the limit*

$$c = \lim_{x \to +0} x^2 V(x)\ (\ge 0).$$

exist. Then the operator $\mathcal{L}u = -u'' + Vu$ *requires a boundary condition at* $x = 0$ *if and only if* $c < 3/4$.

We shall elucidate this result using the model example $V(x) = cx^{-2}$. In this case $\mathcal{L}_{\min} \ge 0$, and so it suffices to study the kernel of $\mathcal{L}_{\max} + I$. The solutions of the equation $u'' - (cx^2 + 1)u = 0$ that belong to $L_2(a, \infty)$, where $a > 0$, have the form $a\sqrt{x}K_p(x)$, where $p^2 = c + 1/4$ (K_p is the Bessel-MacDonald function). The solutions have the asymptotics $\alpha' x^{-p+1/2}$ as $x \to 0$, which implies that they belong to $L_2(\mathbb{R}_+)$ only for $p < 1$, that is, for $c < 3/4$. It follows that $\mathcal{L}_{\min}$ is essentially self-adjoint exactly for $c \geq 3/4$.

Below we state one of the results concerned with the many-dimensional case (see Reed and Simon 1975, Vol. 2).

Theorem 2.9. *Let $V = \bar{V} \in L_{2,\mathrm{loc}}(\mathbb{R}^n \setminus \{0\})$ and let*

$$V(x) \geq -n(n-4)/4x^2 + d,$$

where $d > -\infty$. Then the operator (2.18) requires no boundary condition at $x = 0$.

In particular, an interesting case is when $V(x) \equiv 0$ in $\mathbb{R}^n \setminus \{0\}$, which corresponds to a δ-shaped potential in $\mathbb{R}^n$. The Laplace operator on $C_0^\infty(\mathbb{R}^n \setminus \{0\})$ is essentially self-adjoint for $n \geq 4$, but it is not essentially self-adjoint for $n \leq 3$. It describes the behaviour of a particle in a potential field of *'radius zero.'* The lack of essential self-adjointness means that the operator $-\Delta$ on $C_0^\infty(\mathbb{R}^n \setminus \{0\})$ requires boundary conditions at 0, that is, the physical description of such a particle must include the interaction between the particle and an impenetrable obstacle at 0. For example, for $n = 1$ a self-adjoint extension of the operator $-d^2/dx^2$ to $C_0^\infty(\mathbb{R}_+)$ can be fixed by specifying a boundary condition of the form $y'(0) + \alpha y(0) = 0$ with $-\infty < \alpha \leq \infty$ at zero. The physical meaning of this condition is that a plane wave with momentum k will change its phase by $\arg((ik - \alpha)/(ik + \alpha))$ when reflected by the obstacle. For $n = 2, 3$ the situation is much more involved (see Colin de Verdière 1982, 1983; Pavlov and Shushkov 1988).

2.8. The Dirac Operator

The *Dirac operator* in $\mathbb{R}^3$, which describes the behaviour of a relativistic particle, serves as another important example. Let $\alpha_1, \ldots, \alpha_4$ be complex Hermitian 4×4-matrices that satisfy the *'anticommutation relations'* $\alpha_j\alpha_k + \alpha_k\alpha_j = 2\delta_{jk}$, and let $V \in L_{2,\mathrm{loc}}(\mathbb{R}^3)$ be a 4×4–matrix-valued function. The Dirac operator is generated by the differential expression

$$\mathcal{L} = \mathcal{L}_0 + V(x) = \sum_{j=1}^{3} \alpha_j D_j + \alpha_4 + V(x) \tag{2.22}$$

in the space of four-component vector-valued functions in $\mathbb{R}^3$. As opposed to the Laplace operator, $\mathcal{L}_0$ is not semi-bounded, which can be seen immediately on applying the Fourier transform.

The essential self-adjointness of $\mathcal{L}$ on $C_0^\infty(\mathbb{R}^3)$ depends only on the local properties of the potential.

Theorem 2.10 (Levitan and Otelbaev 1977). *For the operator (2.22) to be essentially self-adjoint it is necessary and sufficient that the Dirac operator* *with the potential* $\chi_B V$, *where* χ_B *is the indicator function of* B, *be essentially self-adjoint for any ball* $B \subset \mathbb{R}^3$.

The essential self-adjointness of the Dirac operator can usually be proved by the methods of perturbation theory.

Example 2.11. The Dirac operator with scalar Coulomb potential. Let $V(x) = c|x|^{-1}$. One can demonstrate (Reed and Simon 1975, Vol. 2) that $\mathcal{L}_{\min}$ is essentially self-adjoint for $|c| < \sqrt{3}/2$. Taking the electric charge and mass of an electron into account, one finds that this condition corresponds to atoms with atomic number $Z < 118$. See (Reed and Simon 1975, Vol. 2; Arai 1983) for a discussion of possible physical consequences.

§3
Defining an Operator by a Quadratic Form

The *variational method*, or, using another terminology, the *method of forms*, is an important technique for defining a self-adjoint operator. The method, which is applicable in the semi-bounded case, is based on the Friedrichs theorem (Theorem 1.5) on the construction of a self-adjoint operator from a quadratic form.

The qualitative and quantitative characteristics of the spectrum of a semi-bounded self-adjoint operator can be well described in terms of its quadratic form. Formula (1.13) can serve as an example. In spectral analysis this often makes it possible to dispense with the explicit description of the domain of the operator and, what is more, even with the explicit description of its 'action.' As we shall see, the method of forms enables one to relax significantly the regularity conditions for the boundary of the domain and the coefficients of the differential expression, as well as the summability conditions for the potential in the Schrödinger operator, etc. Moreover, this method is also applicable when the minimal operator is not defined (that is, does not transform C_0^∞ into L_2), and so the extension scheme presented in §2 does not work. If the minimal operator exists, is lower semi-bounded, and not essentially self-adjoint, then the method of forms distinguishes one concrete self-adjoint realization of $\mathcal{L}$, namely, the Friedrichs extension of $\mathcal{L}_{\min}$.

The construction of a differential operator from a quadratic form is closely related to the problems of classical variational calculus. The differential expression $\mathcal{L}$ corresponding to the form appears on the left-hand side of the Euler equation. The latter should be understood in the sense of an 'integral identity.' The boundary conditions for $\mathcal{L}$ can be divided into two types: those involved in the description of the domain of the form (the 'main' conditions), and the 'natural' conditions in the sense of variational calculus.

3.1. Examples

We shall present a number of typical examples.

Example 3.1. Boundary value problems for a second-order elliptic operator (see Example 2.6). Let $X \subset \mathbb{R}^n$ be an arbitrary domain and let

$$l[u] = \int_X \Big(\sum_{1 \le i,j \le n} a_{ij} D_{x_i} \overline{D_{x_j} u} + a_0(x)|u|^2 \Big)\, dx. \tag{3.1}$$

The quadratic functional (3.1) corresponds formally to the differential expression (2.8), that is, it can be obtained by multiplying (2.8) by $\bar{u}$ and using the integration-by-parts formula with the boundary terms ignored. We assume that $a_{ij} = \bar{a}_{ji}$ and $a_0 = \bar{a}_0$ are measurable, bounded,[2] and satisfy (2.9). If $c \in \mathbb{R}$ is large enough, then the form $l[u] + c\|u\|^2$ defines the metric of $H^1(X)$, and so (3.1) is a closed form on $H^1(X)$. Let $\mathcal{L}_N$ *(the operator of the Neumann problem)* be the corresponding self-adjoint operator in $L_2(X)$. If X is bounded and its boundary as well as the coefficients a_{ij} are smooth enough, then $\mathcal{L}_N = \mathcal{L}_{\text{II}}$, where $\mathcal{L}_{\text{II}}$ is the operator (2.14). The description of the operator given in (2.14) is no longer valid in the general case.

The form (3.1) is also closed and semi-bounded on any closed subspace of $H^1(X)$, and, in particular, on $\mathring{H}^1(X)$. In the smooth case the operator $\mathcal{L}_D = \mathrm{Op}\,(l \restriction \mathring{H}^1(X))$ coincides with (2.13).

Analogously, let $F \subset \partial X$ be a closed set and let $\mathring{H}^1(X; F)$ be the closure in $H^1(X)$ of the set of functions vanishing in a neighbourhood of F. We set $\mathcal{L}_{D(F)} = \mathrm{Op}\,(l \restriction \mathring{H}^1(X; F))$. Roughly speaking, the functions from the domain of this operator satisfy the Dirichlet condition on F and the Neumann condition on $\partial X \setminus F$. If F is 'very small' (of capacity zero), then $\mathcal{L}_{D(F)} = \mathcal{L}_N$.

Now, for $u \in H^1(X)$, we set

$$l_\sigma[u] = l(u) + \int_{\partial X} \sigma(x)|u|^2\, dS(x), \tag{3.2}$$

where l is the form (3.1) and $\sigma = \sigma \in L_\infty(\partial X)$. We assume that the boundary is compact and of class C^2. Thus, by the embedding theorem, the boundary integral in (3.2) is a continuous quadratic functional in $H^1(X)$. In the smooth case the operator $\mathcal{L}_\sigma = \mathrm{Op}\,(l_\sigma \restriction H^1(X))$ coincides with (2.15).

In general, under the above assumptions for a_{ij} and the domain (which are weaker compared with Example 2.6), one cannot construct a self-adjoint realization of $\mathcal{L}$ by extending the minimal operator $\mathcal{L} \restriction C_0^\infty$, since in this case it may no longer be true that $\mathcal{L} : C_0^\infty \to L_2$.

The following example illustrates the difficulties connected with the analytic description of the domain of the operator $\mathcal{L}_D$ defined variationally.

---

[2] One can relax the condition $a_0 \in L_\infty(X)$.

Example 3.2. The operator $(-\Delta)_D$ in a domain with corners. Let (r, φ) be the polar coordinates in $\mathbb{R}^2$ and let $X = \{(r, \varphi) : 0 < r < 1, 0 < \varphi < \pi/\alpha\}$, where $\alpha > 1/2$. Any function u that belongs to $H^2 \cap \overset{\circ}{H}{}^1$ outside a neighbourhood of 0 and equals $r^\alpha \sin \alpha\varphi$ near 0 is contained in $D(\Delta_D)$, but $u \in H^2(X)$ only if $\alpha \geq 1$. Therefore, if $\alpha < 1$, we have $D(\Delta_D) \not\subset H^2(X)$. A similar effect occurs for the Neumann problem. For a problem of the type $\mathcal{L}_{D(F)}$ the effect is also present for $\partial X \in C^\infty$. For more details on operators in domains with corners, edges, and the like, see (Kondrat'ev and Oleinik 1983).

Example 3.3. The Dirichlet and Neumann problems for a polyharmonic operator. Once again, let $X \subset \mathbb{R}^n$ be an arbitrary domain and let

$$l[u] = \int_X |\nabla_r u|^2 \, dx = \sum_{|\alpha|=r} \frac{r!}{\alpha!} \int_X |D^\alpha u|^2 \, dx. \tag{3.3}$$

The operator $(-\Delta)^r$ corresponds formally to the quadratic functional (3.3), which is closed and non-negative on $H^r(X)$. If $\partial X \in C^\infty$, then the self-adjoint operator $\mathrm{Op}\,(L \restriction H^r(X))$ coincides with $(-\Delta)^r_D$ (see Example 2.7). In analogy with the case $r = 1$, we denote by $(-\Delta)^r_N$ the operator corresponding to the form $l \restriction H^r(X)$. Here all the boundary conditions are the natural ones. In the case $r > 1$ their exact description is quite involved even for $\partial X \in C^\infty$.

Example 3.4. The operators of the Dirichlet and Neumann problems for degenerate elliptic differential expressions. Let $X \subset \mathbb{R}^n$ be a bounded domain with smooth boundary ∂X and let $\rho = \rho(x)$ be the *regularized distance* from x to ∂X, that is, a smooth positive function on X equal to $\mathrm{dist}\,(x, \partial X)$ in a neighbourhood of ∂X. The form

$$l_\alpha[u] = \int_X \rho^\alpha \Big(\sum_{1 \leq i,j \leq n} a_{ij}(x) D_{x_i} u \overline{D_{x_j} u} + |u|^2 \Big) \, dx, \tag{3.4}$$

where $a_{ij} = \bar{a}_{ij} \in L_\infty(X)$ satisfy (2.9), determines the metric of the weighted Sobolev space $H^1_\alpha(X)$ (Nikol'skij 1977). The form $l_\alpha \restriction H^1_\alpha(X)$ is non-negative and closed. We denote by $\mathcal{L}_{\alpha,N}$ the self-adjoint operator in $L_2(X)$ that corresponds to this form.

The operator $\mathcal{L}_{\alpha,D}$ corresponds to the same form (3.4) considered in $\overset{\circ}{H}{}^1_\alpha(X)$ (the closure of $C_0^\infty(X)$ in $H^1_\alpha(X)$). If $\alpha \geq 1$, then $\overset{\circ}{H}{}^1_\alpha = H^1_\alpha$, that is, $C_0^\infty(X)$ is dense in H^1_α. In this case $\mathcal{L}_{\alpha,N} = \mathcal{L}_{\alpha,D}$.

The differential expression

$$\mathcal{L}_\alpha u = \sum_{1 \leq i,j \leq n} D_{x_j} \left(\rho^\alpha a_{ij}(x) D_{x_i} u \right) + \rho^\alpha u$$

corresponds to the quadratic functional (3.4). The domains $D(\mathcal{L}_{\alpha,D})$ and $D(\mathcal{L}_{\alpha,N})$ are much harder to describe analytically than for $\alpha = 0$.

Example 3.5. The Dirichlet problem for non-elliptic differential expressions with constant coefficients. Let $\mathcal{L}(\xi)$ be a real polynomial in $\mathbb{R}^n$ such that

$$\inf_{\xi \in \mathbb{R}^n} \mathcal{L}(\xi) > -\infty.$$

With the aid of the Fourier transform, one can easily verify that $\mathcal{L} = \mathcal{L}(D)$ is a lower semi-bounded symmetric operator on $C_0^\infty(\mathbb{R}^n)$, and so also on $C_0^\infty(X)$, where $X \subset \mathbb{R}^n$ is an arbitrary domain. We denote by $\mathcal{L}_D(X)$, or simply by $\mathcal{L}_D$, the Friedrichs extension of $\mathcal{L} \restriction C_0^\infty(X)$. In general, it is scarcely possible to describe the boundary conditions explicitly unless $\mathcal{L}$ is elliptic.

3.2. The Schrödinger Operator and Its Generalizations

Here we present examples connected with the Schrödinger operator (2.18).

Example 3.6. Let

$$V = \bar{V} \geq c > -\infty, \quad V \in L_{1,\mathrm{loc}}(\mathbb{R}^n). \tag{3.5}$$

On multiplying the differential expression (2.18) by $\bar{u}$ and integrating formally by parts, we obtain the quadratic functional

$$a[u] = \int_{\mathbb{R}^n} \left(|\nabla u|^2 + V|u|^2\right) dx, \tag{3.6}$$

which is semi-bounded and closed on the natural domain $d[a] = H^1 \cap L_{2,|V|}$. The corresponding self-adjoint operator A is taken as the realization of the differential expression (2.18).

In the case in question one can give a more or less explicit description of the domain of A (Kato; see Reed and Simon 1975, Vol. 2):

$$D(A) = \{u \in L_2(\mathbb{R}^n) : Vu \in L_{1,\mathrm{loc}}, -\Delta u + Vu \in L_2\}.$$

We shall discuss this example from one more point of view. Let $a_0[u] = \int |\nabla u|^2 dx$ and $b[u] = \int V|u|^2 dx$. Then $d[a_0] = H^1$, $d[b] = L_{2,|V|}$, and the form $a = a_0 + b$ turns out to be closed on the natural domain $d[a_0] \cap d[b]$. The operator $A_0 = -\Delta$ corresponds to a_0 and $B : u \mapsto Vu$ corresponds to b. By definition, the operator equality $A = A_0 + B$ entails, in particular, the equality $D(A) = D(A_0) \cap D(B)$, which means that the operator is separated in the case in question, and which fails to be satisfied in general.

In similar cases it is customary to say that $A = A_0 + B$ in the sense of forms.

The condition $V \in L_{1,\mathrm{loc}}(\mathbb{R}^n)$ is not necessary for the Schrödinger operator (2.18) to admit a realization as the self-adjoint operator in $L_2(\mathbb{R}^n)$ that corresponds to (3.6). It is only important that the natural domain of (3.6) be dense in $L_2(\mathbb{R}^n)$. Thus one can consider potentials $V \in L_{1,\mathrm{loc}}(\mathbb{R}^n \setminus F)$, where F is closed and $\mathrm{mes}_n F = 0$. This being the case, $d[a] = H^1(\mathbb{R}^n \setminus F) \cap L_{2,|V|}$.

Here is a typical example: $V(x) = c|x|^{-\alpha}$, $c > 0 \ \forall\ \alpha > 0$. The *'Schrödinger operator with a δ-shaped potential,'* which corresponds to the form

$$a[u] = \int_{\mathbb{R}^1} |u'|^2 dx + |u(0)|^2, \quad u \in H^1(\mathbb{R}^1),$$

serves as another example.

In analogy with Example 3.6, one can consider the following

Example 3.7. Generalized Schrödinger operator. Let the 'potential' V satisfy (3.5). We consider the quadratic functional

$$a_r[u] = \int_{\mathbb{R}^n} \left(|\nabla_r u|^2 + V(x)|u|^2\right) dx \tag{3.7}$$

on $H^r \cap L_{2,|V|}$. The functional (3.7) is closed and semi-bounded in $L_2(\mathbb{R}^n)$. The operator $\mathrm{Op}(a_r)$ can be taken as a realization of the differential expression

$$\mathcal{L}u = (-\Delta)^r u + Vu. \tag{3.8}$$

3.3. Non–Semi-Bounded Potentials

The semi-boundedness of the Schrödinger operator is not necessarily connected with that of the potential V. It is only necessary that the negative part of (3.6) be dominated by the positive part. Below we denote by $V_\pm$ the positive and negative parts of V, that is, $2V_\pm = |V| \pm V$.

Example 3.8. The (generalized) Schrödinger operator with weakly non–semi-bounded potential. Let $V_+ \in L_{1,\mathrm{loc}}(\mathbb{R}^n)$ and let V_- satisfy the condition

$$\int V_- |u|^2 dx \le a \int |\nabla_r u|^2 dx + b \int |u|^2 dx \quad \forall\, u \in H^r(\mathbb{R}^n) \tag{3.9}$$

for some $a \in (0,1)$ and $b \geq 0$. Under these conditions, the functional (3.7) is semi-bonded and closed on the domain $H^1 \cap L_{2,V+}$. As before, the corresponding self-adjoint operator in $L_2(\mathbb{R}^n)$ is taken as the realization of the differential expression (3.8). In particular, for $r = 1$ we obtain a realization of the Schrödinger operator (2.18).

From (3.9) one can deduce concrete conditions for semi-boundedness. This approach enables one to consider potentials with stronger negative local singularities than the approach described in Theorem 1.1. Let

$$\gamma = 1 \ (n < 2r), \quad \gamma > 1 \ (n = 2r), \quad \gamma = n/2r \ (n > 2r). \tag{3.10}$$

By the embedding theorems, (3.9) is satisfied if $V_- \in L_\infty + L_\gamma$. In particular, for $n \geq 2$ this enables one to define the Schrödinger operator with the potential $-c|x|^{-\alpha}$, where $c > 0$ and $\alpha < 2$. The potential $-c|x|^{-2}$ can be included in consideration on the basis of the *Hardy inequality*

$$\frac{(n-2)^2}{4} \int_{\mathbb{R}^n} \frac{|u(x)|^2}{|x|^2}\, dx < \int_{\mathbb{R}^n} |\nabla u|^2 dx, \quad 0 \neq u \in C_0^\infty(\mathbb{R}^n). \tag{3.11}$$

The constant factor multiplying the integral on the left-hand side is sharp. It follows from (3.11) that the form (3.5) with $V(x) = -c|x|^{-2}$ is semi-bounded and closed on $H^1(\mathbb{R}^n)$ for $n \geq 3$ and $4c < (n-2)^2$.

3.4. Weighted Polyharmonic Operator

Here we shall consider examples of operators that can be defined by variational triples (see Sect. 1.10). In our examples

$$a[u] = a_{r,\varepsilon}[u] = \int_X (|\nabla_r u|^2 + \varepsilon |u|^2)\, dx, \quad r \geq 1, \quad \varepsilon \geq 0, \tag{3.12}$$

$$b[u] = b_p[u] = \int_X |u|^2 p\, dx, \tag{3.13}$$

under various assumptions concerning the domain $X \subset \mathbb{R}^n$, the real-valued function p, and the character of the boundary conditions defining the space d. If T is the operator determined by the triple $\{d; a_{r,\varepsilon}, b_p\}$, then the equality $Tf = u$ implies that u satisfies the equation (in the sense of an integral identity)

$$(-\Delta)^r u + \varepsilon u = pf. \tag{3.14}$$

We shall therefore talk of a 'weighted polyharmonic operator.'

Example 3.9. The Dirichlet problem in a bounded domain. Let $X \subset \mathbb{R}^n$ be an arbitrary bounded domain and let $d = \overset{\circ}{H}{}^r(X)$. Then (3.12) defines a norm in d for any $\varepsilon \geq 0$. Let

$$p \in L_\gamma(X) \quad \text{with } \gamma \text{ defined by (3.10).} \tag{3.15}$$

Then, by the embedding theorem, the form (3.13) is bounded (and compact) in d. We shall refer to $T\{\overset{\circ}{H}{}^r_1(X); a_{r,\varepsilon}, b_p\}$, where $\varepsilon \geq 0$, as the *operator of the Dirichlet problem for equation (3.14).*

Example 3.10. The Neumann problem. Let $X \subset \mathbb{R}^n$ be a bounded domain with Lipschitzian boundary and let (3.15) be satisfied. We shall call $T\{H^r(X); a_{r,\varepsilon}, b_p\}$, where $\varepsilon > 0$, the *operator of the Neumann problem for equation (3.14).*

The definition of the operator of the Neumann problem for $\varepsilon = 0$ is more involved, since (3.12) no longer defines a metric in H^r for $\varepsilon = 0$ (it is degenerate on the space of polynomials of degree less than r). Here d must be a subspace of finite deficiency in H^r on which (3.12) is not degenerate. For a reasonable choice of d (depending on the 'weight' p in (3.13)), the relationship between $T\{d; a_{r,0}, b_p\}$ and (3.14) can be preserved. For details, see (Birman and Solomyak 1972, 1973).

Example 3.11. The Steklov problem. Let $X \subset \mathbb{R}^n$ be a domain with boundary of class C^2 and let

$$\sigma = \bar{\sigma} \in L_\infty(\partial X), \quad \int_{\partial X} \sigma \, dS \neq 0,$$

$$d = \left\{ u \in H^1(X) : \int_{\partial X} u\sigma \, dS = 0 \right\},$$

$$a[u] = \int_X |\nabla u|^2 \, dx, \quad b[u] = \int_{\partial X} \sigma |u|^2 \, dS.$$

To the triple $\{d; a, b\}$ there corresponds the *operator of the Steklov problem*

$$\Delta u = 0 \text{ in } X, \quad \left. \frac{\partial u}{\partial n} \right|_{\partial X} = \sigma f. \tag{3.16}$$

Example 3.12. The Dirichlet problem in an unbounded domain. The passage to an unbounded X involves two new difficulties. One of them, which is purely technical, is that the boundedness conditions for (3.13) with respect to the a-metric are, in general, more complex than those for a bounded X. The other difficulty, which arises only for $\varepsilon = 0$, has a more fundamental nature. The point is that $\overset{\circ}{H}{}^r(X)$ is not necessarily complete in the metric (3.12) with $\varepsilon = 0$, and so it cannot be taken as d. Its completion is not always a space of functions. For instance, for $X = \mathbb{R}^1$ and $r = 1$, we obtain the quotient space of $\mathcal{L}_2^1 = \{u \in L_{2,\mathrm{loc}}(\mathbb{R}^1) : u' \in L_2(\mathbb{R}^1)\}$ relative to the subspace of constants. The form (3.13) is meaningless in this space. The difficulties do not arise, for example, if $2r < n$ and $\mathbb{R}^n \setminus X$ is bounded. In this case the completion of $\overset{\circ}{H}{}^r(X)$ in the $a_{r,0}$-metric is a function space (we designate it by $\overset{\circ}{\mathcal{H}}{}^r(X)$). Namely,

$$\overset{\circ}{\mathcal{H}}{}^r(X) = \left\{ u \in H^r_{\mathrm{loc}}(X) : u|_{\partial X} = 0, \int_X |x|^{-2r} |u(x)|^2 \, dx < \infty \right\}. \tag{3.17}$$

If $p \in L_{n/2r}(X)$, then the form (3.13) is bounded in $\overset{\circ}{\mathcal{H}}{}^r(X)$, and so $\{\overset{\circ}{\mathcal{H}}{}^r(X); a_{r,0}, b_p\}$ is a variational triple. The operator that corresponds to this triple is what we take as the operator of the Dirichlet problem for equation (3.14).

§4
Examples of Exact Computation of the Spectrum

There are two basic sources of differential operators that admit exact computation of the spectrum. One of them consists of invariant operators on Lie groups or homogeneous spaces, in which case the spectral characteristics can be computed by means of representation theory. Special functions and separation of variables constitute the other source (which is, in fact, closely related to the former one). Many examples of the latter kind can be found in standard textbooks on mathematical physics.

Examples of operators whose spectrum can be computed explicitly are important because of many reasons. In particular, they make it possible to observe relationships which are sometimes rather difficult to discover in the general situation. Quite often the analysis of such examples constitutes the first step in proving general assertions – see, for example, the estimates of the spectrum of regular elliptic differential operators in Sect. 5.2 and the computation of the asymptotics of the spectrum by the variational method in §11. Moreover, operators with known spectrum serve as the starting point when the methods of perturbation theory are used – see, for example, §8.

Here we shall present several examples to be used later on. The examples are completely elementary and their analysis requires no direct reference to representation theory. More involved examples, in which the connections with representations of groups are essential, are presented in §16.

4.1. Operators with Constant Coefficients on $\mathbb{R}^n$ and on a Torus

Example 4.1. Scalar differential operators on $\mathbb{R}^n$. Let $\mathcal{L} = \mathcal{L}(D)$ be a self-adjoint differential operator with constant coefficients in $L_2(\mathbb{R}^n)$ considered on the the domain (2.17). The Fourier transform turns $\mathcal{L}$ into the operator of multiplication by the real symbol $\mathcal{L}(\xi)$, which realizes the spectral theorem (Theorem 1.4), and, consequently, leads to the complete spectral analysis of the operator.

The spectrum $\sigma(\mathcal{L})$ is purely continuous (if $\operatorname{ord}\mathcal{L} > 0$) and coincides with the closure of the set of values of $\mathcal{L}(\xi)$, where $\xi \in \mathbb{R}^n$. For any Borel set $\delta \subset \mathbb{R}$ the spectral projection $E^{\mathcal{L}}(\delta)$ is given by the formula $E^{\mathcal{L}}(\delta) = \Phi^{-1} X_{\mathcal{L}^{-1}(\delta)} \Phi$, where Φ is the Fourier transform and X_e is the multiplication operator by the indicator function of $e \subset \mathbb{R}^n$.

In particular, the spectrum of $(-\Delta)^r \restriction H_r^2(\mathbb{R}^n)$ coincides with the half-axis $[0, \infty)$ for any $r > 0$.

The Fourier transform makes it also possible to carry out the spectral analysis of differential operators with constant coefficients defined on vector-valued functions. In particular, for the unperturbed Dirac operator (that is, for the operator (2.22) with $V(x) \equiv 0$) the spectrum coincides with the set $(-\infty, -1] \cup [1, +\infty)$.

Let us present one more proof of the assertion in Example 4.1. We use the fact that $\lambda \in \mathbb{R}$ belongs to the spectrum of a self-adjoint operator A if and only if there exists a sequence $\{u_k\}_1^\infty$ such that $u_k \in D(A)$, $\|u_k\| = 1$, and $\|Au_k - \lambda u_k\| \to 0$.

Fix a function $\varphi \in C_0^\infty(\mathbb{R}^n)$ such that $\|\varphi\|_{L_2} = 1$. Denote by $s(\mathcal{L})$ the set of values of $\mathcal{L}(\xi)$ and assume that $\lambda \in s(\mathcal{L})$, that is, $\lambda = \mathcal{L}(\xi)$ for some $\xi \in \mathbb{R}^n$. Consider the functions $u_k(x) = k^{-n/2}\exp(i\xi \cdot x)\varphi(k^{-1}x)$, where $k = 1, 2, \dots$. Then $\|u_k\| = 1$ and, as can be verified by direct computation, $\|\mathcal{L}u_k - \lambda u_k\| \to 0$. It follows that $\lambda \in \sigma(\mathcal{L})$, that is, $s(\mathcal{L}) \subset \sigma(\mathcal{L})$, which implies that $\overline{s(\mathcal{L})} \subset \sigma(\mathcal{L})$.

The set $s(\mathcal{L})$ is either equal to the whole axis $\mathbb{R}$ (in which case $\sigma(\mathcal{L}) = \mathbb{R}$) or it is a half-axis. For definiteness, let $\overline{s(\mathcal{L})} = [\lambda_0, +\infty]$. Changing to the Fourier transforms, we find that

$$(\mathcal{L}u, u) = \int \mathcal{L}(\xi)\, |\tilde{u}(\xi)|^2\, d\xi \geq \lambda_0 \|u\|^2$$

for $u \in C_0^\infty(\mathbb{R}^n)$. As a consequence, there are no points of the spectrum to the left of λ_0.

Since this proof makes little use of the Fourier transform, it can be easily adapted to analyse the following example.

Example 4.2. The operator $\mathcal{L}_D(X)$ (see Example 3.5) in an unbounded domain. We assume that the domain $X \subset \mathbb{R}^n$ contains balls of arbitrarily large radius. Then the spectrum of $\mathcal{L}_D(x)$ coincides with the closure of the range of $\mathcal{L}(\xi)$.

Indeed, to prove that $\overline{s(\mathcal{L})} \subset \sigma(\mathcal{L})$, it suffices to consider suitable shifts of the functions u_k used above. The proof of the inclusion in the opposite direction remains unchanged.

Example 4.3. Scalar differential operator on the torus $\mathbb{T}^n$. In what follows by the *torus* $\mathbb{T}^n$ we shall understand the manifold $\mathbb{R}^n/(2\pi\mathbb{Z})^n$ with the *flat metric*, that is, the Riemannian metric induced from $\mathbb{R}^n$. A differential operator $\mathcal{L} = \mathcal{L}(D)$ with constant coefficients is self-adjoint in $L_2(\mathbb{T}^n)$ if and only if $\mathcal{L}(\xi)$ is a real symbol.

The functions $\exp ij \cdot x$, where $j \in \mathbb{Z}^n$, are the eigenfunctions of $\mathcal{L}$. Since the system $\{\exp ij \cdot x\}$ is complete in $L_2(\mathbb{T}^n)$, it follows that $\sigma(\mathcal{L})$ is a pure point spectrum. The eigenvalues are equal to $\mathcal{L}(j)$, where $j \in \mathbb{Z}^n$. If $|\mathcal{L}(\xi)| \to \infty$ as $|\xi| \to \infty$, then $\mathcal{L}$ is an operator with discrete spectrum.

Example 4.4. Differential operator in the space of vector-valued functions on a torus. Let $\mathcal{L} = \mathcal{L}(D)$ be a self-adjoint operator on vector-valued functions of dimension k. Its symbol $\mathcal{L}(\xi)$, where $\xi \in \mathbb{R}^n$, is a Hermitian $k \times k$-matrix. Let $\{\lambda_l(\xi)\}$ and $\{f_l(\xi)\}$, $l = 1, 2, \dots, k$, be the eigenvalues (with multiplicity taken into account) and eigenvectors of $\mathcal{L}$, the latter being orthonormalized in $\mathbb{C}^k$. The vector-valued functions $\{f_l(j)\exp ij\cdot x\}$, where $j \in \mathbb{Z}^n$ and $l = 1, 2, \ldots, k$, form a complete system of eigenfunctions of $\mathcal{L}$ in $(L_2(\mathbb{T}^n))^k$. It follows that $\mathcal{L}$ has pure point spectrum; $\sigma_p(\mathcal{L})$ consists of all the eigenvalues $\lambda_l(j)$. If, for $|\xi|$ large enough, $\mathcal{L}(\xi)$ is an invertible matrix and $\|\mathcal{L}^{-1}(\xi)\| \to 0$ as $|\xi| \to \infty$, then $\sigma(\mathcal{L}) = \sigma_p(\mathcal{L})$.

4.2. The Factorization Method

Let W be a differentiable function on $\mathbb{R}^1$. We consider two differential expressions

$$l_\pm u = \pm\frac{du}{dx} + W(x)u. \tag{4.1}$$

Then

$$l_+l_-u = -\frac{d^2u}{dx^2} + (W^2 + W')u, \quad l_-l_+u = -\frac{d^2u}{dx^2} + (W^2 - W')u. \tag{4.2}$$

If $l_+l_-u = \lambda u$ and $v = l_-u \not\equiv 0$, then $l_-l_+v = l_-l_+l_-u = \lambda v$. Hence it is seen that the 'formal' eigenvalues $\lambda \neq 0$ (for the time being we do not assume that $u, v \in L_2$) are the same for the two operators (4.2). Sometimes this simple argument enables one to find the point spectrum of a differential operator.

Example 4.5. Harmonic oscillator. The operator

$$(\mathcal{L}u)(x) = -u'' + x^2u, \quad D(\mathcal{L}) = \{u \in H^2(\mathbb{R}^1) : x^2u \in L_2\} \tag{4.3}$$

is self-adjoint in $L_2(\mathbb{R}^1)$. If we set $W(x) = x$ in (4.1), then

$$\mathcal{L} = l_+l_- - I = l_-l_+ + I.$$

The equation $l_+u = 0$ has a solution $u_0(x) = \exp(-x^2/2) \in L_2$. We set $u_k = l_-^k u_0$ for $k \in \mathbb{N}$. Then

$$l_-l_+u_1 = l_-l_+l_-u_0 = l_-(2I + l_-l_+)u_0 = 2u_1.$$

Repeating this argument, we can find by induction that $l_-l_+u_k = 2ku_k$. To within normalization, u_k are the Hermite functions $H_k(x)$. Since the Hermite system is complete in $L_2(\mathbb{R}^1)$, the numbers $2k$ exhaust the spectrum of l_-l_+. It follows that the spectrum of the operator (4.3) consists of simple eigenvalues $\lambda_k = 2k - 1$, where $k \in \mathbb{N}$.

The example below can be reduced to the last one.

Example 4.5'. Many-dimensional harmonic oscillator. Let B be a positive definite $(n \times n)$-matrix. The operator

$$\mathcal{L}u = -\Delta u + (Bx, x)u, \quad D(\mathcal{L}) = \{u \in H^2(\mathbb{R}^n) : |x|^2u \in L_2\}$$

is self-adjoint in $L_2(\mathbb{R}^n)$. By an orthogonal coordinate transformation in $\mathbb{R}^n$ diagonalizing B, one can reduce the operator to the form

$$\sum_{1\le j\le n}\left(-\frac{\partial^2}{\partial y_j^2}+\mu_j^2 y_j^2\right),$$

where μ_j^2 are the eigenvalues of B. It follows that the spectrum of $\mathcal{L}$ consists of the eigenvalues

$$\lambda_k=\sum_{1\le j\le n}\mu_j(2k_j-1),\quad k=(k_1,\dots,k_n)\in\mathbb{N}^n.$$

In this notation the multiplicity of the eigenvalues is automatically taken into account. The corresponding eigenfunctions are

$$\prod_{1\le j\le n}H_{k_j}(\mu_j y_j).$$

Example 4.6. The operator

$$\mathcal{L}_a u=-u''-\frac{a(a+1)}{\cosh^2 x}u,\quad a>0,\quad D(\mathcal{L}_a)=H^2(\mathbb{R}^1) \tag{4.4}$$

is self-adjoint (which follows, for example, from Theorem 2.4). The essential spectrum of this operator coincides with $[0,\infty)$ (see Theorem 6.2 below) and we are concerned with computing the negative eigenvalues.

We denote by $l_\pm^{(a)}$ the operators (4.1) for $W(x)=a\tanh x$. We have

$$l_-^{(a)}l_+^{(a)}=\mathcal{L}_a+a^2I,\quad l_+^{(a)}l_-^{(a)}=\mathcal{L}_{a-1}+a^2I.$$

Since $u_a(x)=(\cosh x)^{-a}\in L_2$ is a solution of the equation $l_+^{(a)}u=0$, it follows that $-a^2\in\sigma_p(\mathcal{L}_a)$. If $a>1$, then $-(a-1)^2\in\sigma_p(\mathcal{L}_{a-1})$ or, equivalently, $2a-1\in\sigma_p(l_+^{(a)}l_-^{(a)})$ with eigenfunction u_{a-1}. Since $(l_-^{(a)}u_{a-1})(x)=(2a-1)(\cosh x)^{-a}\sinh x\in L_2$, it follows that $2a-1\in\sigma_p(l_-^{(a)}l_+^{(a)})$ and $-(a-1)^2\in\sigma_p(\mathcal{L}_a)$. Next, by induction, we find that if $n-1<a\le n$, where $n\in\mathbb{N}$, then $-(a-k)^2\in\sigma_p(\mathcal{L}_a)$ for $k=0,\dots,n$. It can be demonstrated that these numbers exhaust the point spectrum of $\mathcal{L}_a$.

4.3. Operators on a Sphere and a Hemisphere

Example 4.7. The Laplace-Beltrami operator on the sphere S^n. The spherical functions form a complete system of eigenfunctions of this operator. The spectrum consists of the eigenvalues $\lambda_k=k(k+n-1)$, where $k\ge 0$, each of which has multiplicity

$$\binom{n+k-1}{n-1} + \binom{n+k-2}{n-1}.$$

Here $\binom{p}{q}$ designates the number of q-element combinations out of p elements. If $p < q$, then, by definition, $\binom{p}{q} = 0$.

Example 4.8. The Dirichlet and Neumann problems for the Laplace-Beltrami operator on the hemisphere $S_+^n = \{x \in \mathbb{R}^{n+1} : |x| = 1,\ x_{n+1} > 0\}$. The odd and even spherical functions with respect to x_{n+1} are the eigenfunctions of $-\Delta_N$ and $-\Delta_D$, respectively. As in Example 4.7, the spectrum of either of these operators consists of the eigenvalues $\lambda_k = k(k+n-1)$ with $k \geq 0$ for $-\Delta_N$ and $k > 0$ for $-\Delta_D$. The multiplicity of λ_k is equal to $\binom{n+k-1}{n-1}$ for $-\Delta_N$ and $\binom{n+k-2}{n-1}$ for $-\Delta_D$.

In Sect. 16.5 we shall considerably generalize this example.

§5
Differential Operators with Discrete Spectrum. Estimates of Eigenvalues

The following assertion, which belongs to the general theory of operators, can, as a rule, serve as the basis for establishing that the spectrum is discrete.

Lemma 5.1. *Let A be a self-adjoint operator in a Hilbert space $\mathfrak{H}$, let $D(A)$ be its domain with the A-metric (1.1), and (in the positive definite case) let $d[a] = D(A^{1/2})$ be the domain of its quadratic form with the a-metric $a[x] = \|A^{1/2}x\|^2$. In order that the spectrum of A be discrete it is necessary and sufficient that either of the embedding operators*

$$I_A : D(A) \to \mathfrak{H}, \quad I_a : d[a] \to \mathfrak{H} \tag{5.1}$$

be compact.

Once it is established that the spectrum is discrete, there arise quantitative questions concerned with the estimates of the eigenvalues and eigenfunctions, their asymptotic behaviour, and the like. §§9–15 are devoted to the asymptotics of the spectrum. Here we shall touch upon the simpler, but, nevertheless, important question concerning the estimates.

5.1. Basic Examples of Differential Operators with Discrete Spectrum

Example 5.1. Regular elliptic operator A (see Theorems 2.1 and 2.2). If $\operatorname{ord} A = m$, then $D(A) \subset H^m(X)$ and the A-metric (1.1) generates the $H^m(X)$-topology on $D(A)$. Since the embedding $H^m \to L_2$ is compact in the case of smooth compact manifolds (with or without boundary), it follows that the spectrum of A is discrete.

Example 5.2. The operator $(-\Delta)^r_D$ *in a domain* $X \subset \mathbb{R}^n$ (see Example 3.3). Lemma 5.1 reduces the question of whether the spectrum of the operator is discrete or not directly to the question concerning the compactness of the embedding $I_0 : \overset{\circ}{H}{}^r(X) \to L_2(X)$. For example, the operator I_0 is compact for any domain X of finite measure, including any bounded domain. A simple necessary condition for the embedding I_0 to be compact, which also admits some domains X of infinite measure, is that only finitely many disjoint cubes with edges of length d can be fitted into X for any $d > 0$. For $2r > n$ this condition is not only necessary, but also sufficient. There is also a similar criterion for $2r \leq n$, but in this case one has to consider cubes that may contain a sufficiently small (in the sense of capacity) portion of the complement of X. For a rigorous formulation, see (Maz'ya 1985).

Example 5.3. The operator $(-\Delta)^r_N$ (see Example 3.3). Here it proves more convenient to consider the form $l[u] + \|u\|^2_{L_2}$ instead of the form $l[u]$ given by (3.3), which does not affect the character of the spectrum. By Lemma 5.1, an equivalent condition for the spectrum to be discrete is that the embedding operator $I : H^r(X) \to L_2(X)$ be compact. The latter requires much stronger conditions for X compared with Example 5.2. For instance, it suffices that X be bounded and its boundary be Lipschitzian. For the precise conditions see (Maz'ya 1985).

Everything that has been said in Examples 5.2 and 5.3 can be automatically carried over to the case of operators with variable coefficients. It is only required that the corresponding quadratic form $a[u]$ (or at least $a[u] + c\|u\|^2_{L_2}$ for c large enough) should define the metric of H^r on $d[a] \subset H^r(X)$. In particular, under the assumptions of Example 3.1, the spectrum of $\mathcal{L}_D$ is discrete if $X \subset \mathbb{R}^n$ is an arbitrary domain of finite measure, while the spectrum of $\mathcal{L}_N$ is discrete if X is a bounded domain with Lipschitzian boundary.

Example 5.4. Degenerate second-order elliptic operators (see Example 3.4). Here the problem can be solved by means the embedding theorems for weighted Sobolev classes. The spectrum is discrete if and only if $\alpha < 2$.

Example 5.5. The Schrödinger operator with increasing potential. Under the assumptions of Example 3.6, let $V(x) \to +\infty$ ($|x| \to \infty$). Without loss of generality, we can assume that $V(x) \geq 1$. Then the form $a[u]$ defined by (3.6) is positive definite in $L_2(\mathbb{R}^n)$. We fix $\varepsilon > 0$ and find $r > 0$ such that $V(x) \geq \varepsilon^{-1}$

outside the ball $B_r = \{x \in \mathbb{R}^n : |x| < r\}$. Since $\{u \in d[a] : a[u] \leq 1\}$ is a bounded set in H^1, it is compact in $L_2(B_r)$. Moreover,

$$\int\limits_{|x|>r} |u|^2\,dx \leq \varepsilon \int V(x)|u|^2\,dx \leq \varepsilon a[u] \leq \varepsilon \quad (a[u] \leq 1).$$

It follows immediately that the embedding operator I_a in (5.1) is compact, which implies that the spectrum of the Schrödinger operator defined by (3.6) in $L_2(\mathbb{R}^n)$ is discrete if $V(x) \to +\infty$ $(|x| \to \infty)$. The same is true for the generalized Schrödinger operator from Example 3.7.

The above discussion rests on the following two facts: the L_2-norm of u outside a sufficiently large ball is small compared with $a[u]$, and the embedding of $d[a]$ in L_2 is compact in a bounded domain. The former fact is true, since the potential tends to $+\infty$. The latter is true, since the potential is locally semi-bounded. Both conditions can, in fact, be relaxed. In particular (see Birman 1959, 1961), in order that the spectrum be discrete for $n = 1$ it suffices that, for any $h > 0$,

$$\varphi_h(x) = \int\limits_x^{x+h} V_+(t)\,dt$$

converge to infinity as $|x| \to \infty$ and $V_-(x)$ be small compared with $\varphi_h(x)$ (for example, $V_- \in L_1 + L_\infty$). If $\inf V(x) > -\infty$, then the condition is also necessary.

Similar criteria for the spectrum to be discrete are also valid for the operator (3.8), but only for $2r > n$. If $2r \leq n$, then, even for a semi-bounded potential, a criterion for the spectrum to be discrete can be stated only in terms of capacity (see Maz'ya 1985).

5.2. Estimates of Eigenvalues

The sharp-order estimates

$$N_\pm(\lambda; A) \leq c(A)\lambda^{n/m}, \quad \lambda \geq 1, \quad n = \dim X, \quad m = \operatorname{ord} A \tag{5.2}$$

are valid for regular elliptic operators. Here $N_\pm(\lambda; A)$ are the distribution functions (1.16) for the positive and negative spectrum of A. For the eigenvalues $\pm\lambda_j^\pm(A)$ themselves, an estimate equivalent to (5.2) has the form

$$\lambda_j^\pm(A) \geq c'(A)j^{m/n}.$$

The estimate (5.2) is most easily verified if operators with constant coefficients on the torus $\mathbb{T}^n$ are considered as an example (see Examples 4.3 and 4.4).

Thus, let $\mathcal{L}$ be an elliptic operator with constant coefficients and with positive principal symbol $\mathcal{L}^0(\xi)$. The eigenvalues of the operator coincide with $\mathcal{L}(j)$, where $j \in \mathbb{Z}^n$. By the uniform ellipticity condition (2.7), $\mathcal{L}^0(j) \geq \gamma |j|^m$. Thus, obviously, $\mathcal{L}(j) \geq \gamma_0 |j|^m - C$ for some $\gamma_0 \in (0, \gamma)$ and $C \geq 0$. It follows that $N_\pm(\lambda; \mathcal{L})$ are finite and bounded from above by the number of points $j \in \mathbb{Z}^n$ inside the ball $|j| < \left(\gamma_0^{-1}(\lambda + C)\right)^{1/m}$. For large λ the number of such points is of the same order as the volume of the ball and we obtain (5.2).

One of the methods of proving (5.2) in the general situation consists in reducing the problem to the case already analysed. The estimate (5.2) is also valid for a wide class of operators defined with the aid of a quadratic form (for instance, under the assumptions of Examples 3.1 and 3.3).

Estimates similar to (5.2) are also valid for the spectral function of a regular elliptic operator. Thus, if $\mathcal{L}$ is lower semi-bounded, then the function (1.14) satisfies the relations

$$e^A(\lambda; x, y) = O\left(\lambda^{n/m}\right) \text{ uniformly on } X \times X, \tag{5.3}$$

$$e^A(\lambda; x, y) = O\left(\lambda^{(n-1)/m}\right)$$
$$\text{uniformly on any compact set in } X \times X \setminus \text{diag}. \tag{5.4}$$

If A is an operator defined on vector-valued functions, then the same estimates are valid for the matrix elements of the kernel (1.15). In the non-semi-bounded case (5.3) and (5.4) are valid for $e^A_\pm(\lambda; x, y)$. As was demonstrated by Agmon (1965), one can deduce (5.3) using only the embedding theorems for Sobolev spaces. The estimate (5.3) implies (5.2), since, for example,

$$N(\lambda; A) = \operatorname{Tr} E^A((-\infty, \lambda)) = \int_X \operatorname{tr} e^A(\lambda; x, x)\, dx$$

for a semi-bounded operator in a domain $X \subset \mathbb{R}^n$, as can be seen from (1.15).

The subsequent development of estimates of the form (5.2) consists in refining the constant $C(A)$. Thus, the estimate

$$N\big(\lambda; (-\Delta)^r_D\big) \leq c_{r,n}\, \lambda^{n/2r} \operatorname{mes} X \quad \forall\, \lambda > 0, \tag{5.5}$$

in which the constant $c_{r,n}$ is independent of X, is satisfied for the operator $(-\Delta)^r_D$ in a bounded domain $X \subset \mathbb{R}^n$ (see Rozenblum 1972b). The estimate (5.5) can only partially be carried over to other boundary value problems. For instance, for the Neumann problem the constant in the estimate depends on X (it worsens as the properties of the boundary deteriorate). Moreover, the estimate itself is violated for small λ, because zero is an eigenvalue of $(-\Delta)^r_N$.

The independence of the estimates of the domain (for the Dirichlet problem) makes it possible to extend (5.5) to the case of unbounded domains. The estimate (5.5) is valid for any open set of finite measure. In particular, this includes the assertion that the spectrum of $(-\Delta)^r_D$ is discrete for such domains (see Example 5.2).

5.3. Estimates of the Spectrum of a Weighted Polyharmonic Operator

(On weighted polyharmonic operators see Sect. 3.4.) We are concerned with the spectra of the Dirichlet and Neumann problems for the equation

$$pu = \lambda\big((-\Delta)^r u + \varepsilon u\big), \quad \varepsilon \geq 0, \tag{5.6}$$

which corresponds to (3.14). We use the variational formulation of the problem, that is, by the spectrum of the Dirichlet or Neumann problem for (5.6) we understand the spectrum of the variational triple $\{d; a_{r,\varepsilon}, b_p\}$, where $a_{r,\varepsilon}$ and b_p are the forms (3.12) and (3.13), and $d = \overset{\circ}{H}{}^r(X)$ (for the Dirichlet problem) or $d = H^r(X)$ (for the Neumann problem). For these problems one can succeed in obtaining estimates of $n_\pm(\lambda)$ (see (1.20)) in terms of the integral norms of the 'weight' p. Such estimates have important applications in the study of spectral asymptotics by the variational method (see §11) as well as in the study of the spectrum of the Schrödinger operator.

Example 5.6. The Dirichlet problem for equation (5.6). Under the assumptions adopted in Examples 3.9 and 3.12 (that is, for bounded $X \subset \mathbb{R}^n$ under the condition (3.15), or unbounded $X \subset \mathbb{R}^n$ with $n > 2r$ for $p \in L_{n/2r}(X)$), the operator T that corresponds to $\{\overset{\circ}{H}{}^r(X); a_{r,\varepsilon}, b_p\}$ is compact.

The estimates

$$n_\pm(\lambda; T) \leq c\|p_\pm\|^{n/2r}_{L_\gamma(X)} \lambda^{-n/2r} \quad \forall\, \lambda > 0 \tag{5.7}$$

are true under the assumptions of Example 5.6. In (5.7) γ is defined by (3.10). The constant c does not depend on p or $\varepsilon \geq 0$. If $2r < n$, then it is also independent of $X \subset \mathbb{R}^n$. If only one of the functions $p_\pm$ belongs to $L_\gamma(X)$, while the other one belongs merely to $L_{1,\mathrm{loc}}(X)$, then (5.7) with the corresponding sign is satisfied. The estimate (5.7) implies (5.5) (for $p \equiv 1$ and $2r < n$).

The estimate (5.7) takes a particularly simple form if $2r < n$. Let us write it down for the most important case when $r = 1$:

$$n_\pm(\lambda; T) = c_n \lambda^{-n/2} \int_X p_\pm^{n/2}\, dx \quad \forall\, \lambda > 0,\ n \geq 3. \tag{5.8}$$

It turns out that the following asymptotics is valid subject the conditions under consideration (see §9):

$$n_\pm(\lambda; T) \sim v_n (2\pi)^{-n} \lambda^{-n/2} \int_X p_\pm^{n/2}\, dx, \quad \lambda \to +0 \tag{5.9}$$

(v_n is the volume of the unit ball in $\mathbb{R}^n$). It follows that the estimate (5.8) is sharp in the sense that its right-hand side contains the same functional of p as the asymptotic formula. There are no similar estimates for $n \le 2$ (or $n \le 2r$ for any $r > 1$).

We can see that, under the assumptions of Example 5.6, the conditions for p ensuring that T is a compact operator are too stringent. Indeed, these conditions imply the very strong estimate (5.8) (and the asymptotics (5.9) for $2r < n$). The conditions for p can be relaxed, but this involves 'localization' of singularities. In the case when the boundary ∂X is smooth, T is a compact operator for $|p(x)| \le c[\operatorname{dist}(x, \partial X)]^{-\beta}$, where $\beta < 2r$. On the estimates in this case see (Birman and Solomyak 1974; Rozenblum 1975).

Example 5.7. The Neumann problem for equation (5.6). Under the assumptions of Example 3.10, the operator T determined by the variational triple $\{H^r(X); a_{r,\varepsilon}, b_p\}$ is compact and the estimate (5.7) remains valid for T. However, in this case the constant c_n in (5.8) depends not only on X, but also on $\varepsilon > 0$. As in Example 5.6, T can also be compact under weaker conditions for p.

From the estimate (5.8) (for the Dirichlet problem) one can easily deduce an important estimate for the spectrum of the Schrödinger operator.

Theorem 5.1. *If $n \ge 3$, then the estimate*

$$N(\lambda; \mathcal{L}) \le c_n \int\limits_{\mathbb{R}^n} (\lambda - V(x))_+^{n/2}\, dx = c_n \Phi(V, \lambda) \tag{5.10}$$

is satisfied for the operator (2.18) whenever the integral in (5.10) is finite, and, in particular, also when only the part of the spectrum contained in $(-\infty, \lambda)$ is discrete.

To prove the theorem, we consider the quadratic form

$$b_\lambda[u] = \int (V(x) - \lambda)|u|^2\, dx$$

in $\mathring{\mathcal{H}}^1(\mathbb{R}^n)$ (see (3.17)) for a fixed $\lambda \in \mathbb{R}$. In general, the form is unbounded. But if $(\lambda - V)_+ \in L_{n/2}$, then it is lower semi-bounded and closed on the domain

$$\left\{u \in \mathring{\mathcal{H}}^1(\mathbb{R}^n) : \int |V(x) - \lambda||u|^2\, dx < \infty\right\}.$$

Let $T_\lambda = \operatorname{Op}(b_\lambda)$. According to (1.13),

$$N(-1; T_\lambda) = \max\dim\left\{F \subset C_0^\infty(\mathbb{R}^n) : \int (V - \lambda)|u|^2\, dx < \int |\nabla u|^2\, dx,\ 0 \ne u \in F\right\}.$$

Applying the same formula (1.13) to the operator (2.18) in $L_2(\mathbb{R}^n)$, we find that

$$N(\lambda; \mathcal{L}) = \max\dim \left\{F \subset C_0^\infty(\mathbb{R}^n) : \int \left(|\nabla u|^2 + V|u|^2\right) dx < \lambda \int |u|^2\, dx,\ 0 \neq u \in F\right\}.$$

It follows that

$$N(\lambda; \mathcal{L}) = N(-1; T_\lambda) = n_-(1; T_\lambda).$$

It remains to apply (5.8).

From (5.10) one obtains the following important estimate of the number of negative eigenvalues of the Schrödinger operator for $\lambda = 0$:

$$N(0; \mathcal{L}) \leq c_n \int V_-^{n/2}\, dx.$$

The estimates (5.8) and (5.10), the latter being a consequence of (5.8), were obtained by Rozenblum (1972a) and rediscovered later by Lieb and Cwikel (see Reed and Simon 1978, Vol. 4).

The asymptotic formula $N(\lambda; \mathcal{L}) \sim (2\pi)^{-n} v_n \Phi(V, \lambda)$ for $\lambda \to \infty$ holds true under some additional conditions for V. Nevertheless, one can easily construct potentials such that the estimate (5.10) fails to be sharp, that is, $N(\lambda; \mathcal{L}) < \infty$ and $\Phi(V, \lambda) = \infty$ for all $\lambda > 0$.

The analogous estimates for $n = 1, 2$ are more involved. We shall present one of the simplest examples for $n = 1$ (e.g., see Reed and Simon 1978, Vol. 4; Berezin and Shubin 1983).

Theorem 5.2. *The estimate*

$$N(\lambda; \mathcal{L}) \leq \int_{\mathbb{R}^1} |x| \big(\lambda - V(x)\big)_+\, dx + 1 \tag{5.11}$$

is satisfied for the Schrödinger operator in $L_2(\mathbb{R}^1)$.

5.4. Estimates of the Spectrum: Heuristic Approach

Estimates of the type (5.8) and (5.10) fit into the following general scheme, which is important because of its heuristic power (cf. the discussion about the spectral asymptotic formulae in §9). Let A be a semi-bounded self-adjoint differential operator in a domain $X \subset \mathbb{R}^n$ and let $A(x, \xi)$ be its symbol. We consider the sets

$$\mathcal{E}(\lambda; A) = \big\{(x, \xi) \in X \times \mathbb{R}^n : A(x, \xi) < \lambda\big\}.$$

Then $N(\lambda; A)$ is bounded from above by $c_1 \mathrm{mes}_{2n} \mathcal{E}(c_2\lambda; A)$. In many cases such estimates are two-sided. They can also be easily modified to cover the case of equations of the type (5.6).

These estimates can be violated if $A(x,\xi)$ or X are not regular enough. In such cases $N(\lambda; A)$ can often be estimated by the number of disjoint unit cubes that can be fitted into $\mathcal{E}(\lambda; A)$. Recently Fefferman (1983) proposed an even more general concept, according to which $N(\lambda; A)$ is comparable with the number of disjoint 'cells' of the form $Q_\delta(x_0) \times Q_{\delta^{-1}}(\xi_0)$ that can be fitted into $\mathcal{E}(\lambda; A)$. Here we denote by $Q_h(z)$ the n-dimensional cube with centre z and side h.

The domain of applicability of this concept has not been determined so far. In (Fefferman and Phong 1980) it was established for the Schrödinger operator with a certain class of potentials.

As an illustration, we shall present the following (earlier) result of Rozenblum (1972b), which, obviously, lies within the framework of Fefferman's concept.

Theorem 5.3 (see Example 5.2). *Let X be a domain in $\mathbb{R}^n$ and suppose that only a finite number $Z(d)$ of pairwise disjoint cubes with edges of length d can be fitted into X for any $d > 0$. Then the estimates*

$$Z\left(c_1\lambda^{-1/2r}\right) \leq N(\lambda) \leq Z\left(c_2\lambda^{-1/2r}\right)$$

are satisfied for $(-\Delta)^r_D$ in X if $2r > n$.

5.5. Estimates of Eigenfunctions

For a fixed x, the spectral function $e^A(\xi; x, x)$ of the Schrödinger operator admits the same estimate (5.3) as for regular boundary value problems. This estimate does not, however, characterize the behaviour of the eigenfunctions $\varphi(x)$ as $x \to \infty$. It turns out that in the case of an increasing potential the eigenfunctions decrease very rapidly. In particular, the following result is true.

Theorem 5.4 (see Reed and Simon 1978, Vol. 4). *Let $V \in L_{\infty,\text{loc}}(\mathbb{R}^n)$ and let $V(x) \to \infty$ as $|x| \to \infty$. Then the estimate*

$$\varphi(x) = O\big(\exp(-a|x|)\big) \quad \forall\, a > 0$$

is satisfied for any eigenfunction $\varphi(x)$. Furthermore, if $V(x) \geq c|x|^{2k}$ for sufficiently large $|x|$, then

$$|\varphi(x)| = O\left(\exp\left(-\frac{(c-\varepsilon)^{1/2}}{n+1}\,|x|^{k+1}\right)\right) \quad \forall\, \varepsilon > 0.$$

§6
Differential Operators with Non-Empty Essential Spectrum

The Schrödinger operator (2.18) with a potential that does not tend to infinity is an important example of a differential operator with non-empty essential spectrum. In this connection there arise many new questions concerning the location of the essential spectrum and its spectral characteristics (in the first place, one is interested in conditions for the absolute continuity of the spectrum; see Sect. 1.6), the location and properties of the discrete part of the spectrum, eigenvalues within the continuous spectrum, and so on. Among various interesting classes of potentials, those that tend to zero as $|x| \to \infty$ as well as periodic ones have to be mentioned first. These classes are most interesting from the viewpoint of applications, and, at the same time, relatively complete results are available for them. Nowadays the theory of the Schrödinger operator with an almost periodic potential is also developing rapidly, however, it is still far from being complete. The latter two types of potentials are presented in §17 and §18. Here we shall restrict ourselves to the case when $V(x) \to 0$, omitting questions that can be regarded more as parts of scattering theory, including the problem of absolute continuity conditions for the essential spectrum. We also postpone to §7 the presentation of certain facts concerned with the spectral theory of the multiparticle Schrödinger operator.

6.1. Stability of the Essential Spectrum under Compact Perturbations of the Resolvent

The location of the essential spectrum can often be determined on the basis of the following result from abstract perturbation theory (see Birman and Solomyak 1980).

Theorem 6.1. *Let A and A_0 be self-adjoint operators in a Hilbert space $\mathfrak{H}$ and let $(A - \lambda I)^{-1} - (A_0 - \lambda I)^{-1}$ be a compact operator for some (and then for all) $\lambda \in \rho(A) \cap \rho(A_0)$. Then*

$$\sigma_{\text{ess}}(A) = \sigma_{\text{ess}}(A_0). \tag{6.1}$$

If A and A_0 are introduced by means of quadratic forms, then it is more convenient to use the following corollary of Theorem 6.1, rather than the theorem itself.

Corollary. *Let a_0 be a closed positive definite form in $\mathfrak{H}$. Let a form a be defined on $d = d[a_0]$ so that the operator generated by the variational triple*

$\{d; a_0, a - a_0\}$ is compact. Then a is semi-bounded and closed on d, and (6.1) is satisfied for $A = \mathrm{Op}\,(a)$ and $A_0 = \mathrm{Op}\,(a_0)$.

If a_0 is merely semi-bounded, then it must be replaced by $a_0[u] + c\|u\|^2$ with c large enough when the corollary is applied. It is important to note that the equality $D(A) = D(A_0)$ does not have to be valid under the assumptions of the theorem or the corollary.

6.2. Essential Spectrum of the Schrödinger Operator with Decreasing Potential

Let $Au = -\Delta u + Vu$ be a self-adjoint Schrödinger operator in $L_2(\mathbb{R}^n)$ defined either as the closure of an operator on C_0^∞ (as in Theorem 2.6) or with the aid of the method of forms (as in Example 3.6), depending on the conditions for V. If

$$V(x) \to 0 \quad (|x| \to \infty), \tag{6.2}$$

then, under certain additional local conditions for V, it turns out that

$$\sigma_{\mathrm{ess}}(A) = [0, \infty). \tag{6.3}$$

One of the simplest exact results along these lines is the following.

Theorem 6.2. *Let $V \in L_{\infty,\mathrm{loc}}$ and let (6.2) be satisfied, that is,*

$$\operatorname*{ess\,sup}_{|x| \geq R} |V(x)| \to 0 \quad (R \to \infty).$$

Then (6.3) is satisfied.

This result can be easily deduced from Theorem 6.1 (or the corollary of this theorem). It also remains valid if local singularities of the potential are admitted, for example, if $V_+ \in L_{\infty,\mathrm{loc}}$ and $V_- \in L_{\gamma,\mathrm{loc}}$, where γ is the exponent defined by (3.10) (for $r = 1$), and (6.2) is satisfied. Alternative conditions for V that imply (6.3) can be found in (Reed and Simon 1978, Vol. 4).

If, in addition, $V(x) \geq 0$ under the assumptions of Theorem 6.2, then the form (3.6) is non-negative, and so is the spectrum. It follows that the spectrum of the Schrödinger operator with a non-negative potential that satisfies the assumptions of Theorem 6.2 coincides with the half-axis $[0, \infty)$.

6.3. Negative Spectrum of the Schrödinger Operator

If $V(x) < 0$ on a set of non-zero measure, then the operator can develop a non-empty negative spectrum, which, by Theorem 6.2, must be discrete. Depending on the properties of the potential, the negative spectrum may turn out to be finite or infinite. We shall present relevant examples.

Example 6.1. The hydrogen atom. We consider the operator

$$\mathcal{L}u = -\Delta u - |x|^{-1}u$$

in $L_2(\mathbb{R}^3)$. Here $V = -V_- \in L_{\gamma,\mathrm{loc}}$ for $\gamma < 3$, which implies that $\sigma_{\mathrm{ess}}(\mathcal{L}) = [0, \infty)$. The negative spectrum is infinite and consists of eigenvalues $\lambda_k = -(4k^2)^{-1}$ of multiplicity k^2. The eigenfunctions can be expressed in terms of the Laguerre polynomials and spherical functions. For details, see, for example, (Courant and Hilbert 1953).

Example 6.2. We consider the operators $\mathcal{L}_D$ (with the condition $u(0) = 0$) and $\mathcal{L}_N$ (with $u'(0) = 0$) in $L_2(\mathbb{R}_+)$ defined by the differential expression

$$\mathcal{L}u = -u'' - V_h u, \quad V_h(x) = h^2 \ (x \le 1), \quad V_h(x) = 0 \ (x > 1).$$

The essential spectrum of either of these operators occupies the half-axis $[0, +\infty)$. The eigenvalues are easily computed: $\lambda_k(\mathcal{L}) = h^2 - p_k^2$, where p_k are the solutions of the equation $\tan p = -p(h^2 - p^2)^{-1/2}$ ($\mathcal{L} = \mathcal{L}_D$) or $\tan p = p^{-1}(h^2 - p^2)^{1/2}$ ($\mathcal{L} = \mathcal{L}_N$). Hence we can see that the negative spectrum is finite:

$$N(0; \mathcal{L}_D) = 0 \quad (h \le \pi/2),$$
$$N(0; \mathcal{L}_D) = l \quad ((l - 1/2)\pi < 2h \le (l + 1/2)\pi, \ l \ge 1),$$
$$N(0; \mathcal{L}_N) = l \quad ((l - 1)\pi < h \le l\pi, \ l \ge 1).$$

The study of the negative spectrum of the Schrödinger operator with a decreasing potential can be reduced to the investigation of the spectrum of an equation of the form (5.6) (for $r = 1$). The abstract scheme of such a reduction was developed by Birman (1959, 1961). We shall state only the simplest result from (Birman 1961).

Theorem 6.3. *Let A and A_0 be semi-bounded self-adjoint operators in a Hilbert space $\mathfrak{H}$ and let $A_0 > 0$. Let $a = \mathrm{QF}(A)$, $a_0 = \mathrm{QF}(A_0)$, and $d[a] = d[a_0] = d$. Let T_ε, where $\varepsilon > 0$, be the operator determined by the variational triple $\{d; a_0[u] + \varepsilon\|u\|^2, a - a_0\}$. Then*

$$N(-\varepsilon; A) = n_-(1; T_\varepsilon). \tag{6.4}$$

If A_0 is positive definite, then (6.4) is also valid for $\varepsilon = 0$.

The proof can be reduced to comparing the formulae below, which follow directly from (1.13) and (1.21) (cf. the proof of Theorem 5.1):

$$N(-\varepsilon; A) = \max\dim\left\{F \subset d : a[u] + \varepsilon\|u\|^2 < 0\right\},$$
$$n_-(1; T_\varepsilon) = \max\dim\left\{F \subset d : a[u] - a_0[u] < -a_0[u] - \varepsilon\|u\|^2\right\}.$$

The assertion concerning the case $\varepsilon = 0$ can be extended to any $A \ge 0$, but the formulation becomes more involved (see Birman 1961).

The equality (6.4) is referred to as the '*Birman-Schwinger principle*,' and so are some generalizations of (6.4). It proved to be an efficient tool in the study of various problems concerned with the spectrum. We shall illustrate the application of (6.4) using the operator (3.8) with $V < 0$ as an example. We shall assume that $V \in L_\infty + L_\gamma$ with γ defined by (3.10). Let $A_0 = (-\Delta)^r \restriction H^{2r}(\mathbb{R}^n)$ and let $A = \operatorname{Op}(a)$, where a is the form (3.7). Then $d[a] = d[a_0] = H^r$. The operator T_ε in (6.4) is determined by the variational triple $\{H^r(\mathbb{R}^n), a_{r,\varepsilon}, b_V\}$, where $a_{r,\varepsilon}$ is the form defined by (3.12) and b_V by (3.13). As a consequence, the study of the negative spectrum of the operator (3.8) has been reduced to the problem of the spectrum of equation (5.6). The case $r = 1$ is of course especially important.

In particular, we can see that an equivalent condition for the negative spectrum of (3.8) to be finite is that $n_-(1; T_\varepsilon)$ can be estimated uniformly with respect to $\varepsilon > 0$. Without assuming that $V(x) \to 0$ ($|x| \to \infty$), we find that the negative spectrum is discrete if and only if $n_-(1; T_\varepsilon) < \infty \ \forall\, \varepsilon > 0$.

One can estimate $n_-(1; T_\varepsilon)$, and consequently also $N(-\varepsilon; A)$, most easily if $2r < n$. In particular, by (5.8),

$$N(-\varepsilon; A) \leq c_n \int [V(x) + \varepsilon]_-^{n/2}\, dx, \quad \varepsilon > 0, \quad n \geq 3. \tag{6.5}$$

As has been mentioned in §5, this estimate is satisfied for any real-valued potential $V \in L_{1,\mathrm{loc}}(\mathbb{R}^n)$. It follows from (6.5) that if $V_- \in L_{n/2}(\mathbb{R}^n)$, where $n \geq 3$, then the negative spectrum of the Schrödinger operator with potential V is finite. For example, this will be the case if $V_- \in L_\infty$ and $V(x) \sim -c|x|^{-\alpha}$, $\alpha > 0$, $\forall\, c > 0$ as $|x| \to \infty$. If $V(x) \sim -c|x|^{-2}$ as $|x| \to \infty$, then one can claim that the negative spectrum is finite only if $4c < (n-2)^2$. The latter follows form the Hardy inequality (3.11).

If $\int V_-^{n/2}\, dx < c_n^{-1}$, where $n \geq 3$, then there is no negative spectrum by virtue of (6.5). This is not so for $n = 1, 2$, in which case the Schrödinger operator with potential V must have negative eigenvalues if $V \leq 0$ and $V \not\equiv 0$. This assertion is particularly easy to verify for $n = 1$: we fix $\varphi \in C_0^\infty$ such that $\varphi(x) = 1$ for $x \in (-1, 1)$, and set $\varphi_k(x) = \varphi(x/k)$. Then

$$\int \left(|\varphi_k'|^2 + V|\varphi_k|^2\right) dx \to \int V\, dx < 0$$

as $k \to \infty$, the negative spectrum being non-empty by (1.13).

Following the terminology used in physics, the operator $-d^2/dx^2$ in $L_2(\mathbb{R}^1)$ has a *resonance* at $\lambda = 0$. For an operator on a half-axis the presence or absence of a resonance at zero is determined by the boundary condition. There is a resonance for the operator of the Neumann problem (the proof is the same as in the case of an axis). This is not so for the operator of the Dirichlet problem as a consequence of another version of Hardy's inequality, which reads

$$4 \int x^{-2}|u|^2\, dx < \int |u'|^2\, dx \quad \forall\, u \in \mathring{H}^1(\mathbb{R}_+).$$

It follows that there are certainly no negative eigenvalues if $V_-(x) < 4x^{-2}$.

Example 6.2 is a good illustration of the above discussion.

For $2r > n$ there are no estimates of the negative spectrum similar to (6.5). The estimates of the most accurate order have been obtained in (Egorov and Kondrat'ev 1987). In the case of the one-dimensional Schrödinger operator the following simple estimates are valid for the operator on an axis or half-axis with the Neumann condition (see Birman 1961):

$$N(-\varepsilon; A) \leq \int |x| \big(V(x) + \varepsilon\big)_- \, dx + 1$$

(cf. (5.11)) and

$$N(-\varepsilon; A) \leq (2\varepsilon)^{-1} \int \big(1 - e^{-2\varepsilon|x|}\big) V_-(x) \, dx + 1.$$

For the operator of the Dirichlet problem on a half-axis the term 1 should be discarded.

From Theorem 6.3 and its analogues one can easily obtain convenient criteria for the negative spectrum of the (generalized) Schrödinger operator to be finite and discrete for all the potentials $\alpha V(x)$ with $\alpha > 0$ at once. Thus, in order that the negative spectrum of each of the operators $A_\alpha = -\Delta + \alpha V(x)$ be discrete it is necessary and sufficient that the negative part $(T_1)_-$ of T_1 be compact. A number of conditions for such operators to be compact were obtained in (Birman 1961). For example, it suffices that the integral

$$\int\limits_{|y-x|<a} \big(1 + |y-x|^{1-n}\big) \, V_-(y) \, dy$$

converge to zero as $|x| \to \infty$ for some a.

The compactness of T_ε follows from the compactness of the embedding of $D(A)$ or $d[a]$ into the weighted space $L_{2,|V|}$. For $V \leq 0$ these compactness conditions are equivalent. Necessary and sufficient conditions for the imbedding to be compact expressed in terms of capacity have been obtained in (Maz'ya 1985). Moreover, if the sign of V is not constant, then V_- can be partially compensated by V_+. Strictly speaking, an equivalent condition for T_ε to be compact is that the embedding of $H^r \cap L_{2,V_+}$ into L_{2,V_-} be compact. In (Maz'ya 1985) there are also compactness criteria for this case. At present there are no quantitative estimates similar to (6.5) applying to such a general situation.

6.4. The Dirac Operator

Since the Dirac operator (2.22) is not semi-bounded, there is no variational formulation of the problem. The spectral properties of the Dirac operator can be studied on the basis of Theorem 6.1, which, however, yields no quantitative estimates of the discrete spectrum.

According to Theorem 6.1, if the operator of multiplication by the potential $V(x)$ is compact as an operator from $(H^1(\mathbb{R}^3))^4$ to $(L_2)^4$, then the essential spectrum of (2.22) coincides with that of the unperturbed operator, that is, with $(-\infty, -1] \cap [1, +\infty)$. In particular, this is the case if $V(x) = O(|x|^{-1-\varepsilon})$ with $\varepsilon > 0$ for large $|x|$, $|\nabla V(x)| \le C$, and $V \in L_{p,\mathrm{loc}}$, where $p > 3$. The latter two conditions ensure that (2.22) is a self-adjoint operator on $(H^1)^4$.

6.5. Eigenvalues within the Continuous Spectrum

From the viewpoint of physics it is natural to expect that the one-particle Schrödinger operator (as well as the Dirac operator) with a potential that decays rapidly enough as $|x| \to \infty$ will have no eigenvalues within the continuous spectrum (see the discussion in Reed and Simon 1978, Sect. XIII.13). There are proofs of this fact based on the following uniqueness theorem (see Reed and Simon 1978, Vol. 4):

Theorem 6.4. *Let $u \in H^2_{\mathrm{loc}}(\mathbb{R}^n)$ and let the estimate*

$$|\Delta u(x)| \le M|u(x)| \tag{6.6}$$

be satisfied almost everywhere on a connected open set $X \subset \mathbb{R}^n$. Then $u \equiv 0$ in X if $u(x) = 0$ in a neighbourhood of a point $x_0 \in X$.

On the basis of Theorem 6.4 it is not difficult to prove that the Schrödinger operator with a compactly supported potential $V(x)$ has no eigenvalues $\lambda > 0$. We assume that there exists a (connected) domain $X_0 \subset \mathbb{R}^n$ such that $\bar{X}_0 = \mathbb{R}^n$ and $V \in L_{\infty,\mathrm{loc}}(X_0)$. In particular, this condition admits any local singularities of the potential. If u is an eigenfunction of $-\Delta + V$ corresponding to an eigenvalue λ, then $\Delta u + \lambda u = 0$ outside the support of V. For $\lambda \ge 0$ this equation has no solutions in L_2, and so u is compactly supported. On the other hand,

$$|\Delta u(x)| \le (\lambda + |V(x)|)\, u(x)|$$

in X_0. It follows that (6.6) is satisfied in any strictly internal subdomain $X \subset X_0$. Applying Theorem 6.4, we find that $u \equiv 0$.

Without the assumption that the potential is compactly supported the proof becomes much more complicated. The simplest conditions under which the Schrödinger operator has no positive eigenvalues are the following: $V \in L_{\infty,\mathrm{loc}}(\mathbb{R}^n)$ and $|x|V(x) \to 0$ as $|x| \to \infty$ (*Kato's theorem*). For more general conditions, see (Reed and Simon 1978, Sect. XIII.13), where a survey of other methods of proof is also given, and, in particular, methods that enable one to establish the property under consideration for multiparticle operators are described.

6.6. On the Essential Spectrum of the Stokes Operator

One more type of differential operators with non-empty essential spectrum arises in the theory of elliptic systems in the sense of Douglis-Nirenberg. We confine ourselves to a simple example. For general results, see (Geymonat and Grubb 1977).

Example 6.3. The Stokes operator. Let $\mathcal{L}$ be a differential expression acting on 4-dimensional vector-valued functions $w = (u, v)$, where $u \in (L_2)^3$ and $v \in L_2$, such that

$$\mathcal{L}w = \begin{pmatrix} -\Delta & -\mathrm{grad} \\ \mathrm{div} & 0 \end{pmatrix} \begin{pmatrix} u \\ v \end{pmatrix}.$$

In a bounded domain $X \subset \mathbb{R}^3$ with smooth boundary we consider the operators

$$\mathcal{L}_D = \mathcal{L} \restriction \left\{ w \in D_{\max}(\mathcal{L}) : u\big|_{\partial X} = 0 \right\},$$

$$\mathcal{L}_N = \mathcal{L} \restriction \left\{ w \in D_{\max}(\mathcal{L}) : \frac{\partial u}{\partial n} + (n, v)\bigg|_{\partial X} = 0 \right\}.$$

For $w \in D_{\max}$ the boundary conditions are both meaningful. It turns out that $\mathcal{L}_D$ and $\mathcal{L}_N$ are self-adjoint operators with two-point essential spectrum:

$$\sigma_{\mathrm{ess}}(\mathcal{L}_D) = \{-1, -1/2\}, \quad \sigma_{\mathrm{ess}}(\mathcal{L}_N) = \{-3/2, -1\}.$$

In other cases the essential spectrum of an elliptic operator in the sense of Douglis-Nirenberg can occupy an interval. For example, this is so for the operator arising in the momentum-free theory of shells (Aslanyan and Lidskij 1974).

§7 Multiparticle Schrödinger Operator

7.1. Definition of the Operator. Centre of Mass Separation

The Schrödinger operator of a system of $N \geq 3$ interacting particles (or $N \geq 2$ particles in an external field) has much more subtle spectral properties than the one-particle operator.

Let $x_j \in \mathbb{R}^n$, where $j = 1, \ldots, N$, be the coordinates of the particles, which have masses μ_j, and let $x = (x_1, \ldots, x_N) \in \mathbb{R}^{nN}$. The case of physical interest is, of course, when $n = 3$. The potentials $W_{kl}(x_k - x_l)$ of the mutual interactions between the particles are real-valued functions. In natural physical situations $W_{kl}(z) \to 0$ as $|z| \to \infty$. In the absence of an external field the Schrödinger operator in $L_2(\mathbb{R}^{nN})$ that describes such a system of particles is generated by the differential expression

$$\mathcal{L} = T + V = -\sum_j (2\mu_j)^{-1}\Delta_j + \sum_{k<l} W_{kl}(x_k - x_l). \tag{7.1}$$

In Example 2.9 we stated conditions for this (and even a slightly more general) operator to be essentially self-adjoint on $C_0^\infty(\mathbb{R}^{nN})$. Formally, (7.1) can be regarded as a special case of the operator (2.18), although the potential

$$V(x) = \sum_{k<l} W_{kl}(x_k - x_l)$$

regarded as a function in $\mathbb{R}^{nN}$ does not tend to zero in some directions as $|x| \to \infty$. This has a decisive effect on the spectral properties of the operator (7.1).

It is customary to *separate the motion of the centre of mass* in (7.1). By this we mean writing the operator in terms of the coordinates (y_0, y') corresponding to the direct decomposition $\mathbb{R}^{nN} = Y_0 + Y'$. Namely,

$$y_0 = \mu^{-1} \sum_j \mu_j x_j, \qquad \mu = \sum_j \mu_j.$$

As Y' we take the subspace spanned by all possible vectors $x_k - x_l$. This decomposition of $\mathbb{R}^{nN}$ enables one to separate the variables in (7.1), that is, to consider the decomposition

$$\mathcal{L} = (-2\mu)^{-1}\Delta_{y_0} \otimes I + I \otimes \mathcal{L}' \tag{7.2}$$

of (7.1) in $L_2(\mathbb{R}^{nN}) = L_2(Y_0) \otimes L_2(Y')$. It follows from (7.2) that the spectrum of $\mathcal{L}$ occupies a half-axis. Of interest is the spectrum of $\mathcal{L}'$, which is often called the *Hamiltonian of the system*. In particular, by the *spectrum of a system of N particles* we understand the spectrum of $\mathcal{L}'$. The operator $\mathcal{L}'$ has the form $\mathcal{L}' = H' + W$, where H' is an elliptic second-order differential operator in $\mathbb{R}^{n(N-1)}$ with constant coefficients (which depend on the choice of y'), and where the potential W can be obtained from

$$\sum_{k<l} W_{kl}(x_k - x_l)$$

by going over to the y' coordinates. If $N = 2$, then $\mathcal{L}'$ coincides with the operator of a single particle in an external field. The problems of a single particle in an external field and a free system of two particles are therefore equivalent. They are referred to as *one-particle* or *two-particle* problems (leading to some terminological ambiguity). On the other hand, the operator (2.20) with $N-1$ particles in an external field can be treated as the limiting case of the Hamiltonian of a free system of N particles in which the mass of one of the particles tends to infinity. For $N \geq 3$ such systems are referred to as *multiparticle* systems.

7.2. Subsystems. Essential Spectrum

The spectral properties of the multiparticle operator are to a large extent determined by the properties of the subsystems. Let Z be a non-empty m-element subset of the set $Z^0 = \{1, \ldots, N\}$ of particle numbers. We denote by $\mathcal{L}_Z$ the operator of the form (7.1) in $L_2(\mathbb{R}^{mn})$, but with sums over $j, k, l \in Z$. For this operator we also separate the motion of the centre of mass and obtain the decomposition

$$\mathcal{L}_Z = -(2\mu_Z)^{-1}\Delta_{y_{Z,0}} \otimes I + I \otimes \mathcal{L}'_Z, \qquad \mu_Z = \sum_{j \in Z} \mu_j$$

in the corresponding coordinates $(y_{Z,0}, y'_Z)$. Here $\mathcal{L}'_Z$ is the Hamiltonian of the subsystem. Under reasonable conditions for W_{kl} (see Theorem 2.4 as well as Example 2.9), $\mathcal{L}_Z$ and $\mathcal{L}'_Z$ are essentially self-adjoint operators on C_0^∞. We retain the same symbols for their self-adjoint realizations. If the subsystem consists of a single particle, then formal centre-of-mass separation leads to an operator $\mathcal{L}'_Z$ in the one-dimensional space $L_2(\mathbb{R}^0)$. In this case it is customary to say that the spectrum of $\mathcal{L}'_Z$ consists of one eigenvalue $\lambda = 0$.

The operator $\mathcal{L}'_Z$ describes the behaviour of the particles from Z under the condition that they do not interact with the particles from $Z^0 \setminus Z$.

Now let $Z^0 = Z_1 \cup Z_2$ be any non-trivial division of the set of particles and let $N_i = \operatorname{card} Z_i$. Below we shall write $Y_{i,0}, \mathcal{L}_i, \ldots$ instead of $Y_{Z_i,0}, \mathcal{L}_{Z_i}$, and so on. The separation $\mathbb{R}^{nN_i} = Y_{i,0} + Y'_i$, $i = 1, 2$ of the centre of mass motion in Z_1 and Z_2 gives rise to the direct decomposition $Y' = \tilde{Y}_0 + Y'_1 + Y'_2$, where $\tilde{Y}_0 = (Y_{1,0} + Y_{2,0}) \cap Y'$ and $\dim \tilde{Y}_0 = n$. To this decomposition there corresponds the tensor factorization

$$L_2(Y') = L_2(\tilde{Y}_0) \otimes L_2(Y'_1) \otimes L_2(Y'_2), \tag{7.3}$$

as well as the following representation of the Hamiltonian of the system:

$$\mathcal{L}' = h \otimes I \otimes I + I \otimes \mathcal{L}'_1 \otimes I + I \otimes I \otimes \mathcal{L}'_2 + \tilde{W}, \tag{7.4}$$

where h is the result of separating the centre of mass for $-(2\mu_1)^{-1}\Delta_{1,0} \otimes I - I \otimes (2\mu_2)^{-1}\Delta_{2,0}$, and where $\tilde{W}$ is the sum of interaction potentials between the particles from different subsystems.

Let us elucidate the meaning of each term in (7.4). The first term, which is a differential operator with constant coefficients in $\mathbb{R}^n$, describes the motion of the centres of mass of the subsystems relative to the common centre of mass regardless of the interaction. Its spectrum occupies the half-axis $[0, \infty)$. The second and third terms are the operators of motion of the subsystems relative to their centres of mass with no interaction between the subsystems taken into account, the latter being included in $\tilde{W}$.

Let λ_i be fixed eigenvalues of the operators $\mathcal{L}'_i$, $i = 1, 2$ with corresponding eigenfunctions φ_i. Then we can distinguish a subspace of $L_2(Y')$ which, in terms of the decomposition (7.3), has the form

$$\mathcal{X}_{\varphi_1,\varphi_2} = L_2(\tilde{Y}_0) \otimes \{\varphi_1\} \otimes \{\varphi_2\}.$$

Here $\{\varphi\}$ designates the linear subspace spanned by φ. Let $P = P_{\varphi_1,\varphi_2}$ be the orthogonal projection onto $\mathcal{X}_{\varphi_1,\varphi_2}$ in $L_2(Y')$. The operator $\mathcal{L}_{\varphi_1,\varphi_2} = P\mathcal{L}'P$ has the form

$$\mathcal{L}_{\varphi_1,\varphi_2} = (h + V_{\varphi_1,\varphi_2}) \otimes I \otimes I + I \otimes \lambda_1 I \otimes I + I \otimes I \otimes \lambda_2 I \tag{7.5}$$

with $V_{\varphi_1,\varphi_2} = P\tilde{W}P$. The spectrum of $\mathcal{L}_{\varphi_1,\varphi_2}$ differs from that of $T_{\varphi_1,\varphi_2} = h + V_{\varphi_1,\varphi_2}$ by the shift by $\lambda_1 + \lambda_2$. It can be studied with the aid of 'one-particle methods.' Under reasonable conditions concerning the regularity of W_{kl} and the rate at which W_{kl} decrease, it turns out that the restriction of $\mathcal{L}'$ to $\mathcal{X}_{\varphi_1,\varphi_2}$ is a weak perturbation of $\mathcal{L}_{\varphi_1,\varphi_2}$ in a suitable sense, and the spectrum of $\mathcal{L}_{\varphi_1,\varphi_2}$ generates a branch of the spectrum of $\mathcal{L}'$, which is called a 'channel' in scattering theory.

In a similar way one can construct a representation of the form (7.4) for the case of a division of Z^0 into any number $p \geq 2$ of subsystems with a fixed eigenvalue λ_i for each of the subsystems. (We set $\lambda_i = 0$ for a subsystem consisting of one particle only.) The set of numbers of the form $\lambda_1 + \ldots + \lambda_p$, where $(\lambda_1, \ldots, \lambda_p)$ ran through all possible sequences of eigenvalues of the subsystems of various divisions, is called the *threshold set*. The threshold set (which will be designated by Λ) is always non-empty. It contains zero, which corresponds to the division into one-particle subsystems.

The following theorem is one of the basic results in the spectral theory of the multiparticle Schrödinger operator.

Theorem 7.1 (Hunziker-Van Winter-Zhislin; see Joergens and Weidmann 1973; Reed and Simon 1978, Vol. 4). *Let the multiplication operators by W_{kl} be compact as operators from $H^1(\mathbb{R}^n)$ to $L_2(\mathbb{R}^n)$. Then $\sigma_{\text{ess}}(\mathcal{L}') = [\lambda_0, \infty)$, where $\lambda_0 = \inf(\lambda \in \Lambda)$.*

Let us explain Theorem 7.1 in the case when λ_0 can be attained by a division into two subsystems. The spectrum of the operator $T = \mathcal{L}' - \tilde{W}$, which includes the first three terms in (7.4), is very easy to describe. If λ_1 and λ_2 are the lower bounds of the spectra of the Hamiltonians $\mathcal{L}'_1$ and $\mathcal{L}'_2$ of the subsystems, then $\sigma(T) = [\lambda_1 + \lambda_2, \infty)$. The meaning of Theorem 7.1 is that if a division corresponds to the least possible sum $\lambda_1 + \lambda_2$, then the essential spectrum cannot be changed by the perturbation $\tilde{W}$.

By and large the operator $\mathcal{L}'$ turns out to be a weak perturbation of the direct sum of operators of the type (7.4) corresponding to various divisions into subsystems.

Each threshold $\lambda \in \Lambda$ generates a branch of the spectrum of $\mathcal{L}'$ consisting of the essential spectrum $[\lambda, \infty)$ and at most a countable set of eigenvalues with a possible single accumulation point at λ. These branches are superimposed on each other, as a result of which it is possible that eigenvalues of $\mathcal{L}'$ will appear within the continuous spectrum, which is not the case for one-particle operators (see Sect. 6.5). Besides, for potentials that satisfy the assumptions of Theorem 7.1 there exist neither positive eigenvalues nor singular continuous spectrum.

7.3. Eigenvalues

The eigenvalues of the operator are also controlled by the threshold set. The following result provides a qualitative description of the eigenvalues.

Theorem 7.2 (see Froese and Herbst 1982; Perry 1984). *Let the potentials W_{kl} be such that the operators $W_{kl}(-\Delta+I)^{-1}$ and $(-\Delta+I)^{-1}((x\nabla)W_{kl})(-\Delta+I)^{-1}$ are compact in $L_2(\mathbb{R}^n)$. Then the eigenvalues of $\mathcal{L}'$ form a nowhere dense and at most countable set with possible accumulation points lying only in the threshold set Λ.*

We remark that since the thresholds can be expressed in terms of the eigenvalues of operators with a smaller number of particles, it follows that Λ has the same properties as the set of eigenvalues.

Whether the spectrum to the left of λ_0 is finite or not depends to a large extent on the structure of the lowest threshold λ_0. Divisions into subsystems that yield λ_0 are referred to as *determining divisions*. We assume that all determining divisions consist of two subsystems (*two-cluster divisions*) and all thresholds corresponding to divisions into three or more subsystems are strictly greater than λ_0 (which implies, in particular, that $\lambda_0 < 0$). In this case the properties of the discrete spectrum below λ_0 depend on the spectra of the operators T_{φ_1,φ_2} that correspond to the determining divisions. This enables one to obtain results on the finiteness or infiniteness of the discrete spectrum (see Joergens and Weidmann 1973; Yafaev 1976; Reed and Simon 1978, Vol 4; Murtazin and Sadovnichij 1988, and references therein). In particular, the spectrum is finite provided the potentials decrease fast enough.

Theorem 7.3. *Let each determining division be a two-cluster division. Let the Fourier transform of each potential W_{kl} belong to $L_1 + L_p$, where $p < n(n-2)^{-1}$. Moreover, let the interaction potentials between particles from different subsystems of each determining division belong to $L_2 \cap L_{3/2}$ for $n = 3$ and $L_{n/2}$ for $n > 3$. Then the discrete spectrum to the left of λ_0 is finite.*

On the other hand, for more slowly decreasing potentials the infiniteness of the discrete spectrum below λ_0 can even be established along with a two-sided estimate of $N(\lambda; \mathcal{L}')$, where $\lambda < \lambda_0$, by the sum of the distribution functions

$N(\lambda - \lambda_0; T_{\varphi_1,\varphi_2})$ over various determining divisions. The existing estimates and asymptotic formulae for the spectrum of a one-particle operator (see §6, §9, and §11) enable one to obtain similar formulae also in the multiparticle case (see Murtazin and Sadovnichij 1988).

For concrete physical systems general results of the type of Theorem 7.3 are often inapplicable and the systems must be studied by individual methods. For atoms and positive ions with nuclei of arbitrary mass as well as molecules with nuclei of infinite mass it has been established that the discrete spectrum is infinite. On the other hand, for negative ions the spectrum is finite (Yafaev, Antonets, and Zhislin; see the references in Murtazin and Sadovnichij 1988). For molecules with nuclei of finite mass the problem remains open.

The situation when not all determining divisions are two-cluster divisions has a different character. Here the most extensively studied case is when no one of the subsystems has eigenvalues (then $\lambda_0 = 0$). In this case the properties of the discrete spectrum are determined by the fine structure of the lower bound $\lambda = 0$ of the spectrum of the subsystems. This point is called a resonance if adding the potential $-\varepsilon(1 + |z|)^{-2}$ to the operator of the subsystem will produce eigenvalues for any $\varepsilon > 0$.

The first results on the discrete spectrum in the presence of a resonance were obtained by Yafaev (1972). Subsequent progress has been made in the works by Zhislin, Vugal'ter, Murtazin, Sadovnichij, and others (see the references in Murtazin and Sadovnichij 1988). We shall state a characteristic result, confining ourselves to the case of smooth compactly supported potentials for the sake of simplicity.

Theorem 7.4. *Let $W_{kl} \in C_0^\infty$ and let $n = 3$. We assume that the set Z^0 of particles can be divided into two subsets, $Z^0 = Z_1 \cup Z_2$, $Z_1 \cap Z_2 = \varnothing$, in such a way that any subsystem with a resonance is contained either in Z_1 or Z_2. Then the discrete spectrum is finite. On the other hand, if $N = 3$ and there are two subsystems with a resonance, or $N = 4$ and there are two three-particle subsystems with a resonance, then the discrete spectrum is infinite* (this property is usually called the *Efimov effect*).

7.4. Refinement of the Physical Model

The Pauli principle, according to which identical fermions are forbidden to occupy the same state, and the indistinguishability of identical bosons in the same state both require that the operator $\mathcal{L}$ be considered not for all functions from L_2, but only for those that are invariant under a fixed representation of the permutation groups for identical particles in $L_2(\mathbb{R}^{nN})$. The functions must be antisymmetric under transposition of fermions and symmetric under transposition of bosons. Moreover, in the natural physical situation in which $W_{kl}(z) = w_{kl}(|z|)$ one often considers the restriction of $\mathcal{L}$ to functions invariant under the action of the group $\mathrm{O}(\mathbb{R}^n)$ or one of its subgroups in $L_2(\mathbb{R}^{nN})$.

Finally, to take the spin of the particles into account, the Schrödinger operator must be considered on a space of vector-valued functions. For such operators the spectral properties are, in principle, similar to those of the operator (7.1), although the proofs (and the statements) are often considerably more complicated.

For more complete information on the theory of the multiparticle Schrödinger operator, see (Joergens and Weidmann 1973; Reed and Simon 1978, Vol. 4; Murtazin and Sadovnichij 1988), as well as (Faddeev and Merkur'ev 1985), where scattering theory for multiparticle problems is discussed.

§8
Investigation of the Spectrum by the Methods of Perturbation Theory

One of the basic methods for studying the spectrum consists in considering the operator A under investigation as a member of a family of operators $A(\varepsilon)$ containing an operator $A_0 = A(0)$ of a simpler structure. Linear families

$$A(\varepsilon) = A_0 + \varepsilon T, \tag{8.1}$$

where T is an operator subordinate to A_0 in some sense, are used most frequently. (If T fails to be subordinate to A_0, the perturbation is often said to be singular.)

Because of the importance of perturbation theory in the study of qualitative and quantitative characteristics of the spectrum, it has been a long time since it developed into an independent branch of operator theory, or, more precisely, into several branches (qualitative perturbation theory; the analytic theory of perturbations of the point spectrum; abstract scattering theory, which can be regarded as the theory of perturbations of the absolutely continuous spectrum; the theory of singular perturbations – this list is far from being complete). The majority of these branches were developed in connection with the questions of the theory of differential equations and quantum mechanics. The preceding sections contain many examples of applications of the methods of perturbation theory in the study of the spectral properties of differential operators (self-adjointness – Theorem 1.1, the location of the essential spectrum – Theorem 6.1, and the like). In the present section we shall touch upon the simplest questions of analytic perturbation theory. For more details on this subject see (Friedrichs 1965; Kato 1966; Reed and Simon 1978, Vol. 4). We shall also discuss some 'typical properties' of the spectrum for certain classes of differential operators.

8.1. The Rayleigh-Schrödinger Series

Analytic perturbation theory investigates the behaviour of the isolated eigenvalues of $A(\varepsilon)$ of finite multiplicity as functions of ε. It is assumed that $A(\varepsilon)$ is an analytic operator-valued function in a neighbourhood of $\varepsilon = 0$. In the case when $A(\varepsilon) \in B(\mathfrak{H})$ analyticity can be defined in the standard way, for example, by means of a convergent power series in ε. For unbounded operators analyticity means that the resolvent $(A(\varepsilon) - \lambda_0 I)^{-1}$ exists (as an operator in $B(\mathfrak{H})$) for some $\lambda_0 \in \mathbb{C}$ and all $\varepsilon \in \mathbb{C}$ such that $|\varepsilon| < \varepsilon_0$, and is an analytic operator-valued function in the above sense. We remark that this definition does not imply that the domain $D(A(\varepsilon))$ is independent of ε. There are also more general definitions of analyticity for operator-valued functions (see Kato 1966).

In particular, let A_0 be a self-adjoint operator and let T be a symmetric operator such that $D(T) \supset D(A_0)$. Then the linear family (8.1) is analytic in a neighbourhood of $\varepsilon = 0$, and, by Theorem 1.1, $A(\varepsilon)$ is a self-adjoint operator on $D(A_0)$ for any real ε.

We now assume that λ_0 is an isolated eigenvalue of A_0 of multiplicity m, where $1 \le m < \infty$. Then, for small $|\varepsilon|$, there are m (not necessarily distinct) single-valued analytic functions $\lambda_j(\varepsilon)$, where $j = 1, \dots, m$, such that $\lambda_j(0) = \lambda_j$, and, provided the multiplicity is taken into account, the spectrum of the operator $A(\varepsilon)$ of the form (8.1) in a neighbourhood of λ_0 consists of the eigenvalues $\lambda_j(\varepsilon)$.

The corresponding power series for $\lambda_j(\varepsilon)$ are called the *Rayleigh-Schrödinger (RS) series*. The series can be computed by integrating the resolvent along a contour. We fix a small $\delta > 0$. For any sufficiently small ε the integral

$$P(\varepsilon) = -(2\pi i)^{-1} \int\limits_{|\lambda - \lambda_0| = \delta} (A_0 + \varepsilon T - \lambda I)^{-1} \, d\lambda \tag{8.2}$$

is a projection operator (generally speaking, not orthogonal) onto the sum of the eigenspaces of $A(\varepsilon)$ corresponding to the eigenvalues $\lambda_j(\varepsilon)$. If λ_0 is a simple eigenvalue (that is, $m = 1$) and u_0 is the corresponding normalized eigenvector, then the formula for the single branch $\lambda(\varepsilon)$ has the form

$$\lambda(\varepsilon) = \lambda_0 + \varepsilon \sum_{j=0}^{\infty} a_j \varepsilon^j \left(\sum_{j=0}^{\infty} b_j \varepsilon^j \right)^{-1}, \tag{8.3}$$

where

$$a_j = \frac{(-1)^{j+1}}{2\pi i} \int\limits_{|\lambda - \lambda_0| = \delta} \left((T(A_0 - \lambda I)^{-1})^{j+1} u_0, u_0 \right) d\lambda,$$

$$b_j = \frac{(-1)^{j+1}}{2\pi i} \int\limits_{|\lambda - \lambda_0| = \delta} \left((A_0 - \lambda I)^{-1} (T(A_0 - \lambda I)^{-1})^{j} u_0, u_0 \right) d\lambda.$$

In particular, $a_0 = (Tu_0, u_0)$ and $b_0 = 1$. Thus

$$\lambda(\varepsilon) = \lambda_0 + \varepsilon(Tu_0, u_0)$$

in the first approximation. The expressions for the subsequent terms of the series (8.3) are much more complicated.

Starting from (8.2), one can also obtain a series for the eigenfunction $u(\varepsilon)$ with $u(0) = u_0$. All the formulae can be generalized to the case of an arbitrary analytic dependence on ε. If $m > 1$, then the coefficients of the series for $\lambda_j(\varepsilon)$ can be found by solving a number of finite-dimensional eigenvalue problems (see Kato 1966).

Example 8.1. The family of Schrödinger operators

$$A(\varepsilon) = -\Delta + V_0 + \varepsilon W, \quad V_0 = \bar{V}_0, \quad W = \bar{W}$$

is analytic, for example, if V_0 satisfies (3.5) and W satisfies the assumptions of Theorem 2.4. Of course, there are also analyticity conditions that are not so crude. For numerous examples involving the evaluation of the eigenvalues with the aid of (8.3), see (Reed and Simon 1978, Vol. 4).

8.2. Typical Spectral Properties of Elliptic Operators

The technique of analytic perturbations makes it possible to study the typical spectral properties of regular elliptic operators. Let $X \subset \mathbb{R}^n$ be a bounded domain with boundary of class C^∞ and let $\lambda_j(X)$ be the eigenvalues of $(-\Delta)_D$ in X. Let X_ε be a domain obtained by a small deformation of X, that is, let X_ε be the image of X under a diffeomorphism $\Gamma_\varepsilon = I + \varepsilon J$ of a neighbourhood of X in $\mathbb{R}^n$. The operator $(-\Delta)_D$ in X_ε is unitarily equivalent to the operator $A_\varepsilon = \Gamma_\varepsilon^{-1} \circ (-\Delta)_D \circ \Gamma_\varepsilon$ in X. To the family A_ε one can apply the theory of analytic perturbations described in Sect. 8.1, which makes it possible to obtain the asymptotics of the eigenvalues $\lambda_j(\varepsilon)$ of A_ε with respect to ε to within terms of arbitrarily high order. The study of this asymptotics leads to the following result.

Theorem 8.1 (Micheletti 1972). *The domains for which $(-\Delta)_D$ has simple spectrum form a set whose complement is of the first category (a typical set) in the space of domains $X \subset \mathbb{R}^n$ with smooth boundary (with the natural topology).*

A similar result is also valid for the Laplace-Beltrami operator.

Theorem 8.2 (Tanikawa 1979; Bando and Urakawa 1983). *Let X be a compact manifold with* $\dim X = n$ *and let X^k, where $n+3 \le k \le \infty$, be the space of Riemannian metrics g of class C^k on X. Then the set of metrics $g \in X^k$ for which the operator $-\Delta_g$ on X has simple spectrum is a typical set in X^k.*

An analogue of Theorem 8.2 is also valid if we confine ourselves only to domains $X \subset \mathbb{R}^n$ invariant under a discrete group of orthogonal transformations of $\mathbb{R}^n$. In particular, the following result is true.

Theorem 8.3 (Driscoll 1987). *Let $\Gamma = \{1, \sigma, \dots, \sigma^{p-1}\}$ be the representation of $\mathbb{Z}_p$ as the group of rotations of the plane $\mathbb{R}^2$ by $2\pi j/p$, where $j \in \{0, \dots, p-1\}$, and let $\mathcal{M}$ be the space of Γ-invariant domains in $\mathbb{R}^2$ with C^∞-boundaries. Then the domains for which all the eigenvalues of the Laplace operator have multiplicity one (in which case, for any eigenfunction u, we have $u \circ \sigma = \exp(2\pi i j/p)u$, where $j \in \{0, \dots, p-1\}$) or two (in which case $u + u \circ \sigma + \dots + u \circ \sigma^{p-1} = 0$) form a typical set.*

Theorem 8.3 supplements the conditional result due to Arnol'd (1972), who, starting from the (unproven but likely) transversality hypothesis, demonstrated that the domains of the type in question form a set with complement of codimension 1 in $\mathcal{M}$. However, even Theorem 8.3 is sufficient to substantiate Arnol'd's argument that the semiclassical approximation (see §14) can yield wrong expressions both for the eigenfunctions and the multiplicities of the eigenvalues in the presence of symmetries.

8.3. The Asymptotic Rayleigh-Schrödinger Series

One has to deal with a more complex situation when $\varepsilon = 0$ does not belong to the analyticity domain of $(A(\varepsilon) - \lambda I)^{-1}$, the function being merely continuous in a certain sense at $\varepsilon = 0$. In this case the RS series diverges for any $|\varepsilon| > 0$. It can, nevertheless, be considered as an asymptotic power series in ε. If $(A(\varepsilon) - \lambda I)^{-1}$ converges in the norm to $(A_0 - \lambda I)^{-1}$ and the intersection $D(A(\varepsilon)) \cap D(A_0)$ is 'rich enough' (for more details see Reed and Simon 1978, Theorem XII.14), then the eigenvalues of $A(\varepsilon)$ can, in fact, be represented asymptotically by the series. In many important examples the series is Borel summable, yielding a convergent series suitable for computation. On the other hand, if the resolvent is merely strongly convergent as $\varepsilon \to 0$, then it is well possible that the eigenvalues cannot be represented asymptotically by the RS series.

Example 8.2. A domain with a small hole. Let X_ε be a domain obtained from $X \subset \mathbb{R}^n$ by removing a small neighbourhood $Y_\varepsilon = \{x : \varepsilon^{-1}(x - w) \in Y \subset \mathbb{R}^n\}$ of a fixed point $w \in X$ and let $A(\varepsilon)$ designate the operator $(-\Delta)_D$ in X_ε. We consider the resolvent of $A(\varepsilon)$ as an operator in X. Then, for $n \geq 4$, the resolvent converges in the norm as $\varepsilon \to +0$ and the RS series is asymptotic. But if $n = 2, 3$, then the convergence is merely strong (this is also the case in any dimension if the Neumann boundary conditions are stated on ∂X_ε). In this case (see Ozawa 1983; Maz'ya et al. 1984) a number of initial terms of the expansion of the eigenvalues $\lambda_j(\varepsilon)$ have been found. For $n = 3$ these terms are power functions, but they differ from the corresponding terms of the RS series. For $n = 2$ they involve the powers of $\log \varepsilon$.

Example 8.3. The even anharmonic oscillator $A(\varepsilon) = -d^2/dx^2 + x^2 + \varepsilon x^{2m}$ *in* $L_2(\mathbb{R}^1)$. The resolvent converges in the norm. The RS series is asymptotic and converges to the eigenvalues of $A(\varepsilon)$ in the sense of Borel (Reed and Simon 1978, Vol. 4; Rauch 1980).

8.4. Singular Perturbations

The case when the RS series is inapplicable even asymptotically is characteristic of the theory of singular perturbations. Here we present a typical example.

Example 8.4. The Schrödinger operator with a small potential. Let $V > 0$, let $V \to 0$ fast enough at infinity, and let $A(\varepsilon) = -\Delta - \varepsilon V$ for $\varepsilon > 0$. As $\varepsilon \to 0$, the eigenvalue $\lambda_j(\varepsilon)$ increases, moves towards zero, and is absorbed by the essential spectrum for some $\varepsilon = \varepsilon_j$. It follows that the 'unperturbed eigenvalue' $\lambda_j(\varepsilon_j)$ is missing, and, along with it, the RS series does not exist either. In the vicinity of ε_j the eigenvalue behaves like an analytic function with an algebraic branch point. The non-real eigenvalues $\lambda_j(\varepsilon)$ corresponding to $\varepsilon < \varepsilon_j$ can be interpreted as the poles $\lambda = \zeta^2$ of the continuation of the resolvent $(A(\varepsilon) - \zeta^2 I)^{-1}$, where $\zeta^2 = \lambda$, with respect to ζ into the domain $\operatorname{Re} \zeta < 0$. It is customary to refer to these eigenvalues as *resonances*. Regarding this point, see (Reed and Simon 1978, Vol. 4) and (Rauch 1980).

8.5. Semiclassical Asymptotics

Let $A(\varepsilon) = A(x, \varepsilon D_x)$ be a family of self-adjoint differential operators with smooth coefficients in $\mathbb{R}^n$. For $\varepsilon = 0$, $A(0)$ is the operator of multiplication by a function $V(x) = A(x, 0)$. The spectrum of $A(0)$ coincides with the image of $V(x)$ and, as a rule, is purely continuous. We assume that the spectrum of $A(\varepsilon)$ is discrete in an interval $\Delta \subset \sigma(A(0))$ for any $\varepsilon > 0$. Then the eigenvalues of $A(\varepsilon)$ in Δ accumulate as $\varepsilon \to +0$ in such a way that every $\lambda \in \Delta$ is an accumulation point of a 'branch' $\lambda(\varepsilon)$ of eigenvalues of $A(\varepsilon)$. There arise the following two problems of singular perturbation theory, which are usually regarded as problems of the theory of *semiclassical asymptotics*.

Problem 1. Given an interval $\Delta_0 \subset \Delta$, to find the asymptotics of $N(\Delta_0; A(\varepsilon))$ (see (1.11)) as well as that of the spectral projection $E^{A(\varepsilon)}(\Delta_0)$ as $\varepsilon \to +0$.

Problem 2. To find the asymptotics of a single continuous branch $\lambda(\varepsilon)$ of eigenvalues and the corresponding eigenfunctions as $\varepsilon \to 0$.

The word 'semiclassical' is justified by examples of quantum-mechanical origin, in which the Planck constant plays the role of ε.

In many situations the results of Problem 1 are the same as those concerned with the asymptotic behaviour of the eigenvalues and eigenfunctions of a fixed differential operator with respect to their number. Such results and the corresponding methods are presented in §§9–15.

On the other hand, Problem 2 can be reduced by scaling to perturbations similar to those considered in Sect. 8.3. We confine ourselves to the most extensively studied (and perhaps most important) example of the Schrödinger operator

$$A(\varepsilon) = -\varepsilon^2\Delta + V(x), \qquad V \in C^\infty(\mathbb{R}^n). \tag{8.4}$$

Example 8.5. Potential well. Let $V(x) \geq 0$ have a single non-degenerate minimum at $x = 0$, let $V(0) = 0$, and let $\liminf_{|x|\to\infty} V(x) > 0$. The spectrum of $A(\varepsilon)$ is discrete in the vicinity of zero. We introduce the *scaling transformation* $U(\varepsilon) : u(x) \mapsto u(\varepsilon x)$. The spectrum of $A(\varepsilon)$ differs form that of $B(\varepsilon) = \varepsilon^{-2}U(\varepsilon)A(\varepsilon)U^{-1}(\varepsilon) = -\Delta + V_\varepsilon(x)$, where $V_\varepsilon(x) = \varepsilon^{-2}V(\varepsilon x)$, only by the factor ε^{-2}. As $\varepsilon \to 0$, the analytic family $B(\varepsilon)$ converges strongly to $B(0) = -\Delta + V^0(x)$, where $V^0(x) = \lim_{\varepsilon\to 0} \varepsilon^{-2}V(\varepsilon x)$ is the quadratic part of $V(x)$ at zero. The operator $B(0)$ represents a many-dimensional harmonic oscillator (see Example 4.5′). With the aid of the RS series, one can therefore find the complete asymptotic expansion of the eigenvalues and eigenfunctions of $A(\varepsilon)$. It turns out that the eigenfunctions of $A(\varepsilon)$ are localized in a potential well: if $u_\varepsilon(x)$ is an eigenfunction of $A(\varepsilon)$, then for any $x \neq 0$, $u_\varepsilon(x) = o(\varepsilon^N)$ as $\varepsilon \to 0$ for any N (a more refined analysis can even reveal that u_ε decreases exponentially).

Example 8.6. Double well. Let the potential $V(x) \geq 0$ have two non-degenerate minima at $x = 0$ and $x = a$ with $V(0) = V(a) = 0$, and let $\liminf_{|x|\to\infty} V(x) > 0$, as in Example 8.5. Here, with the aid of scaling with centre at a, we can obtain one more family of operators $C(\varepsilon)$, which converges strongly to $C(0) = -\Delta + V^a(x)$ as $\varepsilon \to 0$, $V^a(x)$ being the quadratic part of $V(x)$ at a. The RS series provides the asymptotics of the sequence of eigenvalues of $A(\varepsilon)$ corresponding to the potential well at a. If the operators $B(0)$ (described in Example 8.5) and $C(0)$ have no common eigenvalues, then the eigenfunctions of $A(\varepsilon)$ turn out to be localized near $x = 0$ or $x = a$ and their asymptotics is given by the RS series for the families $B(\varepsilon)$ and $C(\varepsilon)$. This is not so if the spectra of $B(0)$ and $C(0)$ have a non-empty intersection. For example, let the potential be symmetric, $V(x) = V(a-x)$. Then the RS series for the lower eigenvalues of $C(\varepsilon)$ and $B(\varepsilon)$ have the same power-function terms, which yields asymptotically a double eigenvalue of $A(\varepsilon)$. There is, however, an exponentially small splitting connected with the presence of the symmetric and antisymmetric eigenfunctions $u^\pm(x,\varepsilon)$ with the corresponding eigenvalues $\lambda^\pm(\varepsilon)$, rather than two eigenfunctions, each localized in one of the wells. The functions $u^\pm$ are localized in the union of the wells. The asymptotic expression

$\log\big(\lambda^-(\varepsilon)-\lambda^+(\varepsilon)\big) \sim -\varepsilon^{-1}d(0,a)$ is valid, $d(0,a)$ being the distance between 0 and a in the metric $V(x)\,dx^2$.

This result admits the following interpretation with the aid of the Schrödinger equation $\partial u/\partial t = iA(\varepsilon)u$. We state the initial condition $u(0,x;\varepsilon) = u^+(x,\varepsilon)+u^-(x,\varepsilon)$, which corresponds to a particle in the well near $x=0$ at $t=0$. The solution has the form $u(t,x;\varepsilon) = \exp(it\lambda^+(\varepsilon))u^+ + \exp(it\lambda^-(\varepsilon))u^-$ and becomes equal to $u(t,x;\varepsilon) = u^+ - u^-$ after a time interval t of order $\exp(\varepsilon^{-1}d(0,a))$, which means that the particle will move underneath the potential barrier into the well at $x=a$, demonstrating the *tunnelling effect*. It takes the same length of time for the particle to return to the well at $x=0$, and so on. It is customary to refer to such particles as *instantons*.

For details on the above problems see (Simon 1983b, 1984a,b; Helffer and Sjöstrand 1986; Combes et al. 1987).

The theory of singular perturbations of boundary value problems (Lyusternik and Vishik 1960; Rauch 1980; Nazarov 1987) and the averaging theory for eigenvalue problems (Sánches-Palencia 1980; Oleinik 1987), which are important for applications, are also close to the problems touched upon in the present section.

§9
Asymptotic Behaviour of the Spectrum. I. Preliminary Remarks

9.1. Two Forms of Asymptotic Formulae

In 'non-pathological' situations the eigenvalues of a self-adjoint differential operator exhibit the proper asymptotic behaviour, which can be described either in terms of the eigenvalues themselves, or in terms of their distribution functions (1.12) and (1.16). In the semi-bounded case the corresponding asymptotic formulae can be written as

$$\lambda_j(A) \sim \Psi(j), \quad j \to \infty, \tag{9.1}$$

$$N(\lambda;A) \sim \Phi(\lambda), \quad \lambda \to \infty, \tag{9.2}$$

where Φ and Ψ are mutually inverse increasing functions on $\mathbb{R}_+$, which are, as a rule, sufficiently regular. Subject to general conditions, formulae (9.1) and (9.2) are equivalent. This is the case, for example, for the power and logarithmic-power asymptotics $\Phi(\lambda) = a\lambda^\alpha$ (in which case $\Psi(j) = (j/a)^{1/\alpha}$) and $\Phi(\lambda) = a\lambda^\alpha \log\lambda$ (in which case $\Psi(j) = (\alpha j/a \log j)^{1/\alpha}$). However, let us say, the relation $\lambda_j(A) \sim Ce^\lambda$, which implies that $N(\lambda;A) \sim \log\lambda - \log C \sim$

$\log \lambda$, is much stronger than the latter, because the information about C is lost.

Sometimes it is possible to find the subsequent terms of the asymptotics of the spectrum. For instance, for the Sturm-Liouville operator with smooth coefficients on a finite interval the eigenvalues admit a complete asymptotic expansion in the powers of j^{-1} (see Naimark 1969). Such expansions can no longer be translated into the language of $N(\lambda; A)$: there are certainly no terms of the form $C\lambda^q$ with $q < 0$ in the asymptotic expansion of $N(\lambda; A)$, because the latter has jumps at $\lambda \in \sigma_p(A)$.

For a partial differential operator there are, in general, no asymptotic expansions of the eigenvalues in the powers of the serial number. The asymptotics of the spectrum of such an operator is usually written in the form (9.2). One can often succeed in refining (9.2) by finding a sharp estimate of the remainder or obtaining the second term of the asymptotics.

9.2. Formulae for the Leading Term of the Asymptotics

We shall present the most frequently used expressions for the leading term of the asymptotics of the spectrum of a differential operator.

Let us emphasize that here we confine ourselves to discussing the form of the expressions, and, as a rule, we refrain from stating the precise conditions under which they can be proved. Some information on these conditions is presented in the subsequent sections, and some in the original articles, to which we refer the reader.

We begin with the case of an operator in a domain $X \subseteq \mathbb{R}^n$. It often proves convenient to express the asymptotics of the spectrum of such an operator by the *Weyl symbol* $A_W(x,\xi)$. For any $u \in C_0^\infty(X)$ the operator A can be represented in terms of the Weyl symbol by

$$(Au)(x) = (2\pi)^{-n} \iint e^{i\xi\cdot(x-y)} A_W\left(\frac{x+y}{2}, \xi\right) u(y)\, dy d\xi.$$

There are recalculation formulae connecting $A_W(x,\xi)$ with the complete symbol $A(x,\xi)$ (see Shubin 1978a). For a differential operator with constant coefficients, $A_W(x,\xi) = A(x,\xi)$. This equality remains valid if the coefficient $a_0(x)$ multiplying u is no longer constant, in particular, for the Schrödinger operator (2.18) and the operator (3.8). An important property of the Weyl symbol is that A is a formally self-adjoint differential operator if and only if all the values of $A_W(x,\xi)$ are real numbers (or Hermitian matrices for a differential operator acting on vector-valued functions). If A_W is lower semi-bounded, then (subject to certain additional assumptions) A is also semi-bounded on $C_0^\infty(X)$.

Let A be a lower semi-bounded scalar self-adjoint differential operator in a domain $X \subset \mathbb{R}^n$, A_W being its Weyl symbol. The asymptotic formula

$$N(\lambda; A) \sim (2\pi)^{-n} \mathrm{mes}_{2n}\{(x,\xi) \in X \times \mathbb{R}^n : A_W(x,\xi) < \lambda\}, \quad \lambda \to \infty \tag{9.3}$$

is valid for a large class of cases.

For a differential operator with constant coefficients formula (9.3) yields

$$N(\lambda; A) \sim (2\pi)^{-n} V_A(\lambda)\, \mathrm{mes}_n X, \quad \lambda \to +\infty, \tag{9.4}$$

$$V_A(\lambda) = \mathrm{mes}_n\{\xi \in \mathbb{R}^n : A(\xi) < \lambda\}. \tag{9.5}$$

The meaning of the asymptotic formulae (9.4)–(9.5) is most easily explained by considering an example involving operators with periodic boundary conditions, or, equivalently, operators on the torus $\mathbb{T}^n$ (see Example 4.3). Indeed, in this case

$$N(\lambda; A) = \mathrm{card}\{k \in \mathbb{Z}^n : A(k) < \lambda\}$$

and formulae (9.4)–(9.5) mean that the number of points with integral coordinates in the domain $A(\xi) < \lambda$ is asymptotically equal to the volume of the domain for large λ. This is indeed the case if the polynomial $A(\xi)$ increases regularly enough as $|\xi| \to \infty$. However, for example, for the polynomial $A(\xi) = \xi_1^2 + \xi_2^2 + \xi_1^4\xi_2^4$ in $\mathbb{R}^2$ this geometric property is not satisfied, and so formulae (9.4)–(9.5) fail to be true.

In accordance with (9.4) and (9.5),

$$N\big(\lambda; (-\Delta)^r\big) \sim (2\pi)^{-n} v_n \lambda^{n/2r} \mathrm{mes}_n X \tag{9.6}$$

for the operator $(-\Delta)^r$ with regular boundary conditions. Here and in what follows v_n designates the volume of the unit ball in $\mathbb{R}^n$. For $r = 1$ this formula (with $n = 2, 3$ and with the Dirichlet and Neumann boundary conditions) was first established by Weyl (1911). In this connection, any differential operator (and also any pseudodifferential operator) for which the asymptotic formula (9.3) is valid is often referred to as an '*operator with the Weyl asymptotics of the spectrum*.'

In particular, the operators $\mathcal{L}_D(X)$ from Example 3.5 have the Weyl asymptotics provided that $X \subset \mathbb{R}^n$ is a bounded set with $\mathrm{mes}_n(\partial X) = 0$ and $S_{\mathcal{L}}(\lambda) = o(V_{\mathcal{L}}(\lambda))$ as $\lambda \to \infty$, $S_{\mathcal{L}}(\lambda)$ being the $(n-1)$-dimensional area of the surface $\{\xi \in \mathbb{R}^n : \mathcal{L}(\xi) = \lambda\}$ (Tulovskij 1971). This class includes, for example, all hypoelliptic differential operators with constant coefficients.

For the operator (3.8) formula (9.3) yields

$$N(\lambda; A) \sim (2\pi)^{-n} v_n \int (\lambda - V(x))_+^{n/2r}\, dx. \tag{9.7}$$

In particular,

$$N(\lambda; A) \sim (2\pi)^{-n} v_n \int (\lambda - V(x))_+^{n/2}\, dx \tag{9.8}$$

for $r = 1$, that is, for the Schrödinger operator. Formulae (9.7) and (9.8) are satisfied under certain general conditions for the potential (for more details see §11).

Now let A be a self-adjoint differential operator in the space of vector-valued functions. On analysing Example 4.4, one can conclude that in 'regular situations' the asymptotic behaviour of the spectrum of A must be determined by the sum of contributions of the form (9.3) from the eigenvalues of the matrix $A_W(x,\xi)$. It proves convenient to write the resulting formula as

$$N(\lambda; A) \sim (2\pi)^{-n} \int\limits_{X \times \mathbb{R}^n} N(\lambda; A_W(x,\xi))\, dx d\xi. \tag{9.9}$$

Here $N(\lambda; A_W(x,\xi))$ is the distribution function of the eigenvalues of the Hermitian matrix $A_W(x,\xi)$. The asymptotic formula (9.9) will also be called the *Weyl asymptotics*.

Subject to general conditions, (9.9) remains valid for differential operators in spaces of vector-valued functions of infinite dimension. In this case it is customary to talk of *differential operators with operator-valued coefficients* or *operator-valued differential operators*. The *Schrödinger operator with an operator-valued potential* is an important special case of such a differential operator.

Let $\mathcal{K}$ be an auxiliary Hilbert space and let $V(x)$, where $x \in \mathbb{R}^n$, be a lower semi-bounded self-adjoint operator in $\mathcal{K}$ with discrete spectrum and 'sufficiently regular' dependence on x. Let $v(x) = \mathrm{QF}(V(x))$. The form

$$a[u] = \int\limits_{\mathbb{R}^n} \left(\|\nabla u\|^2_{\mathcal{K}} + v(x)[u]\right) dx$$

is semi-bounded and closed on the natural domain in $\mathfrak{H} = L_2(\mathbb{R}^n, \mathcal{K})$. The operator $A = \mathrm{Op}(a)$ can be taken as a realization of the differential expression $-\Delta + V(x)$. This scheme can also be extended to the case when $\mathcal{K}$ itself is a function of x. For an operator A of the form in question formula (9.9) can be proved in a quite general setting (Kostyuchenko and Levitan 1967; Robert 1982). It yields

$$N(\lambda; A) \sim (2\pi)^{-n} v_n \sum_j \int\limits_{\mathbb{R}^n} \left(\lambda - \lambda_j(x)\right)_+^{n/2} dx. \tag{9.10}$$

Here $\{\lambda_j(x)\}$ are the eigenvalues of $V(x)$ (with the multiplicity taken into account).

9.3. The Weyl Asymptotics for Regular Elliptic Operators

For regular elliptic operators the asymptotic behaviour of the Weyl symbol $A_W(x,\xi)$ as $|\xi| \to 0$ is the same as that of the principal symbol $A^0(x,\xi)$. This makes it possible to replace A_W by A^0 in (9.3) and (9.9). Thus the formula

$$N(\lambda; A) \sim (2\pi)^{-n} \int\limits_{X\times\mathbb{R}^n} N\big(\lambda; A^0(x,\xi)\big)\, dx d\xi \tag{9.11}$$

is equivalent to (9.9). The substitution $\xi \mapsto \lambda^{1/m}\xi$, where $m = \operatorname{ord} A$, reveals (in view of the homogeneity of the principal symbol) the power-function nature of the asymptotics. Formula (9.11) is often written in the 'localized' form

$$N(\lambda; A) \sim a_0 \lambda^{n/m}, \quad m = \operatorname{ord} A, \tag{9.12}$$

$$a_0 = (2\pi)^{-n} \int\limits_X \omega_0(x)\, dx, \tag{9.13}$$

$$\omega_0(x) = \int\limits_{\mathbb{R}^n} N\big(1; A^0(x,\xi)\big)\, d\xi. \tag{9.14}$$

In particular, for operators acting on scalar functions,

$$\omega_0(x) = \operatorname{mes}_n\big\{\xi \in \mathbb{R}^n : A^0(x,\xi) < 1\big\}. \tag{9.15}$$

This yields

$$N(\lambda; A) \sim (2\pi)^{-n} v_n \lambda^{n/2} \int\limits_X \frac{dx}{\sqrt{\det\{a_{ij}(x)\}}} \tag{9.16}$$

for any second-order operator of the form (2.8) with real coefficients. Formula (9.16) was first established by Carleman (1936).

Formula (9.11) can be modified to cover the case of operators on manifolds. In this respect it is preferable to (9.9) (for elliptic differential operators), since the Weyl symbol fails to be an invariant for operators on manifolds. Let A be a semi-bounded self-adjoint elliptic operator on the sections of a finite-dimensional Hermitian bundle $\mathcal{E}$ over a compact manifold X (with or without boundary) with a fixed smooth positive density $d\mu$ on X. The analogue of (9.11) reads

$$N(\lambda; A) \sim (2\pi)^{-n} \int\limits_{T^*X} N\big(\lambda; A^0(z)\big)\, dz. \tag{9.17}$$

Here dz is the symplectic volume element. In local coordinates $z = (x, \xi)$ and $dz = dx d\xi$.

We can see that the leading term of the asymptotics is independent of the choice of $d\mu$. Moreover, the above formulae indicate that the leading term is independent of the boundary conditions on ∂X in a regular situation. In analogy with (9.11), formula (9.17) can be written in the form (9.12)–(9.13) with

$$\omega_0(x) = \int\limits_{T^*_x X} N\big(1; A^0(x,\xi)\big)\, d\xi. \tag{9.14'}$$

One can compute $\omega_0(x)$ in any local coordinate system. The rules of transformation of the principal symbol under a change of variables imply that $\omega_0(x)\, dx$ is a density. It follows that the expression (9.13) for a_0 is an invariant.

If X is a Riemannian manifold, $\mathcal{E}$ is the trivial linear bundle, and A is the Laplace-Beltrami operator on X, then we find from (9.12)–(9.14′) that

$$N(\lambda; A) \sim (2\pi)^{-n} v_n \lambda^{n/2} \operatorname{vol} X,$$

$\operatorname{vol} X$ being the Riemannian volume of the manifold. This is clearly consistent with (9.16).

Formulae of the type (9.9) and (9.11) are also valid for a large class of non–semi-bounded operators:

$$N_\pm(\lambda; A) \sim (2\pi)^{-n} \int\limits_{X\times\mathbb{R}^n} N_\pm(\lambda; A_W(x,\xi))\, dx d\xi, \tag{9.9$_\pm$}$$

$$N_\pm(\lambda; A) \sim (2\pi)^{-n} \int\limits_{T^*X} N_\pm(\lambda; A^0(z))\, dz, \tag{9.17$_\pm$}$$

and the like.

Along with the asymptotic behaviour of the spectrum one can study that of the spectral function (see Sect. 1.11) of a differential operator. For a differential operator acting on vector-valued functions one is usually concerned with the matrix trace of the spectral function. The asymptotic behaviour of the spectral function can be described with the aid of $\omega_0(x)$, the latter being defined by (9.14), (9.15), or (9.14′), depending on the circumstances. For a semi-bounded elliptic operator A acting on scalar functions in a domain $X \subset \mathbb{R}^n$,

$$e^A(\lambda; x, x) \sim (2\pi)^{-n} \omega_0(x) \lambda^{n/m}, \quad m = \operatorname{ord} A, \tag{9.18}$$

where $\omega_0(x)$ is defined by (9.15). In particular, for a second-order operator with real coefficients,

$$e^A(\lambda; x, x) \sim (2\pi)^{-n} v_n \big(\det\{a_{ij}(x)\}\big)^{-1/2} \lambda^{n/2}.$$

This formula, which was established by Carleman (1936), was the first result on the asymptotic behaviour of the spectral function of a partial differential operator. An analogue of (9.18) is also valid for systems, in which case the left-hand side is replaced by $\operatorname{tr} e^A(\lambda; x, x)$, while $\omega_0(x)$ on the right-hand side is given by (9.14). For an operator on a manifold X with a fixed smooth density $d\mu$,

$$\operatorname{tr} e^A(\lambda; x, x) \sim (2\pi)^{-n} \omega_0(x) \frac{dx}{d\mu} \lambda^{n/m} \tag{9.19}$$

with $\omega_0(x)$ defined by (9.14′). We call to mind that $\omega_0(x)dx$ is a density on X, and so $\omega_0(x)\, dx/d\mu$ is a function on X. According to (1.18),

$$N(\lambda; A) = \int\limits_X \operatorname{tr} e^A(\lambda; x, x)\, d\mu.$$

In the case of a compact manifold without boundary it follows that the spectral asymptotic formula (9.12) can be obtained by integrating the corresponding formula for the spectral function. But if $\partial X \neq \emptyset$, then the asymptotics (9.18) fails to be uniform near ∂X. However, even in this case one can succeed in deriving (9.12) by integrating the asymptotics of $\operatorname{tr} e^A(\lambda; x, x)$ over an inner subdomain $X' \subset X$ and using the uniform estimate (5.3) in $X \setminus X'$. To obtain asymptotic formulae for the spectral function that are uniform near ∂X (and also for $x \neq y$) it is necessary to invoke a more advanced apparatus; the formulae themselves have a much more complex form (see Babich 1979; Ivrii 1984).

We also remark that formulae (9.18), (9.19), and the like, have a local character and are valid even for unbounded domains X. It follows that they are also valid when the spectrum fails to be discrete.

9.4. Refinement of the Asymptotic Formulae

In the present section, unless mentioned otherwise, the boundary of the domain as well as the coefficients appearing both in the equation and the boundary conditions are assumed to be infinitely differentiable.

The following estimate of the *remainder* in the spectral asymptotic formulae has been established for regular self-adjoint elliptic operators under general assumptions:

$$R(\lambda; A) \stackrel{\text{def}}{=} N(\lambda; A) - a_0 \lambda^{n/m} = O\big(\lambda^{(n-1)/m}\big). \tag{9.20}$$

For various successively more complex classes of operators the estimate (9.20) was obtained in (Hörmander 1968; Seeley 1978, 1980; Ivrii 1984, 1986a; Vasil'ev 1986). To date, the estimate has been proved for arbitrary elliptic operators on compact manifolds without boundary and, in the case of a single equation, also for the boundary value problem operators. For systems of equations the class of boundary value problems considered so far is incomplete. Nevertheless, there is no reason to doubt that (9.20) is valid for any regular elliptic operator.

The analysis of Examples 4.7 and 4.8 indicates that the estimate (9.20) cannot be improved, in general. Indeed, the eigenvalues $\lambda_k \sim k^2$ of the Laplace operator on the sphere S^n have multiplicity of order $O(k^{n-1}) = O(\lambda_k^{(n-1)/2})$ (the order being sharp). This clearly implies that (9.20) cannot be improved for this operator. This is also the case for the operators $-\Delta_D$ and $-\Delta_N$ on a hemisphere.

The analysis of the 'sharp' spectral asymptotic expressions constitutes a significant achievement of the past 10 to 15 years (this was written in 1988). Under certain geometric conditions, such asymptotic expressions have the form

$$N(\lambda^m; A) = a_0 \lambda^n + a_1 \lambda^{n-1} + o(\lambda^{n-1}), \quad m = \operatorname{ord} A. \tag{9.21}$$

(Here it proves more convenient to deal with $N(\lambda^m; A)$, rather than $N(\lambda; A)$). There are also similar formulae for the spectral function. The first general results in this direction were obtained by Duistermaat and Guillemin (1975) for 'scalar' pseudodifferential operators on a compact manifold without boundary. In particular, let X be a Riemannian manifold and let A be the Laplace-Beltrami operator. We consider the *geodesic flow* Φ_t on the co-spherical bundle S^*X (this bundle can be identified with the unit sphere bundle over the Riemannian manifold). The flow Φ_t moves every point $z = (x, \xi) \in S^*X$ by a distance t along the geodesic starting at that point. We say that $z_0 \in S^*X$ is a *periodic* point if $\Phi_T z_0 = z_0$ for some $T \neq 0$. A periodic point z_0 is said to be *absolutely periodic* if $\operatorname{dist}(\Phi_T z, z)$ is an infinitesimal function of order $O\big((\operatorname{dist}(z, z_0))^\infty\big)$. According to the result obtained by Duistermaat and Guillemin, if the $(2n-1)$-dimensional measure of the set of absolutely periodic points is equal to zero, then (9.21) is satisfied with $a_1 = 0$ (for $m = 2$).

If there are 'many' absolutely periodic points, then the asymptotic formula (9.21) is no longer valid, in general. This can be seen by considering as an example the Laplace operator on a sphere, in which case the eigenvalues have anomalously high multiplicity.

The fundamental results for boundary value problems were obtained by Ivrii (1980, 1984), who proved (9.21) for operators of order $m = 2$ (see Example 2.6). For such operators

$$a_1 = \pm\frac{1}{4}(2\pi)^{-(n-1)} v_{n-1} \operatorname{vol}_{n-1}(\partial X), \tag{9.22}$$

where $\operatorname{vol}_{n-1}(\partial X)$ is the $(n-1)$-dimensional volume of the boundary with respect to the Riemannian metric on X for which A is the Laplace-Beltrami operator. The minus sign in (9.22) corresponds to the Dirichlet problem, while the plus sign corresponds to the Neumann problem or the third boundary value problem. The hypothesis that the asymptotic formulae (9.21)–(9.22) are valid for the Laplace operator on a planar domain was stated by Weyl as early as in 1912 (Weyl 1912). Since then it has been known as the *Weyl hypothesis*. In this connection, it became customary to refer to the sharp asymptotic formula (9.21) as the *two-term Weyl asymptotics* also in the general case. (The reader should be warned that some terminological confusion may arise: Sects. 9.2 and 9.3 were concerned with the one-term Weyl asymptotics!)

The geometric condition under which (9.21)–(9.22) are valid is analogous to the Duistermaat-Guillemin condition. One only needs to consider the billiard flow with the ordinary reflection law on the boundary, rather than the geodesic flow. In (Ivrii 1980) the condition was imposed upon all periodic trajectories, rather than the absolutely periodic ones only. Afterwards, it was established (Safarov and Vasil'ev 1988) that the two conditions are equivalent: the $(2n-1)$-dimensional Lebesgue measure of the set of those periodic points that fail to be absolutely periodic is always equal to zero.

Later the domain of problems for which the asymptotic expression (9.21) can be proved was substantially extended by Ivrii (1982, 1985, 1986b,c,d) and

Vasil'ev (1984, 1986). They also demonstrated that the estimates (9.20) and asymptotic expressions (9.21) remain the same if some standard singularities of the boundary (such as edges, conic points, and the like) are admitted. Sharp asymptotics of different type than (9.21) have been studied in (Safarov 1986) and (Gureev and Safarov 1988). A more detailed presentation of this subject can be found in §13 and §14.

With the aid of the hyperbolic equation method, one can sometimes succeed in obtaining the complete asymptotic expansion of the spectral function $e^A(\lambda; x, x)$ for a differential operator in $\mathbb{R}^n$ or in the exterior of a bounded domain in $\mathbb{R}^n$. This is connected with the fact that, subject to certain additional conditions, the *Schwartz kernel* $U(t; x, y)$ of the evolution operator $\exp(it\sqrt{A})$ is decreasing fast enough as a function of t and one can apply the inverse Fourier transform formula directly, instead of the appropriate Tauberian theorem (see Sect. 13.1). Results of this kind are contained in articles by Buslaev, Popov, Shubin, Vainberg, and others. References to these articles can be found, for example, in the survey article (Vainberg 1988); in particular, see (Buslaev 1975; Bardos et al. 1982; Vainberg 1987).

The following *non-trapping condition* plays a key role in the proof that $U(t; , x, y)$ is a decreasing function: it is required that for any $R > 0$ there exist $T_R > 0$ such that any geodesic originating at (x, ξ) with $|x| \leq R$ lies outside the set $\{|x| \leq R\}$ for $|t| > T_R$. For an operator in the exterior of a bounded domain one has to consider billiard trajectories, rather than geodesics.

9.5. Spectrum with Accumulation Point at 0

There is a wide range of problems for which the spectrum has an accumulation point at $\lambda = 0$, rather than ∞ (for example: pseudodifferential operators of order $-\alpha < 0$, including integral operators with a polar kernel; the Schrödinger operator with $V(x) < 0$ and $V(x) \to 0$ as $|x| \to \infty$, in which case one is concerned with the asymptotic behaviour of the negative spectrum; equation (5.6) and its generalizations). For such operators one can study the behaviour of the distribution functions (1.19) and (1.20) as $\lambda \to 0$, and also talk of the 'Weyl asymptotics,' even though the meaning of this expression depends on the form of the operator. For example, for the Schrödinger operator with a negative potential it is obvious that $n_-(-\lambda; A) = N(\lambda; A)$ if $\lambda < 0$. In this case, subject to certain conditions for $V(x)$ (see Theorem 11.3 below), formula (9.8) remains unchanged, except that now $\lambda \to -0$. (For example, the spectrum of the operator from Example 6.1 is consistent with this asymptotics; see (Rozenblum 1977; Tamura 1981, 1982a)).

Now let A be a classical self-adjoint pseudodifferential operator of order $-\alpha < 0$ acting on the sections of a Hermitian vector bundle $\mathcal{E}$ over a compact manifold X with or without boundary. As always, we assume that there is a fixed smooth positive density $d\mu$ on X. Let $A^0(z)$, where $z \in T^*X$, be the principal symbol of A. In this case the formula analogous to $(9.17_\pm)$ reads

$$n_\pm(\lambda; A) \sim (2\pi)^{-n} \int\limits_{T^*X} n_\pm(\lambda; A^0(z))\, dz$$
$$= (2\pi)^{-n}\lambda^{-n/\alpha} \int\limits_{T^*X} n_\pm(1; A^0(z))\, dz. \qquad (9.23_\pm)$$

This formula (as well as its generalization to the case of 'anisotropically homogeneous' symbols) has been obtained in (Birman and Solomyak 1977b).

Expressions of the form $(9.23_\pm)$ are also valid for many problems of the form $Bu = \lambda Au$ with an elliptic operator A. The integrand $n_\pm(\cdot)$ corresponds to the finite-dimensional problem $B^0(z)f = \lambda A^0(z)f$. In particular, for the spectrum of a weighted polyharmonic operator, i.e., for the equation

$$pu = \lambda\big((-\Delta)^r u + \varepsilon u\big), \qquad (9.24)$$

the asymptotic formula

$$n_\pm(\lambda) \sim v_n(2\pi)^{-n}\lambda^{-n/2r} \int\limits_X p_\pm^{n/2r}\, dx \qquad (9.25)$$

is valid under the conditions of Examples 3.9, 3.10, and 3.12.

For more details on this subject see (Birman and Solomyak 1970, 1972, 1973).

9.6. Semiclassical Asymptotics

We shall briefly discuss the results of Problem 1 in Sect. 8.5. Let $A(\varepsilon) = A(x, \varepsilon D)$ be a family of self-adjoint differential operators in a domain $X \subset \mathbb{R}^n$ with discrete spectrum in an interval $\delta \subset \mathbb{R}$. After simple transformations, the heuristic formula (9.3) yields

$$N\big(\delta, A(\varepsilon)\big) \sim (2\pi\varepsilon)^{-n}\mathrm{mes}_{2n}\big\{(x,\xi) : A_W(x,\xi) \in \delta\big\}, \quad \varepsilon \to 0. \qquad (9.3')$$

This formula has been proved for many classes of problems, in particular, for those differential operators in $\mathbb{R}^n$ for which $A_W(x,\xi) \to \infty$ as $|x| + |\xi| \to \infty$ (under certain regularity conditions for the symbol); see (Helffer and Robert 1981, 1982b; Tamura 1984). The form of the expression suggests that only the behaviour of the coefficients of the differential operator in the set $A_W^{-1}(\delta)$ (and possibly in a small neighbourhood of this set) is essential for (9.3′) to be true. This can, however, be confirmed only for individual classes of operators. In particular, for the Schrödinger operator $A(\varepsilon) = -\varepsilon^2\Delta + V(x)$ in $\mathbb{R}^n$, formula (9.3′), which takes the form

$$N\big(\lambda; A(\varepsilon)\big) \sim (2\pi\varepsilon)^{-n} v_n \int \big(\lambda - V(x)\big)_+^{n/2}\, dx, \quad \varepsilon \to 0$$

for $\delta = (-\infty, \lambda)$, is always valid for $n \geq 3$ provided that the integral on the right-hand side, which is computed over the region in which the integrand is independent of V, i.e., $V(x) > \lambda$, is finite (see Rozenblum 1972a; Birman and Solomyak 1972, 1973). Lower-order terms and sharp estimates of the remainder have been found under additional conditions for the operator in (9.3′); see (Helffer and Robert 1981, 1982b; Tamura 1982b).

9.7. Survey of Methods for Obtaining Asymptotic Formulae

It is customary to divide the methods used for computing the asymptotics of the spectrum into two groups: 'variational methods,' which go back to the works of Weyl (Weyl 1911) and Courant (Courant and Hilbert 1953), and 'Tauberian methods' going back to (Carleman 1936). The relatively new 'approximate spectral projection method' put forward by Shubin and Tulovskij (1973) occupies an intermediate position. Here we shall briefly characterize these methods. For a more detailed presentation and the results obtained with the aid of these methods see §§11-15.

The variational technique, which is applicable in the semi-bounded case, rests on the consecutive use of formulae of the type (1.13) for $N(\lambda; A)$. On the basis of these formulae, using suitably chosen subspaces F, one can succeed in obtaining two-sided estimates for $N(\lambda; A)$ such that the upper and lower estimates approach each other asymptotically as $\lambda \to \infty$. As a rule, the choice of F is connected with dividing the domain into cubes and 'freezing the coefficients' in each cube. It should be noted that the concept of localization is present in some form in each of the methods for computing the asymptotics of the spectrum.

The variational method has the advantage of being elementary. It is not so sensitive to the smoothness of the coefficients, the boundary of the domain, and the like, as the other methods. For many types of operators (the Laplace operator, the system of elasticity theory, the Schrödinger operator, elliptic operators with a degeneracy of the ellipticity condition, the Cohn-Laplace operator, and so on) the spectral asymptotics has been obtained for the first time by means of the variational approach. On the other hand, the variational method has failed to produce (at least so far) any sharp estimates of the type (9.20) of the remainder, or, what is more, any sharp asymptotic formulae of the type (9.21). For more details on the variational method see §11.

We proceed to the characterization of Tauberian methods. Let A be a semi-bounded self-adjoint operator with discrete spectrum in a Hilbert space $\mathfrak{H}$, and let $\{\lambda_j\}$ and $\{e_j\}$, where $j \in \mathbb{N}$, be the eigenvalues and normalized eigenvectors of A. Let φ be a bounded Borel function on $\mathbb{R}$. We form the operator $\varphi(A)$ (see (1.6)). If $\varphi(s)$ is decreasing fast enough as $s \to +\infty$ (the rate at which the function must decrease depends on the growth order of λ_j), then $\varphi(A)$ is a trace class operator and

$$\operatorname{Tr}\varphi(A) = \sum_j \varphi(\lambda_j) = \int \varphi(\lambda)\, dN(\lambda; A). \tag{9.26}$$

It follows that the information on $\operatorname{Tr}\varphi(A)$ contains some information on the behaviour of $N(\lambda; A)$. If the behaviour of $\operatorname{Tr}\varphi(A;t)$ is known for a sufficiently rich family of functions $\varphi(\lambda;t)$, it turns out to be possible to determine the asymptotic behaviour of $N(\lambda; A)$ as $\lambda \to \infty$. It is customary to refer to any theorem that provides such a possibility as a *Tauberian theorem*, from which the name of the method is derived.

To apply the Tauberian technique one has to compute $\operatorname{Tr}\varphi(A;t)$ (as well as some relatively crude estimates of $N(\lambda; A)$) independently. If A is a differential operator, then in many cases one can succeed in performing such a computation for the functions

$$\begin{aligned}
\varphi(A;t) &= \exp\left(it\sqrt{A}\right), \quad t \in \mathbb{R},\\
\varphi(A;t) &= \exp\left(-At\right), \quad t > 0,\\
\varphi(A;t) &= (A + tI)^{-1}, \quad t > t_0.
\end{aligned}$$

The computation of $\varphi(A;t)$ is connected with solving the equation $u_{tt} + Au = 0$ or $u_t + Au = 0$ in the first two cases, and $Au + tu = f$ in the third case. In this connection, one can talk of the hyperbolic equation method, the parabolic equation method, and the resolvent method. The resolvent method was proposed by Carleman (1936). The hyperbolic equation method was proposed by Levitan (1952). Each of the two methods can be extended to non-semi-bounded operators A. In the case of the resolvent method this involves the change to complex numbers t. The parabolic equation method, which was put forward by Minakshisundaram and Plejel (1949), is applicable only in the semi-bounded case.

During the most recent years the greatest achievements in the field of spectral asymptotics (see Sect. 9.4) have been connected with the hyperbolic equation method. When using this method, one has to bear in mind that $\exp\left(it\sqrt{A}\right)$ is not a trace class operator and $\sigma(t) = \operatorname{Tr}\exp\left(it\sqrt{A}\right)$ must be regarded as a distribution on $\mathbb{R}$. The singularities of $\sigma(t)$ are localized only for $\operatorname{ord} A = 2$ (and also for $\operatorname{ord} A = 1$ if $\exp(itA)$ is considered instead of $\exp\left(it\sqrt{A}\right)$). What makes it possible to find the sharp asymptotics of the spectrum of A is the information on the singularities of $\sigma(t)$. To obtain this information one can use the powerful technique of Fourier integral operators and the theory of propagation of singularities for hyperbolic equations. §13 and §14 are devoted to a detailed presentation of this subject.

An important feature of the parabolic equation method is that $\exp(-At)$, where $t > 0$, is a trace class operator if no one of the eigenvalues $\lambda_j(A)$ grows faster than a power function. In particular, this is the case for any regular elliptic operator. The function

$$\theta_A(t) = \operatorname{Tr}\exp(-At) = \sum_j \exp\left(-\lambda_j(A)t\right) \tag{9.27}$$

is called the *θ-function* of A. A remarkable (and rather unexpected) property of the θ-function of any regular elliptic operator (and many other classes of operators) is that it admits a complete asymptotic expansion as $t \to +0$. In the simplest situation the expansion has the form

$$\theta_A(t) \sim \sum_{k=0}^{\infty} c_k t^{-\alpha_k}, \quad \alpha_0 > \alpha_1 > \dots, \quad \alpha_k \to -\infty \tag{9.28}$$

(in more complex cases the expansion can also contain terms of the form $t^{-\alpha} \log^p t$). This can be interpreted as the 'smoothing of irregularities' in the behaviour of $\lambda_j(A)$ in the course of computing the sum in (9.27).

If A is a regular elliptic operator of order m in a domain $X \subset \mathbb{R}^n$, then $a_k = (n-k)/m$ and $c_0 = \Gamma(\alpha_0 + 1)a_0$ in (9.28), a_0 being the coefficient from (9.12). The asymptotic formula (9.12) can be obtained from (9.28) with the aid of Karamata's Tauberian theorem (see Taylor 1981), although the information on the subsequent terms of the asymptotics and even on the estimate of the remainder is lost when the theorem is applied. Nevertheless, the expansion (9.28) is interesting in its own right. The coefficients in (9.28) are called the *Plejel-Minakshisundaram coefficients*. If A is the Laplace-Beltrami operator on a Riemannian manifold X, then these coefficients contain extensive information on the geometry of X (see §12).

If A is a regular elliptic operator of order m on an n-dimensional manifold, then, by the estimate (5.2) of sharp order, $(A+tI)^{-1}$ is a trace class operator only if $m > n$. If $m \le n$, one must consider either the resolvent of A^k with $mk > n$, or the powers of the resolvent of A when using the resolvent method. The asymptotic formula (9.11) was established with the aid of the resolvent method for a very large class of elliptic differential operators. Then qualified (but not of sharp order) estimates of the remainder were found (Agmon 1968). Afterwards, having refined the method, Métivier (1982) used it to obtain the estimate (9.20) (which was first established in the framework of the hyperbolic method). The study of the resolvent for complex values of λ plays an important role for the estimates of $R(\lambda; A)$.

By investigating the behaviour of the resolvent at large t, one can study the analytic properties of the *ζ-function* of A:

$$\zeta_A(z) = \operatorname{Tr}(A^{-z}) = \sum_j (\lambda_j(A))^{-z}. \tag{9.29}$$

Seeley (1967) was the first to carry out such a study in the case of regular elliptic operators. In this case the series (9.29) converges for $\operatorname{Re} z > n/m$ ($m = \operatorname{ord} A$ and $n = \dim X$). It turns out that the sum of the series can be extended to a meromorphic function on the whole complex plane. The poles of the function are simple[3] and lie at $z_k = \alpha_k = (n-k)/m$, where

[3] For a more general class of differential operators the ζ-function can have multiple poles.

$k = 0, 1, \ldots$. The residue at α_k is equal to $c_k/\Gamma(\alpha_k)$, where c_k are the coefficients from (9.28). Thus Ikehara's Tauberian theorem (see Shubin 1978a) enables one to obtain the asymptotics (9.12), but produces no sharp estimates of the remainder. For more details on the parabolic equation method and the resolvent method see §12.

Finally, we shall explain the basic concept of the approximate spectral projection method. If $\varphi(\lambda) = \chi_t(\lambda)$, where χ_t is the indicator function of the interval $(-\infty, t)$, then (9.26) yields

$$N(t; A) = \operatorname{Tr} \chi_t(A). \tag{9.30}$$

Therefore, if there existed an independent method of computing $\chi_t(A)$, we would immediately have a 'working formula' for $N(t; A)$.

Let A be a self-adjoint pseudodifferential operator in $\mathbb{R}^n$ with the Weyl symbol $A_W(x, \xi)$. For a large class of functions φ one can claim that $\varphi(A)$ can be approximated well enough by a pseudodifferential operator with the Weyl symbol $\varphi(A_W(x, \xi))$. Had the class of functions contained χ_t, a computation based on (9.30) would have led us directly to (9.3). Unfortunately, this is not so. One must therefore approximate χ_t by smooth functions and estimate the resulting discrepancy. The approximation technique used in this case shares certain common features with the variational technique. For more details see §15.

§10
Asymptotic Behaviour of the Spectrum. II. Operators with 'Non-Weyl' Asymptotics

10.1. The General Scheme

Classes of differential operators for which the spectral asymptotic expressions (9.3), (9.9), or (9.11) are no longer valid have also been intensively studied in recent years. As a rule, the violation of the asymptotics is connected with the fact that the expression on the right-hand side of the formula in question becomes infinite. To date, quite a lot of information has been collected to that effect. Below we present the basic examples of such operators.

In the majority of cases the resulting asymptotic formulae can be arranged into a general scheme. Let us consider an operator $A = A^*$ on a manifold X. With such an operator one can usually associate the following objects: 1) a conic submanifold $Z \subset T^*X$ with density dz defined on it; 2) a family of Hilbert spaces $\mathfrak{H}(z)$ parametrized by $z \in Z$; 3) a family of self-adjoint operators $\mathcal{A}(z)$ in $\mathfrak{H}(z)$ with discrete spectrum. The spectral asymptotic formulae for A read

$$N(\lambda; A) \sim c \int_Z N(\lambda; \mathcal{A}(z))\, dz, \tag{10.1}$$

and, in the non–semi-bounded case,

$$N_\pm(\lambda; A) \sim c \int_Z N_\pm(\lambda; \mathcal{A}(z))\, dz, \tag{10.1$_\pm$}$$

where c is a constant. It usually turns out that Z is a symplectic submanifold of T^*X and dz is the corresponding invariant volume form. Moreover, if $\dim Z = 2d$, then $c = (2\pi)^{-d}$. It is appropriate to call the family of operators $\mathcal{A}(z)$ the *operator symbol* of the problem.

The formulae from §9 also fit into the above scheme. In these formulae $Z = T^*X$, $\mathfrak{H}(z) = \mathbb{C}^k$ (k being the vector dimension of the problem), and $\mathcal{A}(z)$ is the Weyl symbol or the principal symbol of A. There is a complete analogy between the form of (10.1) and (9.9), (9.11). It follows that the 'non-Weyl' asymptotics turns out to be a Weyl asymptotics after all, but the meaning of the symbol is different!

In the 'non-Weyl' case the evaluation of the spectral asymptotics can be reduced, at least at the Hermitian level, to finding the above-mentioned characteristics $(Z, dz, \mathfrak{H}(z), \mathcal{A}(z))$ of the operator and the constant c. A general scheme designed to this end was proposed by Levendorskij (1988a); see also Sect. 15.4. As a rule, the construction of the operator symbol $\mathcal{A}(z)$ constitutes the most complex problem.

Very often it turns out that $Z = T^*\tilde{X}$, where $\tilde{X}$ is a submanifold of X. In this case one can usually associate with A an operator-valued differential operator or pseudodifferential operator $\tilde{A}$ on $\tilde{X}$ such that $N(\lambda; A) \sim N(\lambda; \tilde{A})$. The operator symbol of $\tilde{A}$ should then taken as $\mathcal{A}(z)$, and (10.1) itself will turn into formula (9.9) or (9.11) for $\tilde{A}$. The description of examples constitutes the remainder of the present section.

10.2. The Operator $-\Delta_D$ in Infinite Horn-Shaped Domains

Let $X' \subset \mathbb{R}^{n-1}$ be a given bounded domain and let $\varphi(t) > 0$, where $t \geq 1$, be a given continuous function such that $\varphi(1) = 1$ and $\varphi(t) \to 0$ as $t \to \infty$. We consider (in the variational setting; see Example 3.1) the operator $A = -\Delta_D$ in the domain

$$X = \{x = (x', t) \in \mathbb{R}^n = \mathbb{R}^{n-1} \times \mathbb{R} : \ t > 1,\ x'/\varphi(t) \in X'\}.$$

If

$$\int_1^\infty \varphi^{n-1}(t)\, dt < \infty,$$

then $\text{mes}_n X < \infty$ and the asymptotics (9.6) is preserved (for $r = 1$). But if the integral is infinite, then, subject to certain additional conditions for φ, the spectrum of A is the same as that of the Schrödinger operator $\tilde{A}y = -d^2y/dt^2 + V(t)y$ on the semi-axis $t > 1$ with $y(1) = 0$ and with the operator-valued potential $V(t)$ equal to the $(n-1)$-dimensional operator $-\Delta_D$ in $X'_t = \varphi(t)X'$. In the case under consideration the spectrum of A has the asymptotics (10.1), where $Z = T^*\tilde{X}$ ($\tilde{X} = \{t > 1\}$), $dz = dtd\xi$, $c = (2\pi)^{-1}$, $\mathfrak{H}(z) = \mathfrak{H}(t,\xi) = L_2(X'_t)$, and $\mathcal{A}(t,\xi) = \xi^2 I + V(t)$. Taking into account that the eigenvalues $\lambda_j(t)$ of the operator $V(t)$ can be expressed in terms of $\lambda_j = \lambda_j(1)$ by the formula $\lambda_j(t) = \varphi^{-2}(t)\lambda_j$, we find that

$$N(\lambda; A) \sim \pi^{-1} \sum_j \int_1^\infty (\lambda - \varphi^{-2}(t)\lambda_j)_+^{1/2}\, dt \tag{10.2}$$

(cf. (9.10)). Let, in particular, $\varphi(t) \sim t^{-\alpha}$. Then $\text{mes}\, X = \infty$ whenever $\alpha \leq (n-1)^{-1}$. If $\alpha < (n-1)^{-1}$, then, in accordance with (10.2),

$$N(\lambda; A) \sim c(n,\alpha)\lambda^{\frac{1}{2}+\frac{1}{2\alpha}} \sum_j \lambda_j^{-1/2\alpha},$$

the series on the right-hand side being convergent precisely for $\alpha < (n-1)^{-1}$. In the 'border' case when $\alpha = (n-1)^{-1}$ we have $N(\lambda; A) \sim c(n)\lambda^{n/2} \log \lambda\ \text{mes}_{n-1} X'$. For more details see (Rozenblum 1972b).

10.3. Elliptic Operators Degenerate at the Boundary of the Domain

Let $A = \mathcal{L}_{\alpha,D}$ or $A = \mathcal{L}_{\alpha,N}$ be the operator from Example 3.4. Then the measure on the right-hand side of (9.15) is finite precisely for $\alpha < 2n^{-1}$. Subject to this condition, the spectrum has the Weyl asymptotics. If $2n^{-1} < \alpha < 2$ (we recall that the spectrum is no longer discrete for $\alpha \geq 2$), then formula (10.1), in which $Z = T^*\partial X$, dz is the symplectic $(2n-2)$-dimensional volume form on Z, and $c = (2\pi)^{-(n-1)}$, is satisfied. For all $z \in Z$ we have $\mathfrak{H}(z) = L_2(\mathbb{R}_+)$. We introduce the following notation to describe the operators $\mathcal{A}(z)$. Let $x \in \partial X$ and let $\nu = \nu(x) = (\nu_1, \ldots, \nu_n)$ be the inner normal unit vector to ∂X at x. We can identify every element $\xi \in T^*_x\partial X$ with a vector $\xi \in \mathbb{R}^n$ such that $\xi \perp \nu(x)$. Then $\mathcal{A}(z) = \text{Op}\,(a_z)$, where $a_z = a_{x,\xi}$ is the form

$$a_{x,\xi}[f] = \int_{\mathbb{R}_+} t^\alpha \sum_{i,j} a_{ij}(x)(\xi_i + \nu_i D_t)f(t)(\xi_j - \nu_j D_t)\overline{f(t)}\, dt$$

in $L_2(\mathbb{R}_+)$ considered on the domain

$$H^1_\alpha(\mathbb{R}_+) = \left\{ f \in H^1_{\text{loc}}(\mathbb{R}_+) : \int\limits_{\mathbb{R}_+} t^\alpha \left(|f'|^2 + |f|^2\right) dt < \infty \right\}$$

in the case of $\mathcal{L}_{\alpha,N}$ and on $\mathring{H}^1_\alpha(\mathbb{R}_+)$ in the case of $\mathcal{L}_{\alpha,D}$. It follows that for operators with strongly degenerate ellipticity (i.e., for $\alpha n > 2$) the asymptotic behaviour of the spectrum depends on the form of the boundary condition.

A similar asymptotic formula is also true for strongly degenerate elliptic operators of order $m > 2$. For $m = 2$ the result can be transformed to the simpler form

$$N(\lambda; A) \sim \frac{v_{n-1}}{(2\pi)^{n-1}} \lambda^\theta w \int\limits_{\partial X} \frac{(a\nu, \nu)^{\frac{n}{2}-\theta}}{\sqrt{\det a}}\, dS, \quad \theta = (n-1)/2\alpha,$$

where $a = a(x) = \{a_{ij}(x)\}$. Here $w = \sum_j \mu_j^\theta$, where μ_j are the eigenvalues of the equation $-(t^\alpha y')' + t^\alpha y = \mu y$ for $t \in \mathbb{R}_+$, subject to the condition $y(0) = 0$ (in the case of $\mathcal{L}_{\alpha,D}$) or $\lim_{t\to 0} t^\alpha y'(t) = 0$ (in the case of $\mathcal{L}_{\alpha,N}$). The series for w converge precisely for $\alpha n > 2$.

As opposed to the previous example, the scheme at hand leaves out the 'border case' $\alpha n = 2$. In this case

$$N(\lambda; A) \sim \frac{n v_n}{(2n-2)(2\pi)^n} \lambda^{n/2} \log \lambda \int\limits_{\partial X} \frac{dS}{\sqrt{\det a}}$$

for each of the two operators.

For more details see (Solomyak and Vulis 1974; Vulis 1976; Birman and Solomyak 1977a) and references therein.

10.4. Hypoelliptic Operators with Double Characteristics

Let A be a scalar self-adjoint differential operator of order m on a compact n-dimensional manifold X without boundary. We assume that the principal symbol A^0 of A is non-negative and vanishes on a submanifold $\Sigma \subset T^*X$ of dimension $2d$, on which it has a zero of order two. We also assume that Σ is symplectic, that is, the canonical 2-form $\omega = \sum_j d\xi_j \wedge dx_j$ is non-degenerate on Σ. It is known (Tréves 1982; Hörmander 1983–1985) that Σ is microlocally embedded in T^*X like $T^*\mathbb{R}^d$ in $T^*\mathbb{R}^n$. In the corresponding canonical coordinates $(y, \eta, t, \tau) \in T^*\mathbb{R}^d \times T^*\mathbb{R}^{n-d}$ we set

$$P_{y,\eta}(t,\tau) = \sum_{|\alpha|+|\beta|=2} D_t^\alpha D_\tau^\beta A^0(y,\eta,t,\tau)\Big|_{t=\tau=0} t^\alpha \tau^\beta + A_1(y,\eta,0,0), \quad (10.3)$$

where A_1 is the symbol of order $m-1$ of A. We assume that the differential equation $P_{y,\eta}(t, D_t)u = 0$ with polynomial coefficients has no non-zero solutions in the Schwartz class $S(\mathbb{R}^{n-d})$ for any $y, \eta \neq 0$. This is the *ellipticity condition with the loss of one derivative.* Under this condition, the spectrum of A is discrete. However, A is not necessarily semi-bounded under the above assumptions.

The results on the asymptotic behaviour of the spectrum are the following. If $m(n-d) > n$, then the Weyl formulae ($9.17_\pm$) are satisfied for $N_\pm(\lambda; A)$. Since the principal symbol A^0 is non-negative, we find from ($9.17_\pm$) that $N_+(\lambda; A) \sim a_0\lambda^{n/m}$ with $a_0 > 0$, and $N_-(\lambda; A) = o(\lambda^{n/m})$.

If $m(n-d) < n$, then formulae ($10.1_\pm$), where $Z = \Sigma$, dz is the symplectic volume form on Σ, and $c = (2\pi)^{-d}$, are satisfied. Moreover, $\mathfrak{H}(z) = L_2(\mathbb{R}^{n-d})$ for all $z = (y, \eta) \in \Sigma$. As $\mathcal{A}(z)$ one must take $\mathcal{A}(z) = \mathcal{A}(y, \eta) = \overline{P_{y,\eta}(t, D_t)}$. We remark that the operators (10.3) are essentially self-adjoint in $L_2(\mathbb{R}^{n-d})$. We also remark that the operators $P_{y,\eta}$ are changed to unitarily equivalent ones under any canonical transformation of variables on Σ, so that the spectra of the operators (appearing in ($10.1_\pm$)) are independent of the choice of canonical coordinates.

For $m(n-d) < n$ it turns out that $N_\pm(\lambda, A)$ has power asymptotics of order $\lambda^{d/(m-1)}$, the coefficients in the two formulae for $N_\pm$ being distinct from zero, in general. As in Sect. 10.3, the scheme does not cover the 'border case' $m(n-d) = n$, in which $N_+(\lambda; A) \sim a_0'\lambda^{n/m}\log\lambda$ and $N_-(\lambda; A) = o(\lambda^{n/m}\log\lambda)$. The expression for a_0' contains the integral of the Hessian of A_0 over Σ. On these results as well as their generalizations to the case of a non-symplectic Σ and characteristics of high multiplicity see (Métivier 1976; Menikoff and Sjöstrand 1978, 1979; Mohamed 1982, 1983; Aramaki 1983; Levendorskij 1988a).

10.5. The Cohn-Laplace Operator

(On the Cohn-Laplace operator see, for example, (Métivier 1981; Trèves 1982)). Let X be a bounded domain in $\mathbb{C}^N \cong \mathbb{R}^{2N}$. We consider the spaces $L_2(X, q)$ of $(0, q)$-forms on X, that is, forms that can be written as $\sum u_J d\bar{z}_J$, where the sum is over all sequences $J = (j_1, \dots, j_q)$ such that $1 \le j_1 < \dots < j_q \le N$, and where $d\bar{z}_J = d\bar{z}_{j_1} \wedge \dots \wedge d\bar{z}_{j_q}$. The notation $|J| = q$ is used below. The linear Cauchy-Riemann operator $\bar{\partial}$ transforms $(0, q)$-forms into $(0, q+1)$-forms such that

$$\bar{\partial}(u_J d\bar{z}_J) = \sum_{j=1}^{N} \frac{\partial u_J}{\partial \bar{z}_j}\, d\bar{z}_j \wedge d\bar{z}_J .$$

Next, we assume that X is defined by the inequality $\varphi(z) = \varphi(x, y) < 0$, where φ is a smooth function such that $\nabla\varphi \ne 0$ on ∂X. For any $z \in \partial X$ we consider the matrix $\sigma_q(z) = \{\sigma_{J,J'}(z)\}$ with $|J| = q$ and $|J'| = q-1$ such that $\sigma_{J,J'} = 0$ for $J' \not\subset J$ and $\sigma_{J,J'} = 2i\varepsilon_J^{k,J'}\partial\varphi/\partial z_k$ for $J = J' \cup \{k\}$, where $\varepsilon_J^{k,J'}$ is the sign of the permutation that turns (k, J') into J. The *Cohn-Laplace operator* is defined in $L^2(X, q)$ by the differential expression

$$\square\Big(\sum_{|J|=q} u_J d\bar{z}_J\Big) = -\frac{1}{2}\sum_{|J|=q} \Delta u_J d\bar{z}_J$$

and the boundary conditions $\sigma_q u\big|_{\partial X} = 0$ and $\sigma_{q+1}\partial u\big|_{\partial X} = 0$.

If $q < N$, then A is not the operator of any regular boundary value problem, since the Shapiro-Lopatinskij condition is violated at every point of the boundary. When A is reduced to a pseudodifferential operator on the boundary (see Tréves 1982), it turns out that the corresponding characteristic manifold $\Sigma \subset T^*\partial X$ intersects each fibre $T^*_z\partial X$ along one ray, the characteristics being exactly of multiplicity two. In a neighbourhood of $z_0 \in \partial X$ we introduce local coordinates such that X is given by the inequality $s = \operatorname{Im} z_n > \psi(z', t)$, where $t = \operatorname{Re} z_n$, $\psi(z_0) = 0$, and $\nabla\psi(z_0) \neq 0$. Let $\mu_j = \mu_j(z)$ be the eigenvalues of the Levi form

$$L(z_0, w) = \sum_{j,k<N} \frac{\partial^2\psi(z_0)}{\partial z_j \partial \bar{z}_k} w_j \bar{w}_k.$$

In order that A be hypoelliptic with the loss of one derivative it is necessary and sufficient that there be at least $N-q$ strictly positive numbers or at least $q+1$ strictly negative ones among $\mu_j(z_0)$ for all $z_0 \in \partial X$. If this condition is met, then

$$N(\lambda; A) \sim N_X(\lambda) + N_{\partial X}(\lambda). \tag{10.4}$$

The term $N_X(\lambda)$ has the same nature as the asymptotics of a regular problem and satisfies a formula analogous to (9.6) with $n = 2N$ and $r = 1$. Thus

$$N_X(\lambda) = \binom{N}{q} \frac{\operatorname{mes} X}{(2\pi)^n\, n!} \lambda^N.$$

The second term in (10.4) reflects the violation of the regularity of the boundary condition and has the form

$$N_{\partial N}(\lambda) \sim \lambda^N \int_Z c(z, \tau)\, dz d\tau, \tag{10.5}$$

where $c(z, \tau)$ is a function on $Z = \partial X \times \mathbb{R}_+$, which can be expressed in terms of $\mu_j(z)$ (see Métivier 1981; Tréves 1982). Formula (10.5) cannot be reduced directly to (10.1), since the characteristic manifold Z fails to be symplectic and the rank of the fundamental form on Z is exactly equal to one.

10.6. The n-Dimensional Schrödinger Operator with Homogeneous Potential

Let the potential $V(x) \geq 0$ in (2.18) be continuous and positively homogeneous of degree $a > 0$:

$$V(tx) = t^\alpha V(x) \quad \forall t > 0.$$

If $V(x) \neq 0$ for $x \neq 0$, then the spectrum is discrete and the asymptotic formula (9.8) is satisfied. The situation is more complex if $V(x) = 0$ on a (conic) subset $M = M(V) \subset \mathbb{R}^n$, which clearly means a very irregular behaviour of the potential at infinity.

We assume that M is a smooth conic manifold and $\dim M = d$. Below we denote by $N_y M$ the normal subspace to M at $y \in M \setminus \{0\}$. We assume that the limit

$$F(y, v) = \lim_{t \to 0} t^{-(a-a_0)} V(y + tv)$$

exists and is positive for some $a_0 \in (0, a)$ and any $y \in M \setminus \{0\}$ and $v \in N_y M$, the convergence being uniform with respect to $|y| = |v| = 1$. Then the spectrum of $-\Delta + V(x)$ is discrete. To describe the asymptotic behaviour of the spectrum we introduce the index $\theta = da_0^{-1}(1 + a/2)$. If $da < na_0$, then the asymptotic expression (9.8) is preserved. But if $da > na_0$, then formula (10.1), where $Z = T^*M$, dz is the symplectic volume element on Z, and $c = (2\pi)^{-d}$, is satisfied. Moreover, if $z = (y, \xi)$ with $y \in M$ and $\xi \in T_y^* M$ in local coordinates, then $H(z) = L_2(N_y M)$ and $\mathcal{A}(z) = \mathcal{A}(y, \xi)$ is the $(n-d)$-dimensional Schrödinger operator

$$\mathcal{A}(y, \xi) = -\Delta_v + |\xi|^2 + F(y, v).$$

We denote by $\lambda_j(y)$ the eigenvalues of $A(y, 0)$. Then, in accordance with (10.1) (cf. (9.10)),

$$\begin{aligned} N(\lambda; A) &\sim (2\pi)^{-d} v_d \sum_j \int_M (\lambda - \lambda_j(y))_+^{d/2} \, dS(y) \\ &= c\lambda^\theta \int_{M \cap S^{n-1}} \sum_j (\lambda_j(y))^{\theta - \frac{d}{2}} \, dS(y), \\ \theta &= da_0^{-1}\left(1 + \frac{a}{2}\right), \qquad c = c(d, a, a_0). \end{aligned}$$

The integral is computed over the $(d-1)$-dimensional surface $M \cap S^{n-1}$.

As in a number of previous examples, the formula does not cover the border case $da = na_0$, in which the asymptotic expression has the form $N(\lambda; A) \sim c\lambda^\theta \log \lambda$.

On these and closely related results as well as some more general ones see (Solomyak 1985; Levendorskij 1988a).

10.7. Compact Operators with Non-Weyl Asymptotic Behaviour of the Spectrum

Some classes of integral (compact) operators also have a non-Weyl asymptotics. As a rule, formulae of the type $(10.1_\pm)$ are preserved for such operators. One only has to replace $N_\pm$ by $n_\pm$ everywhere in the formulae (cf. $(9.23_\pm)$).

In particular, the above remarks apply to integral operators whose kernels have singularities only on the boundary of the domain. We shall describe the simplest ('model') result in this direction. For more details see (Laptev 1981; Grubb 1986). Let $X \subset \mathbb{R}^n$ be a bounded convex domain with smooth boundary and let

$$\rho_\alpha(x,y) = \left(\min_{v \in \partial X} \left(|x-v| + |v-y|\right)\right)^\alpha, \quad \alpha > -n.$$

In $L_2(X)$ we consider the integral operator T_α with kernel $\rho_\alpha(x,y)$. It turns out that the spectrum of T_α has the same asymptotics as that of a certain operator-valued pseudodifferential operator $\tilde{T}_\alpha$ in $L_2(\partial X; L_2(\mathbb{R}_+))$. For any $x \in \partial X$ and $\xi' \in T_x^* \partial X$ the operator symbol $\tilde{t}_\alpha(x,\xi')$ of this pseudodifferential operator is an integral operator in $L_2(\mathbb{R}_+)$:

$$\left(\tilde{t}_\alpha(x,\xi')f\right)(t) = \int_0^\infty \omega_\alpha\left(|\xi'|, t+s\right) f(s)\, ds. \tag{10.6}$$

Here $\omega_\alpha(|\xi'|,t)$ is the Fourier transform (in the distribution sense) of $(|y|^2 + t^2)^{\alpha/2}$ with respect to $y \in \mathbb{R}^{n-1}$, i.e., the so-called Bessel-MacDonald kernel. Expressions of the form $(10.1_\pm)$ (with $N_\pm$ replaced by $n_\pm$) can be applied to compute the asymptotics of the spectrum of $\tilde{T}_\alpha$. As a result, we find that

$$n_\pm(\lambda; T_\alpha) \sim v_{n-1}(2\pi)^{-(n-1)} \lambda^{-\frac{n-1}{n+\alpha}} \sum_k \left(\mu_k^\pm\right)^{\frac{n-1}{n+\alpha}} \tag{10.7}$$

as $\lambda \to +0$, $\pm\mu_k^\pm$ being the positive and negative eigenvalues of the operator (10.6) for $|\xi'| = 1$. (There is no negative spectrum for $\alpha < 0$.) Formula (10.7) should be compared with those in Sects. 10.2 and 10.3.

§11
Variational Technique in Problems on Spectral Asymptotics

11.1. Continuity of Asymptotic Coefficients

A characteristic feature of those works on the spectral asymptotics in which the variational technique is used is that the standard variational methods (of the type of (1.12) and Theorem 1.6) are combined with the ideas of perturbation theory.

The following important (even though simple) assertion was proved and used with success by Weyl as early as in his first work on the asymptotic behaviour of the spectrum of the Laplace operator (Weyl 1911).

Lemma 11.1. *Let T and V be compact self-adjoint operators in a Hilbert space $\mathfrak{H}$ such that $n_\pm(\lambda; T) \sim a_\pm \lambda^{-\alpha}$ with $a_\pm \neq 0$ and $n_\pm(\lambda; V) = o(\lambda^{-\alpha})$ as $\lambda \to +0$. Then $n_\pm(\lambda; T+V) \sim a_\pm \lambda^{-\alpha}$.*

The proposition below, which also has many applications in works on the spectral asymptotics, is close to Lemma 11.1. We denote by Σ_q, where $q > 0$, the linear space of compact operators T in $\mathfrak{H}$ whose s-numbers (see Sect. 1.12) satisfy the relation $s_n(T) = O(n^{-1/q})$, or, equivalently, $n(s; T) = O(s^{-q})$ as $s \to 0$. The function $\sup_s\big(sn^{1/q}(s;T)\big)$ is a *quasinorm*,[4] which can be used to introduce a topology in Σ_q in the usual way. If $T \in \Sigma_q$ is a self-adjoint operator, then the functionals

$$\Delta_q^\pm(T) = \limsup_{\lambda \to +0} \lambda^q n_\pm(\lambda; T), \quad \delta_q^\pm(T) = \liminf_{\lambda \to +0} \lambda^q n_\pm(\lambda; T) \tag{11.1}$$

are finite. It is clear that the existence of the correct power asymptotics of $n_\pm(\lambda; T)$ is equivalent to the condition

$$\Delta_q^\pm(T) = \delta_q^\pm(T) \neq 0$$

for some $q > 0$.

Lemma 11.2 (Birman and Solomyak 1972, 1973, 1974). *The functionals (11.1) are continuous on the set of self-adjoint operators $T \in \Sigma_q$.*

In (Rozenblum 1974) there are also similar assertions for operators such that $n_\pm(\lambda; T)$ exhibits a non-power asymptotic behaviour.

---

[4] A *quasinorm* Q on a linear space $\mathcal{X}$ is a functional that satisfies all the conditions for a norm but the triangle inequality, which is replaced by the following weaker condition: there is a constant $c \geq 1$ such that $Q(x+y) \leq c(Q(x) + Q(y))$ for all $x, y \in \mathcal{X}$.

11.2. Outline of the Proof of Formula (9.25)

A suitable analytic apparatus is necessary to apply Lemma 11.2. The problem concerned with the spectrum of a weighted polyharmonic operator (see (9.24)) is one of the simplest problems for which the variational technique proves effective. The estimate (5.8) serves as such an apparatus for this problem.

Below $a_{r,\varepsilon}$ and b_p designate the quadratic forms (3.12) and (3.13). Let us fix a domain $X \subset \mathbb{R}^n$ and the parameters r and ε. We denote by $\Phi_D = \Phi_{X,D}$ the mapping that assigns the operator $T\big(H^r(X), a_{r,\varepsilon}, b_p\big)$ (see Sect. 1.10) to p. We attach a similar meaning to $\Phi_N = \Phi_{X,N}$. It follows from (5.7) that Φ_D and Φ_N are continuous as functions from $L_\gamma(X)$ to $\Sigma_{n/2r}$, subject to the conditions of Examples 5.6 and 5.7. Hence, by Lemma 11.2, it follows that the set of all p such that the asymptotic formula (9.25) is satisfied is closed in $L_\gamma(X)$.

Another consequence of (5.7) and Lemmas 11.1 and 11.2, which is somewhat harder to prove, is that

$$\Delta^{\pm}_{n/2r}(\Phi_D p) = \Delta^{\pm}_{n/2r}(\Phi_N p), \quad \delta^{\pm}_{n/2r}(\Phi_D p) = \delta^{\pm}_{n/2r}(\Phi_N p) \tag{11.2}$$

for any bounded domain with Lipschitz boundary and $p \in L_\gamma(X)$. It is important that (11.2) has an a priori character: the existence of the 'correct' asymptotic formulae, i.e., equalities of the form $\Delta^{\pm} = \delta^{\pm}$, is not required in advance.

Finally, one more consequence is that the functionals $\Delta^{\pm}_{n/2r}$ and $\delta^{\pm}_{n/2r}$ depend continuously on the domain X in a certain sense in the case of the Dirichlet problem.

Having made these remarks, one can prove the asymptotic formula (9.25) without difficulty. The proof can be divided into two stages.

Analysis of a model problem. Let $p = \text{const}$ and let $X = (0, 2\pi)^n$. On identifying the opposite edges of the cube X, we obtain a torus $\mathbb{T}^n$. Let $\tilde{H}^r(X)$ be the space consisting of functions on X, which turns into $H^r(\mathbb{T}^n)$ when X turns into $\mathbb{T}^n$. The eigenvalues of the operator $\tilde{T} = T(\tilde{H}^r(X); a_{r,\varepsilon}, b_p)$ are equal to $p(|j|^{2r} + \varepsilon)^{-1}$, where $j \in \mathbb{Z}^n$. An elementary computation (of the same kind as in Sect. 9.2) leads to formula (9.25) for the spectrum of $\tilde{T}$. Since $\mathring{H}^r(X) \subset \tilde{H}^r(X) \subset H^r(X)$, it follows by Theorem 1.4 that $\Delta^{\pm}_{n/2r}(\tilde{T})$ and $\delta^{\pm}_{n/2r}(\tilde{T})$ are enclosed between the same quantities for the operators $\Phi_D p$ and $\Phi_N p$. Now, it follows from (11.2) that the asymptotic formula (9.25) already established for $\tilde{T}$ can be extended to the operators of the Dirichlet and Neumann problems on X. The same is true for a cube of arbitrary dimensions.

Passage to the general case. To start with, suppose that X can be divided into finitely many cubes X_j, where $j = 1, \dots, j_0$, in each of which $p = p_j = \text{const}$. From the variational principle (1.17) one can easily find that

$$\sum_j n_\pm(\lambda; \Phi_{X_j,D}p_j) \le n_\pm(\lambda; \Phi_{X,D}p)$$
$$\le n_\pm(\lambda; \Phi_{X,N}p) \le \sum_j n_\pm(\lambda; \Phi_{X_j,N}p_j).$$

Here the first and last terms are already known to have the same asymptotic behaviour as $\lambda \to +0$. It follows that the two middle terms also exhibit this behaviour. This yields (9.25) for either of the operators $\Phi_{X,D}p$ and $\Phi_{X,N}p$ in the case under consideration. Due to the above-mentioned continuous dependence of $\Delta^\pm_{n/2r}$ and $\delta^\pm_{n/2r}$ on p and X, the result can be automatically extended to the case of $\Phi_{X,D}p$ for any X and p that satisfy the conditions of Example 5.6. If X is a bounded domain with Lipschitz boundary, then, by (11.2), the same formula is also valid for the spectrum of $\Phi_{X,N}p$.

Now, the asymptotic formula (9.25) is valid for the spectrum of any weighted polyharmonic operator, subject to the conditions of Examples 5.6 and 5.7. In particular (for $p \equiv 1$) we find that the spectrum of $(-\Delta)^r_D$ exhibits the asymptotic bahaviour (9.6) in any domain $X \subset \mathbb{R}^n$ if $2r < n$ and in any bounded domain if $2r \ge n$. In fact, the boundedness condition can also be relaxed for $2r \ge n$ (see Rozenblum 1972b). In doing so one can use the estimate (5.4).

The main disadvantage of this version of the variational technique, which is based on assertions of the type of Lemma 11.2, is that it produces no qualified estimates of the remainder of the asymptotics.

11.3. Other Applications of the Variational Method

Various results can be obtained in the way described above. We shall state some of them. One group of results is concerned with operators with discontinuous leading coefficients. We confine ourselves to the case of second-order operators.

Theorem 11.1. *Let X be any domain of finite measure in the case of the Dirichlet problem or any bounded domain with Lipschitz boundary in the case of the Neumann problem. Let l be the form (3.1) with real coefficients that satisfy (2.9). Then the asymptotic formula (9.16) is valid for the spectra of the corresponding operators $\mathcal{L}_D$ and $\mathcal{L}_N$ (see Example 3.1).*

Some of the assumptions of the theorem can be relaxed. One can admit coefficients $a_{ij} \in L_{1,\mathrm{loc}}(X)$, relax the condition that X must be of finite measure for $n > 2$, and permit a not too strong degeneracy of ellipticity (not necessarily localized near the boundary). The theorem can also be extended to higher-order equations and systems of 'variational type.' For more details see (Birman and Solomyak 1972, 1973).

The variational technique has been used to obtain the results of Sect. 10.3 (operators with degeneracy of ellipticity at the boundary of the domain)

(Solomyak and Vulis 1974) as well as their generalizations to other types of degenerate elliptic operators (Vulis 1976). Various alternative versions of the variational method have been used to derive the formulae presented in Sects. 10.2, 10.5, 10.6, and 10.7.

For the Schrödinger operator with an increasing potential V formula (9.8) was first obtained with the aid of the elementary version of the variational method under quite general assumptions. Afterwards, the Tauberian technique was applied to problems of this kind (Titchmarsh 1946; Kostyuchenko 1966, 1968), which made it possible to extend considerably the class of admissible potentials. Under still more general assumptions, formula (9.8) was proved in (Rozenblum 1974), where the variational method was used again, but now in the spirit of Sect. 11.2.

We shall present the basic result of (Rozenblum 1974). The following notation is adopted below. If V is a real-valued function in $\mathbb{R}^n$, then

$$\sigma(\lambda, V) = \mathrm{mes}_n\{x \in \mathbb{R}^n : V(x) > \lambda\}. \tag{11.3}$$

For any unit cube $Q \subset \mathbb{R}^n$ we set $V_Q = \int_Q V\,dx$ and denote by $\omega_1(V, t; Q)$ the continuity modulus of V in the metric of $L_1(Q)$.

Theorem 11.2. *Let $V(x) \geq 1$ satisfy (2.21) and let*
a) $\sigma(2\lambda, V) \leq c_1\sigma(\lambda, V)$ for any sufficiently large λ;
b) there exist a number $\beta \in [0, 1/2]$ and a continuous increasing function $\eta(t)$, where $t \in [0, \sqrt{n}]$, such that $\eta(0) = 0$ and

$$\omega_1(V, t; Q) \leq \eta(t)t^{2\beta}V_Q^{1+\beta}$$

for any unit cube $Q \subset \mathbb{R}^n$. Then formula (9.8) is valid for the Schrödinger operator with potential V.

Theorem 11.2 does not cover potentials that grow slower than any power function (condition a) is violated). For such potentials V it turns out that (9.8) is also valid under certain regularity conditions for the growth of V. Regarding this point, see (Bojmatov 1974).

If V is sufficiently regular, then qualified estimates of the remainder can also be obtained in the asymptotic formula for $N(\lambda)$ (see Tamura 1982b). These results can be generalized to a wide class of operators in $\mathbb{R}^n$ whose Weyl symbol $A_W(x, \xi)$ tends to infinity as $|x| + |\xi| \to \infty$ (Hörmander 1979b, 1979c; Helffer and Robert 1982a; Helffer 1984). The Tauberian technique was used in these articles.

For negative potentials that tend to zero as $|x| \to \infty$ the following result was also obtained by the variational technique (Rozenblum 1977). We recall that in this case we are concerned with the asymptotic behaviour of the negative spectrum as $\lambda \to -0$.

Theorem 11.3. *Let $n \geq 3$ and let $V(x) < 0$ be a potential such that $V(x) \to 0$ as $|x| \to \infty$ and the following conditions are satisfied:*

a) $\sigma(\lambda/2, V) \leq c_2\sigma(\lambda, V)$ *for any* $\lambda < 0$ *whose modulus is small enough* (*σ is defined by (11.3)*);

b) $|\nabla V(x)| = o\big(|V(x)|^{3/2}\big)$ *as* $|x| \to \infty$.

Then the asymptotic formula (9.8) is true for $\lambda \to -0$.

In this case one can also find qualified estimates of the remainder in the asymptotic expression for smooth V (see (Tamura 1981, 1982a) and (Ivrii 1986b,c,d), where the Tauberian technique is used).

As has already been mentioned, the variational approach usually fails to produce any good estimates of the remainder in the asymptotic formulae. The case of regular boundary value problems for elliptic operators with constant coefficients is one of the exceptions. It was Courant (Courant and Hilbert 1953) who obtained the estimate $O\big(\lambda^{(n-1)/2}\ln\lambda\big)$ of the remainder. It differs only by the logarithmic factor from the sharp estimate (9.20), which was proved later.

A special version of the variational technique suitable for estimating the remainders in the asymptotic formulae was developed by Métivier (1977). In this connection an important role is played by the two-parameter (Λ and the spectral parameter) estimates of the spectrum of a variational triple $\{d_\Lambda, a, b\}$ such that $d_\Lambda = \{u \in H^r(X) : \mathcal{L}u = \Lambda u\}$ ($\mathcal{L}$ is a given elliptic differential expression of order $2r$), $a[u] = \|u\|^2_{H^r}$, and $b[u] = \|u\|^2_{L_2}$; cf. the problems in Sect. 11.4. The technique was used in (Métivier 1977) to obtain 'Courant type' estimates for operators with constant leading coefficients. Estimates of the form $R(\lambda; A) = O(\lambda^{(n-\theta)/2r})$, where θ is determined by the Hölder index, were obtained for operators with variable (Hölderian) leading coefficients. Such estimates had been obtained earlier in the framework of the resolvent method (see Agmon 1968) under somewhat stronger assumptions concerning the regularity of the coefficients and the boundary of the domain. Subsequently, Métivier's technique was extended to problems of the form $Bu = \lambda Au$ with an elliptic operator A and an arbitrary differential operator B (with variable sign) (Kozlov, V.A. 1983).

An important result of (Métivier 1977) is the proof that the decomposition

$$N(\lambda; A) \sim \Phi_1(\lambda) + \Phi_2(\lambda) \tag{11.4}$$

is valid in quite a general setting, Φ_1 being the Weyl principal term of the asymptotics of A, and Φ_2 being determined by the contribution from a neighbourhood of the boundary. Examples of irregular domains have been constructed, in which Φ_2 turns out to be the leading term in the decomposition (11.4) for $-\Delta_N$.

11.4. Problems with Constraints

Problems with differential constraints (in other words, variational problems on subspaces) constitute a distinctive class of variational problems. We shall present a number of examples.

Example 11.1. The Stokes problem on the spectrum of a variational triple $\{d; a, b\}$ with

$$d = \left\{ u = \{u_j\}_1^3 \in \left(H^1(X)\right)^3 : \operatorname{div} u = 0, \int\limits_X u_j \, dx = 0, \ j = 1, 2, 3 \right\}$$

(X denotes a bounded domain with Lipschitz boundary in $\mathbb{R}^3$) and with

$$a[u] = \int\limits_X \sum_j |\nabla u_j|^2 \, dx, \quad b[u] = \int\limits_X \sum_j |u_j|^2 \, dx.$$

Example 11.2. The Steklov problem on the spectrum of a triple with

$$d = \left\{ u \in H^1(X) : \Delta u = 0, \int\limits_{\partial X} \sigma u \, dS = 0 \right\},$$

$$a[u] = \int\limits_X |\nabla u|^2 \, dx, \quad b[u] = \int\limits_{\partial X} \sigma |u|^2 \, dS.$$

Here $X \subset \mathbb{R}^n$ is a bounded domain with smooth boundary. It is assumed that $\int_{\partial X} \sigma \, dS \neq 0$. The present example is, however, not so typical, since the 'constraint' $\Delta u = 0$ arises as a natural condition in the sense of variational calculus and can be excluded from the original formulation of the problem (see Example 3.11 and (3.16)).

Example 11.3. Let

$$d = \left\{ u \in H^1(X) : \Delta u = 0, \int\limits_X u \, dx = 0 \right\},$$

$$a[u] = \int\limits_X |\nabla u|^2 \, dx, \quad b[u] = \int\limits_X |u|^2 \, dx.$$

The spectrum of the variational triple $\{d; a, b\}$ coincides with that of the operator $\Delta_D^{-1} - \Delta_N^{-1}$ in X.

In the general situation one has to deal with variational problems on vector-valued functions of dimension $k \geq 1$ satisfying scalar differential constraint equations $\mathcal{L}_s u = 0$, where $s = 1, \ldots, l$ with $l \leq k$, and possibly some orthogonality relations. One can talk of an *incomplete system of constraints* for

$l < k$, and a *complete system of constraints* for $l = k$. The articles (Métivier 1978) and (Birman and Solomyak 1982b) are devoted to problems of the former kind. Admitted in (Birman and Solomyak 1982) are also pseudodifferential constraint equations. The Weyl asymptotics is characteristic for such problems. The main difference between the case at hand and the formulae in §9 is that the operator $A^0(z) = A^0(x, \xi)$, the spectrum of which appears in (9.11) and (9.23), now acts on the $(k-l)$-dimensional subspace $\{f \in \mathbb{C}^k : \mathcal{L}_s^0(x,\xi)f = 0,\ s = 1, \ldots, l\}$ of $\mathbb{C}^k$, rather than on the whole space $\mathbb{C}^k$.

In particular, in Example 11.1, this subspace can be written as

$$\{f = \{f_j\}_1^3 \in \mathbb{C}^3 : \ \xi_1 f_1 + \xi_2 f_2 + \xi_3 f_3 = 0\}. \tag{11.5}$$

If one 'removes' the constraint $\operatorname{div} u = 0$ from this example, then the problem reduces to an orthogonal sum of three specimens of the operator $-\Delta_N^{-1}$. Then, according to (9.6) (for $n = 3$ and $r = 2$), $n_+(\lambda) \sim 2v_3(2\pi)^{-3}\lambda^{-3/2}\operatorname{mes} X$. The constraint (11.5) reduces the vector dimension of the problem from three to two. Because of this,

$$n_+(\lambda) \sim \frac{2}{3}\, 3v_3(2\pi)^{-3}\lambda^{-3/2}\operatorname{mes} X = (3\pi^2)^{-1}\lambda^{-3/2}\operatorname{mes} X$$

in Example 11.1.

Problems with a complete system of constraints were considered in (Birman and Solomyak 1979, 1982a). In these articles it was established that, subject to the ellipticity condition for the system of constraints, the spectrum of a problem of this kind coincides asymptotically with that of a pseudodifferential operator of negative order on dX. Explicit formulae can be given for the principal symbol of this operator. Then the evaluation of the asymptotics for the spectrum of the problem under consideration can be reduced to the application of (9.23).

In particular,

$$n_\pm(\lambda) \sim (2\pi)^{-(n-1)} v_{n-1} \int_{\partial X} (\sigma_\pm(x))^{n-1}\, dS\ \lambda^{-(n-1)}$$

in the case of Example 11.2, and

$$n_-(\lambda) = 0, \quad n_+(\lambda) \sim (2\pi)^{-(n-1)} v_{n-1} S\ (2\lambda)^{-(n-1)}$$

in the case of Example 11.3. (S denotes the $(n-1)$-dimensional area of ∂X.)

Example 11.2 can also be considered in the framework of the theory of 'equations with the spectral parameter in the boundary condition.' A very general method for analysing such problems was developed by Kozhevnikov (1973). Spectral asymptotic formulae for the so-called 'singular Green operators' were found in (Grubb 1986), providing one more approach to the study of the spectrum of any problem with a complete system of constraints.

§12 The Resolvent and Parabolic Methods. Spectral Geometry

As has already been mentioned in Sect. 9.7, Tauberian methods for studying the asymptotic behaviour of the spectrum rest on the investigation of the functions

$$\varphi(A, z) = \int \varphi(\lambda, z)\, dE^A(\lambda)$$

of an operator A for a suitable family of functions $\varphi(\lambda, z)$. In the present section we shall consider the Tauberian methods corresponding to the functions $(\lambda - z)^{-1}$, $(\lambda^l - z)^{-1}$, λ^{-z}, and $e^{-\lambda z}$. A common property of these methods is that the corresponding transformation

$$\Phi(A, z) = \operatorname{Tr} \varphi(A, z) = \int \varphi(\lambda, z)\, dN(\lambda; A)$$

of the spectral distribution function 'smooths out' the measure $dN(\lambda; A)$ for each of the methods (the effect is particularly strong for λ^{-z} and $e^{-\lambda z}$). Because of this, $\Phi(A, z)$ turns out to be very regular. It can describe explicitly a large number of 'average' spectral characteristics of A. On the other hand, $\Phi(A, z)$ fails to 'notice' any anomalies in the distribution of the eigenvalues (such as, say, eigenvalues of high multiplicity). As a consequence, for the time being the methods in question cannot be used to identify situations in which the second-order term of the asymptotics exists and is a power function. In this respect the methods are inferior to the hyperbolic equation method; see §13.

The concrete realization of each of the methods under consideration can be divided into two stages. We first construct a sufficiently accurate explicit approximation of $\varphi(A, z)$ (in doing so we use the corresponding differential equation) and find the asymptotics of $\Phi(A, z)$. Then we apply the appropriate Tauberian theorem.

12.1. The Resolvent Method

The resolvent method uses the function $\varphi(\lambda, z) = (\lambda^l - z)^{-1}$ or $\varphi(\lambda, z) = (\lambda - z)^{-l}$. The positive integer l is chosen in such a way that $\varphi(A, z)$ is a trace class operator. In particular, if A is the operator of a regular elliptic boundary value problem in a domain $X \subset \mathbb{R}^n$, then we take $\varphi = (\lambda^l - z)^{-1}$ with $ml > n$, where $m = \operatorname{ord} A$. Below we assume for simplicity that $m > n$. Thus we can set $l = 1$.

Let $G(z;x,y)$ be the Green function of A, that is, the kernel of the resolvent $(A-zI)^{-1}$. Subject to the conditions at hand, G is a continuous function, and, in accordance with (1.17),

$$\Phi(A,z) = \operatorname{Tr}(A-zI)^{-1} = \int_X G(z;x,x)\,dx \tag{12.1}$$

in the scalar case. It follows that the problem can be reduced to the study of the asymptotic behaviour of the Green function for large $|z|$. To obtain the first approximation of $G(z;x,y)$ one can resort to the 'coefficient freezing' procedure, i.e., discard the lower terms in the differential expression $A(x,D)$, fix $x_0 \in X$, and solve the equation $A^0(x_0,D)u - zu = f$ with constant coefficients over the whole space $\mathbb{R}^n$ with the aid of the Fourier transform. This yields

$$u(x) = (2\pi)^{-n} \iint \left(A^0(x_0,\xi) - z\right)^{-1} f(y) e^{i\xi\cdot(x-y)}\,dy\,d\xi.$$

Then the coefficients can be 'unfrozen' and the kernel

$$G_0(z;x,y) = (2\pi)^{-n} \int_{\mathbb{R}^n} \left(A^0(x,\xi) - z\right)^{-1} e^{i\xi\cdot(x-y)}\,d\xi \tag{12.2}$$

taken as an approximation of $G(z;x,y)$. In the regular case this approximation turns out to be sufficient for finding the desired asymptotics. The error $G - G_0$ can be estimated with the aid of a relatively simple argument based on Sobolev's embedding theorems (see Agmon 1965). As a result, for the function (12.1), which is the *Stieltjes transform* of the measure $dN(\lambda;A)$, i.e.,

$$\Phi(A,z) = (SN)(z) \stackrel{\text{def}}{=} \int (\lambda - z)^{-1}\,dN(\lambda;A), \tag{12.3}$$

we obtain the power asymptotics

$$(SN)(z) \sim c(-z)^{n/m-1}, \quad z \to -\infty.$$

Next, we can apply the Hardy-Littlewood Tauberian theorem (see Titchmarch 1946; Levitan and Sargsyan 1970).

Theorem 12.1. *Let $f(\lambda)$ be a non-negative non-decreasing function on $\mathbb{R}^1$ and let $(Sf)(z) \sim (Sg)(z)$ as $z \to -\infty$, where $g(\lambda) = c\lambda^\alpha$ with $\alpha > 0$. Then $f(\lambda) \sim g(\lambda)$ as $\lambda \to +\infty$.*

The above argument proves (9.12)–(9.14'). Since $G(z;x,x)$ is the Stieltjes transform of the spectral function $e^A(\lambda;x,x)$, in the same way one can also obtain the asymptotics (9.18), which is uniform on compact sets $K \subset X$.

The above method of deriving the spectral asymptotic formulae was put forward by Carleman (1936) for second-order operators. Subsequently, it was extended to larger classes of operators in the works of a number of authors.

The most complete result (and perhaps the clearest exposition) is due to Agmon (1965).

A suitable modification of Agmon's estimates made it possible to extend the method to operators with coefficients of moderate smoothness (cf. the survey (Birman and Solomyak 1977a)), the corresponding conditions being much more restrictive than in the case of the variational method.

The construction can also be carried over to elliptic operators in the sense of Douglis-Nirenberg and the boundary value problems for such operators; see (Kozhevnikov 1973; Grubb 1978, 1986).

In problems with the power asymptotic behaviour of the spectrum the approach based on considering the ζ-function of A, that is, the *Mellin transform* of $N(\lambda; A)$, is practically equivalent to the resolvent method. If A is a positive elliptic differential operator, then A^{-z} is a trace class operator for $\operatorname{Re} z > n/m$. If the power asymptotics of $G(z; x, x)$ is known, then it is possible to find the asymptotics of $\zeta(A, z) = \operatorname{Tr}(A^{-z})$ as $z \to n/m$ on the basis of (9.32). Ikehara's Tauberian theorem makes it possible to reconstruct the asymptotics of $N(\lambda; A)$ from that of $\zeta(A, z)$ (see Shubin 1978a). The ζ-function can also be applied in a similar way to other problems on the leading term of the asymptotics. One usually resorts to the ζ-function in this manner when taking the powers of A or its resolvent presents difficulties due to any reason.

To obtain the required estimate of $G - G_0$ when considering the Schrödinger operator $A = -\Delta + V(x)$ with $V(x) \to +\infty$ as $|x| \to \infty$, it is assumed that V satisfies (2.21), $|\nabla V| = O(V^{3/2-\delta})$ with $\delta > 0$ (i.e., the smoothness condition is somewhat stronger than in Theorem 11.2), and $V(x) = O\big(\exp\big(2^{-1}|x - y|\sqrt{V(y)}\big)\big)$. The asymptotics does not, in general, exhibit the power behaviour any more. Thus, in order to prove the Weyl formula (9.8), one must extend Theorem 12.1 to other functions (and kernels $(\lambda - z)^{-l}$ with $l > 1$) that may appear in (12.3). In this case it is advisable to take a power of the resolvent, rather than A itself. Such a generalization was obtained by Keldysh in 1951. Below we present the *Keldysh theorem* in a refined form due to Korenblyum. We use the notation

$$(S_r f)(z) = \int_0^\infty (\lambda - z)^{-r-1}\, df(\lambda).$$

Theorem 12.2 (see Kostyuchenko and Sargsyan 1979). *Let $f(\lambda)$ and $g(\lambda)$ be non-negative non-decreasing functions defined for $\lambda > 0$ and equal to zero in a neighbourhood of $\lambda = 0$. Moreover, let $f(\lambda)$ be differentiable, let $f(\lambda) \to \infty$ as $\lambda \to \infty$, and let*

$$\alpha f(\lambda) < \lambda f'(\lambda) < \beta f(\lambda), \quad \alpha, \beta > 0$$

for any sufficiently large λ. This being the case, if $(S_r f)(z) \sim (S_r g)(z)$ as $z \to -\infty$, r being the integral part of β, then $f(\lambda) \sim g(\lambda)$ as $\lambda \to +\infty$.

In this way one can prove the Weyl formula for a large class of operators of the form $A = A_0 + A_1 + V$, where A_0 is an elliptic operator in $\mathbb{R}^n$ with sufficiently smooth bounded coefficients, $V(x)$ is a potential with unbounded growth at infinity, and the operator A_1 with possibly unbounded coefficients is small compared with A_0+V in a certain sense (see Kostyuchenko 1968). For the operator-valued Schrödinger operator formula (9.10) can also be proved with the aid of the presented method (Kostyuchenko and Levitan 1967).

Going over to non-semi-bounded operators (the Dirac operator and non-semi-bounded elliptic systems can serve as examples), one has to consider the resolvent $(A - zI)^{-1}$ for complex numbers z. The corresponding two-sided Tauberian theorem, which generalizes Theorem 12.2 (see Kostyuchenko and Sargsyan 1979, Chap. X), makes it possible, subject to a number of regularity conditions for the measure μ on $\mathbb{R}^1$, to find the asymptotics of $\mu((0,\lambda))$ and $\mu((-\lambda,0))$ as $\lambda \to \infty$ from that of the Stieltjes transform $\int(\lambda - z)^{-r-1}\,d\mu(\lambda)$ as $z \to \infty$ along a ray. In this manner one can prove formulae of the type $(9.9_\pm)$ and $(9.17_\pm)$ (in particular, see Kostyuchenko and Sargsyan 1979).

12.2. The Case of Non-Weyl Asymptotic Behaviour of the Spectrum

Another method of constructing the resolvent is required in the case of an operator with a non-Weyl asymptotic behaviour of the spectrum (see §10), since the approximation (12.2) of the resolvent is no longer accurate enough.

For problems to which one can apply the scheme presented in Sect. 10.1 a method of constructing the resolvent is also suggested by the scheme. Let A be an operator with the same asymptotic behaviour of the spectrum as a differential (or pseudodifferential) operator $\mathcal{A}$ on a submanifold $\tilde{X} \subset X$ with symbol $\mathcal{A}(y,\eta)$, where $(y,\eta) \in T^*\tilde{X}$, $\mathcal{A}(y,\eta)$ being an operator in an auxiliary Hilbert space $\mathfrak{H}(y,\eta)$. Then the desired approximation of the resolvent is given by the pseudodifferential operator on $\tilde{X}$ with the operator-valued symbol $(\mathcal{A}(y,\eta) - zI)^{-1}$ (this is usually the most complex technical point in the proof). Accordingly,

$$\operatorname{Tr}(A - zI)^{-1} \sim \iint\limits_{T^*\tilde{X}} \operatorname{Tr}\left(\mathcal{A}(y,\eta) - zI\right)^{-1} dy d\eta. \tag{12.4}$$

In many special cases the operator-valued symbol $\mathcal{A}(y,\eta)$ turns out to be a differential operator having sufficiently simple structure with polynomial (or power function) coefficients in a Euclidean space. It is now relatively easy to find the asymptotics of $\operatorname{Tr}(A - zI)^{-1}$ by reducing the task to the use of various homogeneity properties of the operator-valued symbol (see Menikoff and Sjöstrand 1978; Aramaki 1983; Mohamed 1983). The distinct forms of the result for the basic non-Weyl and border cases follow automatically from (12.4).

A considerably more subtle technique of constructing the resolvent is required in the study of operators for which no operator-valued pseudodifferential operator $\mathcal{A}$ can be constructed. Of this kind are, for example, hypoelliptic operators with multiple characteristics and a non-symplectic characteristic manifold (which cannot, therefore, be represented in the form $T^*\tilde{X}$). The spectral asymptotic formulae obtained for such operators do not fit into the framework of (10.1), ($10.1_\pm$); see (Menikoff and Sjöstrand 1979; Mohamed 1983).

12.3. Refinement of the Asymptotic Formulae

For a regular elliptic operator one can also obtain further terms of the expansion of $\operatorname{Tr}(A - zI)^{-l}$ in the powers of z as $z \to \infty$ along a ray on the complex plane. This can be achieved quite easily if there is no boundary, namely, the asymptotic series for the resolvent can be obtained by applying an elementary successive approximation procedure starting from the kernel (12.2). For a boundary value problem one also needs an initial approximation of the Green function near the boundary. Using another terminology, this is related to the feasibility of the analytic continuation of the ζ-function of A as a meromorphic function on the whole complex plane with simple poles on the real axis. The poles were studied in (Seeley 1967, 1969). Generalizations to the case of hypoelliptic operators and operators degenerate at the boundary were obtained in (Smagin 1976) and (Karol' 1981), respectively.

This additional information on the asymptotic behaviour of the trace of the resolvent does not, however, lead to any further terms of the asymptotics of $N(\lambda; A)$. The reason is that since $R(\lambda; A) = N(\lambda; A) - a_0\lambda^{n/m}$ is not a monotone function, neither the Tauberian Theorem 12.1 nor any of its generalizations can be applied to the Stieltjes transform of $R(\lambda; A)$ (even though the Stieltjes transform may have a power asymptotics).

To obtain an estimate of the remainder for $N(\lambda; A)$, one must study the asymptotic behaviour of $(A - zI)^{-1}$ in close proximity to the spectrum of A, for example, along the parabolas $|\operatorname{Im} z| = (\operatorname{Re} z)^\gamma$, where $\gamma < 1$. Here one can apply the *Plejel-Malliavin Tauberian theorem* (see Agmon 1968).

Theorem 12.3. *Let $f(\lambda) \geq 0$ be a non-decreasing function on the semi-axis $\mathbb{R}_+$, let $(Sf)(z)$ be the Stieltjes transform of the measure df, and let Γ_λ be a contour in $\mathbb{C}^1$ that connects the points $z_\pm = \lambda \pm i\mu$ and does not intersect $\mathbb{R}_+$. Then*

$$\left| \frac{1}{2\pi i} \int\limits_{\Gamma_\lambda} (Sf)(z)\, dz - f(\lambda) \right| \leq c\mu \big| (Sf)(z_+) \big|.$$

It follows that the closer one can approach the spectrum when studying the asymptotic behaviour of the resolvent, the more accurate results on the distribution of the spectrum will be obtained. On the other hand, technical difficulties arising in the construction of the resolvent increase as one approaches the spectrum (see Agmon 1968). Having found the asymptotics of the resolvent along the parabola $|\mathrm{Im}\, z| = (\mathrm{Re}\, z)^{1/m}$, Métivier (1982) proved a sharp-order estimate of the remainder for an elliptic operator of order m.

12.4. The Parabolic Equation Method

This method rests on the study of

$$\theta_A(t) = \mathrm{Tr}\, e^{-At} = \sum_j e^{-t\lambda_j(A)} = \int e^{-t\lambda}\, dN(\lambda; A).$$

It is customary to call $\theta_A(t)$ the *θ-function of A*. It is defined for all $t > 0$ whenever A is a lower semi-bounded self-adjoint operator with discrete spectrum for which $N(\lambda; A)$ grows no faster than a power function. In particular, by (5.2), this condition is satisfied for regular elliptic operators. By (5.10) and (5.11), the condition is also satisfied for the Schrödinger operator with a potential $V(x) \geq c|x|^{\varepsilon}$, where $\varepsilon > 0$.

If the spectrum of A is infinite, then it is clear that $\theta_A(t) \to \infty$ as $t \to +0$. Let us assume that the power asymptotics

$$\theta_A(t) \sim ct^{-\gamma}, \quad t \to +0 \tag{12.5}$$

is known. The asymptotics of $N(\lambda; A)$ can be derived from relations of the type (12.5) with the aid of the following theorem.

Theorem 12.4 (Karamata's Tauberian theorem; see (Taylor 1981)). *Let $f(\lambda)$ be a non-decreasing non-negative function on $\mathbb{R}_+$ that grows no faster than a power function and let the asymptotic formula*

$$\int_0^\infty e^{-t\lambda}\, df(\lambda) \sim ct^{-\gamma}, \quad \gamma > 0, \quad t \to +0$$

be valid. Then $f(\lambda) \sim c(\Gamma(\gamma+1))^{-1}\lambda^{\gamma}$ as $\lambda \to +\infty$.

Let, for example, $A = A(x, D)$ be a semi-bounded self-adjoint elliptic operator acting on vector-valued functions of dimension k in a bounded domain $X \subset \mathbb{R}^n$. Theorem 12.4 reduces the problem of finding the leading term of the asymptotics of $N(\lambda; A)$ as $\lambda \to \infty$ to the search for the asymptotics of

$$\theta_A(t) = \mathrm{Tr}\, U(t) = \int_X \mathrm{tr}\, U(t; x, x)\, dx$$

as $t \to +0$. Here $U(t; x, y)$ is the Green function of the parabolic equation (system)

$$\frac{\partial u}{\partial t} + A(x, D)u = 0 \tag{12.6}$$

and $\operatorname{tr} U$ is the trace of the matrix U as an operator in $\mathbb{C}^k$. Moreover, applying Theorem 12.4 to $f(\lambda) = \operatorname{tr} e^A(\lambda; x, x)$, where $e^A(\lambda; x, y)$ is the spectral function of A, one can find the asymptotics of $f(\lambda)$ as $\lambda \to +\infty$ from that of $U(t; x, x)$ as $t \to +0$.

By analogy with Sect. 12.1, in regular situations a sufficiently accurate approximation of $U(t; x, y)$ can be obtained by freezing the coefficients. We fix $x_0 \in X$. Let $U^{x_0}(t; x, y)$ be the fundamental solution of the Cauchy problem for the system

$$\frac{\partial u}{\partial t} + A^0(x_0, D)u = 0. \tag{12.7}$$

The matrix-valued function U^{x_0} admits the explicit representation

$$U^{x_0}(t; x, y) = (2\pi)^{-n} \int_{\mathbb{R}^n} e^{i\xi\cdot(x-y)} \exp\big(-tA^0(x_0, \xi)\big)\, d\xi. \tag{12.8}$$

In many cases (e.g., for all regular problems) even the matrix $U_0(t, x, y) = U^y(t, x, y)$ is a sufficient approximation of $U(t; x, y)$. Then Theorem 12.4 yields the Weyl asymptotics of $N(\lambda; A)$ and $e^A(\lambda; x, x)$ on compact sets in X.

For the Schrödinger operator one can take as U_0 the fundamental solution of the parabolic equation with frozen complete symbol. The same conditions are imposed upon V as in the case when the resolvent method is applied. Then the Tauberian Theorem 12.4 extended in a suitable way beyond the class of power functions leads to a proof of the Weyl formula (9.8) (see Kostyuchenko 1968).

Another modification of the method under consideration, which was proposed in (Menikoff and Sjöstrand 1978, 1979) and (Sjöstrand 1980), made it possible to obtain the asymptotics of the spectrum of any hypoelliptic operator with double characteristics (cf. Sect. 10.4), but without the condition that the characteristic manifold should be symplectic. The kernel $U_0(t; x, y)$ was constructed in the form of the Fourier integral operator

$$U_0(t; x, y) = \int e^{i(\varphi(t;x,\xi) - \xi\cdot y)} a(t, x, \xi)\, d\xi$$

with a complex-valued phase function. The phase φ is chosen in a special way: it takes into account the behaviour of the two leading symbols of A in a neighbourhood of the degeneracy manifold. A similar technique was applied in (Stanton and Tartakoff 1984) to study an operator A with double characteristics in a space of vector-valued functions arising when the Cohn-Laplace operator is reduced to the boundary of the domain; see Sect. 10.5.

To date, no one has succeeded in extending the parabolic equation method to the case of characteristics of multiplicity greater than two.

12.5. Complete Asymptotic Expansion of the θ-Function

Let us go back to the regular case. Starting from the 'zero-order approximation' (12.8), one can easily construct the complete asymptotic expansion of the Green function of equation (12.6). If $B^{x_0} = B^{x_0}(x, D) = A(x, D) - A^0(x_0, D)$, then (12.6) and (12.7) imply that

$$\begin{aligned} U(t; x, y) &= U_0(t; x, y) + \int_0^t \int_X U(s; x, z) B^y(z, D_z) U_0(t-s; z, y)\, ds dz \\ &= U_0 + U * \Phi, \end{aligned}$$

where $*$ is the convolution with respect to s and z, and where $\Phi = B^y(z, D_z)$ $U_0(s; z, y)$. By iterating, we obtain the formal sum

$$U = U_0 + \sum_{k \geq 1} U_0 * \underbrace{\Phi * \ldots * \Phi}_{k}. \tag{12.9}$$

This expansion converges to $U(t; x, y)$ uniformly on any compact set in $\mathbb{R}_+ \times X \times X$. It also turns out to be an asymptotic expansion as $t \to +0$ (see Minakshisundaram and Plejel 1949; McKean and Singer 1967).

If A is an elliptic operator on a compact manifold X without boundary, then the initial approximation U_0 of the fundamental solution can be constructed with the aid of a partition of unity by pasting together the kernels (12.8) found for separate coordinate neighbourhoods. Then, as above, the expansion (12.9) can be obtained for $U(t; x, y)$.

As follows easily from (12.8), the leading term of the asymptotics of the *local θ-function* $\theta_A(t, x) \stackrel{\text{def}}{=} \operatorname{tr} U(t; x, x)$ has the form

$$\theta_A(t, x) \sim \Gamma\left(\frac{m}{n} + 1\right) (2\pi)^{-n} t^{-n/m} \omega_0(x), \quad t \to +0,$$

where $\omega_0(x)$ is defined by (9.14).

In other words, the leading term of the asymptotics of $\theta_A(t)$ can be obtained by integrating a density over X, the latter being a local characteristic of the operator. This result can be refined considerably.

Theorem 12.5 (Minakshisundaram-Plejel). *The function $\theta_A(t)$ admits the asymptotic expansion*

$$\theta_A(t) \sim (2\pi)^{-n} t^{-n/m} \sum_{j=0}^{\infty} a_{2j} t^{2j/m}, \quad t \to +0 \tag{12.10}$$

with $a_{2j} = \int_X \omega_{2j}(x)\, dx$, where $\omega_{2j}(x)$ can be expressed in terms of the values of the coefficients of A and their derivatives of order up to $4j$ at x.

Starting from the explicit representation (12.8), one can derive (12.10) from the expansion (12.9).

In the case of an operator A acting on a manifold X with boundary, the right-hand side of (12.10) must be supplemented with yet another similar series determined by the boundary conditions for A (see (Minakshisundaram and Plejel 1949); a generalization to the case of operators on fibre bundles was given in (Greiner 1971)). After such a modification the expansion of the Green function will converge uniformly on compact sets in $\mathbb{R}_+ \times \bar{X} \times \bar{X}$. Theorem 12.5 can be generalized to boundary value problems in the following way.

Theorem 12.6. *For a regular elliptic operator A on a manifold X with boundary $\theta_A(t)$ admits the asymptotic expansion*

$$\theta_A(t) \sim (2\pi)^{-n} t^{-n/m} \sum_{j=0}^{\infty} a_{2j} t^{2j/m} + t^{-n/m} \sum_{j=1}^{\infty} b_j t^{j/m} \tag{12.11}$$

as $t \to +0$, where the coefficients a_{2j} are the same as in (12.10), and where each b_j is the integral over ∂X of a density, which can be expressed in terms of the coefficients of the differential expression and the boundary conditions along with their derivatives of order up to $2j$.

The expansions (12.10) and (12.11), which are called the *Minakshisundaram-Plejel expansions*, do not even ensure the existence of the second term of the asymptotics of $N(\lambda; A)$. Nevertheless, the coefficients a_{2j} and b_j are important characteristics of the distribution of the spectrum and can be studied in their own right.

12.6. Spectral Geometry

In what follows we shall confine ourselves to second-order operators. Let A be the Laplace-Beltrami operator on a Riemannian manifold X without boundary (see, for example, Sect. 2.2). In local coordinates the densities ω_{2j} can be expressed in terms of the components of the metric tensor $g = \{g_{ik}\}$ and its derivatives. It follows that formula (12.10) establishes a connection between the spectral, i.e., global characteristics of A and the local geometric characteristics of the manifold. Formula (12.11) plays the same role in the case of the Laplace-Beltrami operator on a manifold with boundary and, for definiteness, with the Dirichlet condition on ∂X. The objective of spectral geometry is the search for the explicit expressions for the coefficients a_{2j} and b_j corresponding to these and other elliptic operators arising on a Riemannian manifold in a natural way.

In principle, any number of Minakshisundaram-Plejel coefficients can be found starting from (12.9) (or the analogue of (12.9) in the case of a boundary value problem). The first term a_0 of the expansion is proportional to the Riemannian volume of X. Next, $b_1 = \pm\sqrt{\pi}S/2$, where S is the $(n-1)$-dimensional volume of ∂X in the Riemannian metric induced from X. The plus sign corresponds to the Neumann problem, while the minus sign corresponds to the Dirichlet problem. Next,

$$a_2 = -\frac{1}{3}\int_X K(x)\,dx, \quad b_2 = -\frac{1}{6}\int_{\partial X} J(x)\,dS,$$

where $K(x)$ is the scalar curvature of X at $x \in X$, and where $J(x)$ is the mean curvature of the boundary, i.e., the trace of the second fundamental form of the boundary multiplied by two.

12.7. Computation of Coefficients

An algebraic approach to the computation of the densities $\omega_j(x)$ and coefficients a_j, b_j was developed in (Patodi 1971) and (Gilkey 1975). It facilitates a significant simplification of the computational procedure by taking into account the a priori homogeneity and symmetry properties of $\omega_j(x)$.

We consider the expansion (12.10) for the Laplace-Beltrami operator on a manifold X without boundary. We fix semi-geodesic coordinates $\{y^\mu\}$ in a neighbourhood of $x_0 \in X$, so that $g_{ik}(x_0) = \delta_{ik}$. Then, by the Cartan theorem, the Taylor expansion of the metric tensor in a neighbourhood of x_0 has the form

$$g(y) = \{g_{ik}(y)\} = \delta_{ik} + g^{(1)} + g^{(2)} + \dots\,, \tag{12.12}$$

where $g^{(j)}$ are universal polynomials of degree j with respect to the coordinates y and the components of the curvature tensor $R = R^i_{\mu\nu k}$ and their covariant derivatives at x_0. By saying that the polynomials are universal we mean that they have the same form for any manifold of arbitrary dimension and any coordinate system. In particular,

$$g^{(1)}_{ik} = \frac{2}{3}\sum_{\mu,\nu} y^\mu y^\nu R_{i\mu\nu k}.$$

The analysis of (12.10) with the substitution (12.12) indicates that the coefficients $\omega_j(x_0)$ must also be universal polynomials of $R(x_0)$ and the covariant derivatives of R:

$$\omega_j(x_0) = Q_j\big(R(x_0), R'(x_0), \dots\big). \tag{12.13}$$

Next, the homogeneity of $\omega_j(x_0)$ under the transformation $g \mapsto sg$ of the metric implies that Q_j in (12.13) must be a homogeneous polynomial of degree $2j$ if the degree associated with the l-th order covariant derivative of R is $2+l$. Finally, one should use the fact that $\omega_j(x_0)$ must be invariant under orthogonal transformations that map distinct semi-geodesic coordinate systems with centre at x_0 into one another. According to the Weyl theorem on the invariants of the orthogonal group, every invariant polynomial (12.13) must be a linear combination with constant coefficients of elementary homogeneous $O(n)$-invariant terms of the form

$$m(R) = \sum{}' R_{\alpha_1} \dots R_{\alpha_r},$$

where $R_\alpha = R_{\alpha^1 \dots \alpha^4, \alpha^5 \dots \alpha^q}$ are the components of the covariant derivative of order $q-4$ of the covariant curvature tensor for any multi-index α of length $q \geq 4$, and where $\sum'$ denotes the sum with respect to all the indices.

As a result, the search for the densities $\omega_j(x)$ can be reduced to the algebraic problem of enumerating all the elementary terms with a prescribed degree of homogeneity and finding all the constant coefficients that multiply these terms. Taking into account the universality of the coefficients to be found, one can solve the latter problem by considering sufficiently many examples, the geometry of which is simple enough, and for which the coefficients can be found explicitly.

In (Patodi 1971) this programme was implemented in the case of the Laplace-Beltrami operator on a manifold with or without boundary. Then it was applied to the Laplace operator on p-forms (Gilkey 1975). Subsequently, Gilkey and Smith (Smith 1981; Gilkey and Smith 1983) extended the method at hand to second-order operators acting on the sections of a vector bundle over a manifold with or without boundary, subject to the condition that the principal part of the operator must coincide with $-\Delta$. In particular, for the latter problem the coefficients a_2, a_4, and a_6 were computed and expressed geometrically. We should like, however, to remark that even for a_6 the dimension of the corresponding space of invariant homogeneous polynomials equals 46 (for large n), so that we arrive very quickly at insurmountable computational problems.

12.8. The Problem of Reconstructing the Metric from the Spectrum

In the case of the Laplace operator on p-forms over an n-dimensional manifold X without boundary the formulae for a_0, a_2, and a_4 read

$$a_0 = \binom{n}{p} \operatorname{Vol} X,$$

$$a_2 = \int_X \left(\frac{1}{6}\binom{n}{p} - \binom{n-2}{p-1} \right) K(x)\, dx,$$

$$a_4 = \int_X \left(C_1 K^2 + C_2 |E|^2 + C_3 |R|^2 \right) dx,$$

where the coefficients C_1, C_2, and C_3 depend on n and p, and where $R = R^i_{\mu\nu k}$ is the curvature tensor, $E = E_{\mu\nu} = R^i_{\mu i \nu}$ is the Ricci curvature, and

K is the scalar curvature (see Berger et al. 1971). In some cases the study of these quantities makes it possible to solve the problem concerned with the characterization of the manifold by the spectrum. In particular, one can determine from the spectrum whether or not the manifold is flat, Einsteinian, Kählerian, and so on. On this subject see the book (Berger et al. 1971) and the later surveys (Singer 1974; Gilkey 1975; Bérard 1986).

The following problem is connected with the above range of questions: To what extent does the spectrum of the Laplace-Beltrami operator determine the Riemannian manifold X? It must be noted right now that, in general, the manifold is not uniquely determined even by the complete system of coefficients a_j, or, more generally, any system of integrals of local densities. For example, let X be diffeomorphic to the sphere S^n and let the metric of $X = X_{\gamma\gamma'}$ have the form $g(x) = g_0(x)(1 + \varphi(\gamma x) + \varphi(\gamma' x))$, where $\varphi(x)$ is a smooth function with a sufficiently small support, g_0 is the standard metric on S^n, and γ, γ' are two elements of the isometry group $O(n+1)$ of S^n. If the supports of $\varphi(\gamma x)$ and $\varphi(\gamma' x)$ do not intersect each other, then the integral of any local density over X is independent of γ, γ'. (This argument belongs to Molchanov (1975)).

On the other hand, some manifolds of special form corresponding to the extremum values of the coefficients a_j or their linear combinations are uniquely characterized by those coefficients. If the coefficients a_0, a_2, and a_4 for the Laplace operator on functions and p-forms over a manifold X are the same as in the case of S^n, then X is isometric to S^n. For $n > 7$ it is not known whether or not S^n can be characterized by the spectrum of $-\Delta$ on the set of functions alone, even though the sphere is uniquely characterized by the spectrum of $-\Delta$ on p-forms with $p = [n/3]$ for such n (see Tanno 1980).

The standard metric on S^n and other compact spaces of constant curvature is *spectrally isolated*, which means that if a metric g is sufficiently close to g_0 and either the spectrum of the corresponding Laplace operator on functions, or at least the coefficients a_0, a_2, and a_4 are the same as for S^n, then $g = g_0$. It follows that such spaces are *spectrally rigid*, by which we mean that there are no non-trivial deformations preserving the spectrum, that is, *isospectral deformations*. Spaces of constant zero or negative curvature are also spectrally rigid (Tanikawa 1979; Guillemin and Kazhdan 1980; Ikeda 1980a). Examples of lens spaces (Ikeda 1980b; 1983) (see §16) indicate that *isospectral*, i.e., having the same spectrum, spectrally isolated manifolds can fail to be isometric (or even diffeomorphic).

12.9. Connection with Probability Theory

In the conclusion we shall describe the probabilistic treatment of a second-order parabolic equation and the related approach to the study of $\theta(t)$.

For a diffusion process Z_t on a smooth manifold X that vanishes on ∂X (see Molchanov 1975) one can introduce the characteristic operator

$$(Af)(x) = \lim_{V \to x} \frac{Ef(Z_{\tau_V}) - f(x)}{E\tau_V},$$

where E is the mathematical expectation, V is a neighbourhood of x that contracts to x, and τ_V is the first exit time from V of the process starting at x. It turns out (Molchanov 1975) that A is a second-order differential operator. The leading coefficients of A define a Riemannian metric on X. The operator A itself takes the form $A = \Delta/2 + \varphi$, where Δ is the Laplace-Beltrami operator on X and φ is a smooth function. Conversely, with any Riemannian manifold X one can associate a diffusion process whose characteristic operator A is equal to $\Delta/2$.

The transition density $p(t; x, y)$ of Z_t satisfies the parabolic equation

$$\frac{\partial p}{\partial t} - Ap = 0, \quad p\Big|_{\partial X} = 0, \quad p(0) = \delta. \tag{12.14}$$

In particular, $p(t; x, x)$ corresponds to the density of the probability that Z will return to the initial point x at time t.

Equation (12.14) and, in particular, the asymptotic behaviour of $p(t; x, y)$ as $t \to 0$ were studied by probabilistic methods in (Kac 1959, 1966) and (Louchard 1969). For the most advanced results in this direction see (Molchanov 1975), where, in particular, a method of computing all the coefficients of the expansion of $p(t; x, x)$ was proposed for a two-dimensional manifold with boundary. The method yields the coefficients of the expansion of $\theta_A(t)$ up to b_5. In the three-dimensional case a modification of this approach produces the coefficients up to a_4 and b_4. In (Simon 1983a) the non-Weyl asymptotics for the Schrödinger operator $-\Delta + |x|^\alpha |y^\beta|$ in $\mathbb{R}^2$ was found with the aid of the probabilistic method.

§13 The Hyperbolic Equation Method

The hyperbolic equation method turns out to be the most effective one in problems connected with the study of fine asymptotic properties of the spectrum. A number of important results on sharp-order estimates of the remainder, on the second term of the spectral asymptotics, and also on the connection between the asymptotic properties of the spectrum and the geometry of bicharacteristics have been obtained in recent years; §14 is devoted to the latter problem.

As has already been mentioned, the method at hand can also be extended to certain non–semi-bounded operators. For a non–semi-bounded operator A one usually introduces the projections $E^A_\pm = E^{\pm A}(0,\infty)$ onto the positive and negative subspaces and considers A in each of the subspaces separately. The wave equation method can be applied when the projections $E^A_\pm$ have 'good' properties (for example, when they are pseudodifferential operators). We shall not dwell on this problem, and, for the sake of simplicity, we always assume that $A \geq \operatorname{const} I$.

13.1. Tauberian Theorem for the Fourier Transform

Various Tauberian theorems for the Fourier transform can be used in the method under consideration. We shall present one of them, namely, that for the case of the exponential transform. For other transforms the results are similar. The formulation to be presented is taken from (Safarov 1986). It is a sharpened version of a number of similar assertions from (Hörmander 1968; Levitan 1971; Shubin 1978a; Ivrii 1980).

Let $Z(\lambda)$ be a non-decreasing function that grows like a power function on $\mathbb{R}^1$. We denote by $h(t)$ the *Fourier transform of the measure* $dZ(\lambda)$:

$$S'(\mathbb{R}^1) \ni h(t) = \int e^{-i\lambda t}\, dZ(\lambda).$$

Let us fix an even real-valued function $\tilde{\rho} \in C_0^\infty(\mathbb{R}^1)$ such that $\tilde{\rho}$ and its Fourier transform ρ satisfy the following conditions: $\tilde{\rho}(\tau) \geq 0$, $\tilde{\rho}(0) = 1$, $\rho(t) \geq 0$, and $\rho(0) > 0$. We set $\tilde{\rho}_T(t) = \tilde{\rho}(t/T)$ for any $T > 0$ and consider the convolution

$$(\tilde{\rho} * dZ)(\lambda) = (\widetilde{\rho h})(\lambda) = (2\pi)^{-1} \int e^{i\lambda t} \rho(t) h(t)\, dt.$$

Since $Z(\lambda)$ is polynomially bounded,

$$(\tilde{\rho} * dZ)(\lambda) \leq c(|\lambda|^\varkappa + 1) \tag{13.1}$$

for some $\varkappa > 0$.

The behaviour of $\tilde{\rho} * dZ$ as $\lambda \to +\infty$ is determined by the singularities of the distribution $h(t)$ in the support of ρ. Since $(\tilde{\rho} * dZ)(\lambda) = (\tilde{\rho} * Z)'(\lambda)$, these singularities also determine the asymptotic behaviour of $\tilde{\rho} * Z$ with accuracy up to $O(1)$. In turn, the Tauberian theorem makes it possible to obtain the asymptotics of $Z(\lambda)$ from the latter.

Theorem 13.1. *Let $\varkappa > 0$ and let (13.1) be satisfied. Then*

$$(\tilde{\rho}_T * dZ)(\lambda - \varepsilon) - c(\varepsilon^{-1}T^{-1}|\lambda|^{\varkappa} + 1) \le Z(\lambda)$$
$$\le (\tilde{\rho}_T * dZ)(\lambda + \varepsilon) + c(\varepsilon^{-1}T^{-1}|\lambda|^{\varkappa} + 1) \qquad (13.2)$$

for every $\varepsilon > 0$*, where the constant depends only on* $\varkappa$ *and* ρ*.*

By (13.1) and (13.2), if $T = 1$, then

$$Z(\lambda) = (\tilde{\rho} * dZ)(\lambda) + O(\lambda^{\varkappa}), \quad \lambda \to +\infty. \qquad (13.3)$$

It follows that if the singularities of $h(t)$ are known in a neighbourhood of $t = 0$, then the asymptotic behaviour of $Z(\lambda)$ can be determined immediately with accuracy up to $O(\lambda^{\varkappa})$. If the singularities of $h(t)$ are known for all t, then one can expect to obtain an asymptotic formula with remainder $o(\lambda^{\varkappa})$ by letting T tend to $+\infty$ in (13.2). In doing so one can neglect any 'minor' singularities, which introduce a contribution of order $o(\lambda^{\varkappa})$ into $(\tilde{\rho}_T * Z)(\lambda)$.

To make the asymptotics of $Z(\lambda)$ more accurate one must know not only the singularities of $h(t)$, but also the behaviour of $h(t)$ as $t \to \infty$ (see Volovoj 1987).

When the asymptotics of the spectrum of a self-adjoint differential operator A is computed, $Z(\lambda)$ is usually replaced by $N(\lambda^m; A) = N(\lambda; B)$, where $m = \operatorname{ord} A$ and $B = A^{1/m}$, or, in more complicated situations, by $N_{\Gamma}(\lambda; B) = \operatorname{Tr}(E^B(-\infty, \lambda)\Gamma)$, where Γ is a suitably chosen bounded operator. When computing the asymptotics of the spectral function, one most frequently takes $Z(\lambda) = e^B(\lambda; x, x)$. In these cases, as a rule, $\varkappa = n - 1$.

We set $U(t) = \exp(-itB)$. The Green function $U(t; x, y)$ of the Cauchy problem

$$u_t + iBu = 0, \quad u(0) = u_0 \qquad (13.4)$$

is the kernel of this operator. The function $U_{\Gamma}(t) = U(t)\Gamma$ solves the operator equation

$$\frac{dU_{\Gamma}}{dt} + iBU_{\Gamma} = 0, \quad U_{\Gamma}(0) = \Gamma. \qquad (13.4_{\Gamma})$$

If B is an operator with discrete spectrum and $P(\lambda)$, where $\lambda \in \sigma(B)$, are the orthogonal projections onto the eigenspaces of B, then (see Example 1.1)

$$U_{\Gamma}(t) = \sum_{\lambda \in \sigma(B)} \exp(-i\lambda t) P(\lambda) \Gamma.$$

Therefore

$$B_{\rho} = \int \rho(t) U_{\Gamma}(t)\, dt = \sum_{\lambda \in \sigma(B)} \tilde{\rho}(\lambda) P(\lambda) \Gamma$$

will be a trace class operator whenever the function $\tilde{\rho}$ decreases fast enough and the eigenvalues of B increase not too slowly (in particular, it is sufficient that $\sum_{\lambda \in \sigma(B)} \tilde{\rho}(\lambda)$ be an absolutely convergent series). For example, if A is a regular elliptic operator and $B = A^{1/m}$, then this will be the case for any $\tilde{\rho} \in C_0^{\infty}(\mathbb{R}^1)$.

It follows that the functional $\rho \mapsto \operatorname{Tr} B_\rho$ defines a distribution σ_Γ on $\mathbb{R}^1$. It is natural to treat this distribution as the trace of $U_\Gamma(t)$:

$$\sigma_\Gamma(t) = \operatorname{Tr} U_\Gamma(t).$$

It is the Fourier transform of the measure $dE(\lambda) = d(\operatorname{Tr} E^B(\lambda)\Gamma)$. By analogy, the Schwartz kernel $U(t; x, y)$ of $U(t)$ is the Fourier transform of the (complex-valued) measure $de(\lambda) = d_\lambda e^B(\lambda; x, y)$. Thus, to construct the asymptotics of $N_\Gamma(\lambda)$ and $e^B(\lambda; x, x)$, it suffices to study the singularities of $\sigma_\Gamma(t)$ and $U(t; x, x)$ (for a fixed x).

This approach can lead to the goal only if equation (13.4) turns out to be hyperbolic in the appropriate sense. Then the singularities of the solutions propagate regularly enough at finite speed. In terms of these singularities one can effectively describe those of $\sigma_\Gamma(t)$ and $U(t, x, x)$. The latter functions have an isolated singularity at $t = 0$. Computations indicate that it is a power singularity, and, in both cases,

$$(\tilde{\rho} * Z)(\lambda) = c_0\lambda^n + c_1\lambda^{n-1} + o(\lambda^{n-1}) \tag{13.5}$$

if the support of $\tilde{\rho}$ is sufficiently small. The one-term asymptotics

$$Z(\lambda) = c_0\lambda^n + O(\lambda^{n-1}), \quad \lambda \to +\infty$$

with a sharp estimate of the remainder can now be obtained immediately from (13.3). If the non-zero singularities of $\sigma_\Gamma(t)$ and $U(t, x, x)$ are 'weaker' than the singularity at zero, then (13.5) is also true for $(\tilde{\rho}_T * Z)(\lambda)$ for all $T > 1$. In this case (13.2) yields the two-term asymptotics (when $T \to +\infty$, $\varepsilon \to 0$, and $\varepsilon T \to +\infty$)

$$Z(\lambda) = c_0\lambda^n + c_1\lambda^{n-1} + o(\lambda^{n-1}), \quad \lambda \to +\infty.$$

The change from A to $B = A^{1/m}$ is just what is needed to ensure hyperbolicity.

Example 13.1. Let A be the operator $-d^2/dx^2$ on the circle S^1. The function

$$\operatorname{Tr} \exp(-itA) = \sum_n \exp(-itn^2)$$

is known as the 'modular function' (see Serre 1970). Its singularities exhaust the set $\mathbb{R}^1$. At the same time the function

$$\operatorname{Tr} \exp(-itA^{1/2}) = \sum_n \exp(-it|n|)$$

(see (14.1)) has singularities only at $2\pi k$, where $k \in \mathbb{Z}$.

In a number of cases it proves more convenient to consider, in place of (13.4_Γ), another hyperbolic equation connected with A. In particular, for a second-order differential operator A it is natural to begin with the wave equation

$$\frac{d^2U_\Gamma}{dt^2}+AU_\Gamma=0,\quad U_\Gamma(0)=\Gamma,\quad \dot{U}_\Gamma(0)=0. \tag{13.6}$$

The solution of this equation reads $U_\Gamma(t)=\cos(tA^{1/2})\Gamma$, which corresponds to the cosine Fourier transform of the spectral function.[5] The advantage of (13.6) as compared to (13.4_Γ) is that (13.6) is a differential problem, while $B=A^{1/2}$ fails to be a differential operator, which gives rise to certain difficulties (especially for a manifold with boundary).

13.2. Outline of the Method

The hyperbolic equation method was first applied by Levitan (1952) to obtain an estimate of the remainder of the form

$$e^A(\lambda;x,x)-\omega_0(x)\lambda^{n/2}=O\big(\lambda^{(n-1)/2}\big) \tag{13.7}$$

for the spectral function of a second-order elliptic operator A in a domain $X\subset\mathbb{R}^n$. The estimate (13.7) fails to remain uniform as x approaches the boundary. At the same time, for an operator on a compact manifold X without boundary the estimate (13.7) is uniform with respect to $x\in X$. On integrating this estimate, one can obtain an estimate of the form (9.20) in the asymptotic formula for $N(\lambda;A)$.

In (Levitan 1952) the Green function $U(t;x,y)$ of equation (13.6) was constructed with the aid of the Hadamard method for small t. This was done using the fact that for $|t|<cd(y)$ (where $d(y)=\operatorname{dist}(y,\partial X)$, and where c is the ellipticity constant of A) the Green function $U(t;x,y)$ 'does not feel the boundary,' that is, coincides with the fundamental solution of (13.6) in the whole space, which follows immediately from the property of finite propagation speed.

Next, Hörmander (1968) considered the case of a scalar elliptic differential operator A of arbitrary order m. To begin with, let X be a compact manifold without boundary. We consider equation (13.4_Γ) with $B=A^{1/m}$ and $\Gamma=I$. It follows that B is a pseudodifferential operator of order one (see Taylor 1981; Tréves 1982; Egorov and Shubin 1988b). Let $V\subset X$ be an arbitrary coordinate neighbourhood. The operator B acts on functions with compact support in V as follows (in local coordinates):

$$(Bu)(x)=(2\pi)^{-n}\int\limits_{\mathbb{R}^n}(1-\vartheta(\xi))e^{i\xi\cdot(x-y)}B(x,\xi)\tilde{u}(\xi)\,d\xi+(Tu)(x).$$

Here T is a *smoothing operator*, that is, an integral operator with an infinitely differentiable kernel. The symbol $B(x,\xi)$ can be expanded into the asymptotic series

[5] The sine Fourier transform was used by Ivrii (see Ivrii 1984).

$$B(x,\xi) \sim B_1(x,\xi) + B_0(x,\xi) + B_{-1}(x,\xi) + \dots , \tag{13.8}$$

where B_j belongs to the space O_j of positively homogeneous functions of degree j in ξ for each $j = 1, 0, -1, \dots$, and where $\vartheta \in C_0^\infty(\mathbb{R}^n)$ is equal to one in a neighbourhood of zero. Although the expansion (13.8) changes as one goes over to another local coordinate system, the principal symbol B_1 is invariant as a function on T^*X (as in the case of differential operators). For our operator $B = A^{1/m}$ the principal symbol is equal to $B_1(x,\xi) = (A_m(x,\xi))^{1/m}$.

The fundamental solution of equation (13.4_Γ) can be constructed with the aid of *Fourier integral operators*. Let Γ be the operator of multiplication by a function $\varphi \in C^\infty(X)$ with small support. We shall seek the kernel $U_\Gamma(t;x,y)$ of $U_\Gamma(t)$ for any small t in the form of the oscillatory integral

$$U_\Gamma(t;x,y) = (2\pi)^{-n} \int\limits_{\mathbb{R}^n} e^{i(\Phi(t,x,\theta)-\theta\cdot y)} q(t,x,y,\theta)(1-\vartheta(\theta))\, d\theta, \tag{13.9}$$

where $\theta \in \mathbb{R}^n$, with an *amplitude* $q(t,x,y,\theta)$ admitting the asymptotic expansion

$$q \sim q_0 + q_{-1} + q_{-2} + \cdots,$$

where $q_j \in O_j$, and with a real-valued *phase function* $\Phi(t,x,\theta) - \theta \cdot y \in O_1$. Here it is assumed that x varies in a small neighbourhood of the support of φ, and the same coordinate system is chosen for x and y. The integral (13.9) is divergent, in general. However, subject to certain non-degeneracy conditions, it can be interpreted as a distribution on $X \times X$, which depends on t as a parameter. We require that all the homogeneous terms should vanish as a result of the formal substitution of (13.9) into (13.4_Γ). In the end, this ensures that $U_\Gamma(t;x,y)$ will differ from the Green function by a smooth term, which is sufficient for our purposes. To compute the result of the substitution one should use the formulae describing the action of the pseudodifferential operator upon a rapidly oscillating exponent (Fedoryuk and Maslov 1976; Taylor 1981; Tréves 1982). We present only the first two terms of the corresponding asymptotic expansion:

$$\begin{aligned} e^{-i\psi(x)\tau} B(x,D_x)\big(e^{i\psi(x)\tau} g(x)\big) &\sim B_1(x,\nabla\psi) g(x)\tau \\ &+ \big(\mathcal{L}(x,D_x) + H(x)\big) g + O(\tau^{-1}), \quad \tau \to \infty, \end{aligned} \tag{13.10}$$

where

$$\mathcal{L}g = \sum_{|\alpha|=1} B_1^{(\alpha)}(x,\nabla\psi) D_x^\alpha g,$$

$$H = B_0(x,\nabla\psi) + i \sum_{|\alpha|=2} B_1^{(\alpha)}(x,\nabla\psi) \frac{D_x^\alpha \psi}{\alpha!}, \quad B_1^{(\alpha)} = \left(\frac{\partial}{\partial\xi}\right)^\alpha B_1.$$

Thus, substituting (13.9) into (13.4) and equating to zero the homogeneous terms in the resulting expansion of the form (13.10), we arrive at the equations

$$\frac{\partial \Phi}{\partial t} + B_1\left(x, \frac{\partial \Phi}{\partial x}\right) = 0 \tag{13.11}$$

and

$$\frac{\partial q_0}{\partial t} + \mathcal{L}q_0 + Hq_0 = 0. \tag{13.12}$$

The former equation is called the *eikonal equation*. The latter is called the *transport equation*. Using the lower-order terms of the expansion (3.10) (which are not presented here), one can obtain the transport equations for $q_{-1}, q_{-2}, \dots,$ which are similar to (3.12), but no longer homogeneous. The initial conditions for (3.11) and (3.12) must be stated in such a way that $U_\Gamma(0) = \Gamma$ to within an operator with an infinitely smooth kernel. For example, it proves convenient to set

$$\Phi(0, x, y, \theta) = x \cdot \theta, \quad q_0(0, x, y, \theta) = \varphi(y),$$
$$q_j(0, x, y, \theta) = 0, \quad j < 0.$$

Then

$$U(0; x, y) = (2\pi)^{-n} \int e^{i\xi\cdot(x-y)} \varphi(y)(1 - \vartheta(\theta))\, d\theta = \varphi(y)\delta(x-y)$$

to within a C^∞-function.

Equation (3.11) can be solved with the aid of the Hamilton-Jacobi method. For fixed y and θ we consider the *Hamilton system*

$$\frac{dx}{dt} = \frac{\partial B_1}{\partial \xi}, \quad \frac{d\xi}{dt} = -\frac{\partial B_1}{\partial x} \tag{13.13}$$

with initial conditions $x(0) = x_0$ and $\xi(0) = \theta$. The integral curves $x(t; x_0, \theta)$ and $\xi(t; x_0, \theta)$ of the system (13.13) in $T'X$ are called the *bicharacteristics* of B (and also of A and of the symbols B_1 and A_m). We consider the surface $S_t = \{(x(t; x_0, \theta), \xi(t; x_0, \theta)), x_0 \in X\} \subset T'X$. Since S_t is close to the surface $S_0 = \{(x, \theta), x \in X\}$ if t is small enough, it follows that it can be projected onto X without singularities (using the natural projection $\pi : T'X \to X$). It follows that the function

$$x_0 \mapsto x(t; x_0, \theta) \tag{13.14}$$

is invertible. We set $x_0 = x_0^t(x)$. Then $\Phi(t, x, y)$ can be found by integration along the bicharacteristic starting from $x_0^t(x)$:

$$\Phi(t, x, \theta) = x_0^t(x, \theta).$$

This being the case,

$$\frac{\partial \Phi(t, x, \theta)}{\partial x} = \xi(t). \tag{13.15}$$

Next, in view of (13.15), the transport equation (13.12) can be transformed to the form

$$\frac{dq_0}{dt} = \frac{\partial q_0}{\partial t} + \frac{dx(t)}{dt}\frac{\partial q_0}{\partial x} = Hq_0,$$

and it can therefore be solved explicitly by integration along the bicharacteristics. In a similar way one can also find the solutions of the remaining transport equations for small t. This completes the formal construction of $U_\Gamma(t;x,y)$ in the form (13.9) for small t. The above discussion can be justified. Indeed, (13.9) yields $U_\Gamma(t;x,y)$ to within a smooth term. For small t the Green function $U(t;x,y)$ can, in turn, be expressed as a finite sum of oscillatory integrals of the form (13.9) to within a smooth additive term.

Now, if the support of ρ is sufficiently small, then the representation (13.9) can be used to compute the integral $\int \tilde{\rho}(t)e^{i\lambda t}U(t;x,x)\,dt$. Hence one can derive the complete asymptotic power series in λ for $(\tilde{\rho}U)(\lambda;x,x)$, which begins with the following terms:

$$(\tilde{\rho}U)(\lambda;x,x) \sim n\omega_0(x)\lambda^{n-1} + (n-1)\omega_1(x)\lambda^{n-2} + \dots . \tag{13.16}$$

By (13.1), this expansion implies the following asymptotic formula for the spectral function with a sharp estimate of the remainder:

$$e^A(\lambda^m;x,x) = e^B(\lambda;x,x) = \omega_0(x)\lambda^n + O(\lambda^{n-1}). \tag{13.17}$$

Example 4.7 indicates that this estimate cannot be improved, in general.

A modification of the above discussion makes it also possible to obtain the estimate $e^B(\lambda;x,y) = O(\lambda^{n-1})$, which is uniform on any compact set in $X \times X \setminus \{x = y\}$.

In the case of a compact manifold without boundary currently under consideration the estimate (13.17) is uniform in $x \in X$ and can be integrated to obtain the estimate (9.20) for $N(\lambda;A)$.

The estimate (13.17) can be carried over to the case of a manifold with boundary. This is connected with the locality of (13.17): if two self-adjoint differential operators A_1 and A_2 on manifolds X_1 and X_2 coincide (have the same coefficients) on an open subset $X_3 \subset X_1 \cap X_2$ and the estimate (13.17) is satisfied for A_1 uniformly on any compact set in X_3, then this is also the case for A_2. However, as one approaches the boundary the estimate fails to remain uniform and can no longer be used directly to obtain (9.20).

13.3. Global Fourier Integral Operators

As before, let Γ be the operator of multiplication by a C^∞ function with small support. The main obstacle in the construction of the kernel $U_\Gamma(t;x,y)$ of the form (13.9) for large t is that the projection of the surface S_t onto X may develop singularities starting from some t_0. The set of singular points $x \in X$ of the projection of S_t is called the *caustic* (see (Babich 1987) and §14). The function (13.14) fails to be invertible near the caustic. Consequently, in this region the eikonal equation cannot be solved by the Hamilton-Jacobi method. Moreover, in general, the eikonal equation has no smooth solutions in a neighbourhood of the caustic.

This difficulty can be overcome in the following way. We note that the Schwartz kernel of the composition of two operators defined by the oscillatory integrals (13.9) has the form

$$\int \exp i\big(\Phi_1\big(t^{(1)}, x, \theta^{(1)}\big) - \theta^{(1)} \cdot z + \Phi_2\big(t^{(2)}, z, \theta^{(2)}\big) - \theta^{(2)} \cdot y\big)$$
$$\times\, q^{(1)}\big(t^{(1)}, x, z, \theta^{(1)}\big) q^{(2)}\big(t^{(2)}, z, y, \theta^{(2)}\big)$$
$$\times\, \big(1 - \vartheta^{(1)}(\theta^{(1)})\big)\big(1 - \vartheta^{(2)}(\theta^{(2)})\big) dz d\theta^{(1)} d\theta^{(2)}.$$

On changing the variables to $\tilde{z} = \big(|\theta^{(1)}|^2 + |\theta^{(2)}|^2\big)^{1/2} z$ and setting $\theta = \big(\theta^{(1)}, \theta^{(2)}, \tilde{z}\big)$, we obtain the oscillatory integral

$$\int \exp i\psi\big(t^{(1)}, t^{(2)}, x, y, \theta\big) q\big(t^{(1)}, t^{(2)}, x, y, \theta\big)\big(1 - \vartheta(\theta)\big)\, d\theta,$$

where $\theta \in \mathbb{R}^{3n}$, in which the phase function is homogeneous in θ and $q \sim \sum_j q_j$ with $q_j \in O_j$. By analogy, on multiplying operators with such kernels, we obtain an operator whose kernel is again given by a similar oscillatory integral (with a greater number of phase variables). Since the operators $U(t)$ form a group, the kernel $U(t; x, y)$ can be represented as a finite sum of oscillatory integrals

$$\int \exp \Psi(t, x, y, \theta) q(t, x, y, \theta)(1 - \vartheta(\theta))\, d\theta \tag{13.18}$$

on any bounded time interval. (Here x and y can vary in different coordinate neighbourhoods, in general.)

The representation (13.18) is inconvenient because the number of phase variables grows with no limit as t increases. By changing the variables in a special way, (13.18) can be transformed into an integral of the same form with $\theta \in \mathbb{R}^n$ (see Hörmander 1968, 1983–1985). Such a reduction can be executed in various ways, and various phase functions will then be obtained. There arises the natural question about the conditions under which two kernels of the form (13.18) corresponding to different phase functions and amplitudes (possibly with a different number of phase variables) will differ by a smooth term. The answer can be given in terms of the Lagrange manifold associated with the phase function Ψ.

We set $C_\Psi^t = \{(x, y, \theta) : \Psi_\theta = 0,\ \theta \neq 0\}$. If Ψ satisfies the appropriate non-degeneracy conditions, then C_Ψ^t is a smooth $2n$-dimensional manifold. We consider the mapping

$$C_\Psi^t \ni (x, y, \theta) \mapsto (x, \Psi_x, y, \Psi_y) \in T'X \times T'X. \tag{13.19}$$

It is a diffeomorphism, which transforms C_Ψ^t onto a conic submanifold $\Lambda_\Psi^t \subset T'X \times T'X$. The latter turns out to be a *Lagrange manifold*, i.e., the symplectic form $dx \wedge d\xi + dy \wedge d\eta$ on $T'X \times T'X$ vanishes on Λ_Ψ^t and $\dim \Lambda_\Psi^t = 2n$ (see Hörmander 1983–1985, Vol. 4). This submanifold serves as the basic invariant object, making it possible to identify integrals of the form (13.18). Namely, the following result is true.

Theorem 13.2. *We denote by $I(\Psi, q)$ the distribution on $X \times X$ defined by the oscillatory integral (13.18). Let Ψ' be another non-degenerate phase function (possibly with a different number of phase variables). If $\Lambda^t_\Psi = \Lambda^t_{\Psi'}$, then there exists an amplitude q' such that $I(\Psi, q) = I(\Psi', q')$ modulo an infinitely smooth function.*

Example 13.2. Let the Lagrange manifold Λ_Ψ be an open subset of the manifold $D = \{(x, \xi), (x, -\xi)\} \subset T'X \times T'X$. Then the integral (13.18) defines the kernel of a pseudodifferential operator on X.

The amplitude q' can be computed explicitly from Ψ, Ψ', and q. Moreover, any finite number of terms in the expansion of q' can be determined by means of finite-order jets of Ψ, Ψ', and the corresponding terms of q. We refrain from stating these very cumbersome formulae. We shall, however, describe the following important construction connected with them. Let q_0 and q'_0 be the leading homogeneous terms of the amplitudes from Theorem 13.2, and let $\iota : C^t_\Psi \to \Lambda^t$ and $\iota' : C^t_{\Psi'} \to \Lambda^t$ be of the form (13.19). We introduce arbitrary coordinates on Λ^t and transfer them with the aid of the functions ι and ι' to C^t_Ψ and $C^t_{\Psi'}$. We denote by ω and ω' the resulting coordinates on C^t_Ψ and $C^t_{\Psi'}$. We set

$$d_\Psi = \left|\det \frac{D(\omega, \Psi_\theta)}{D(x, y, \theta)}\right|, \quad d_{\Psi'} = \left|\det \frac{D(\omega', \Psi'_{\theta'})}{D(x, y, \theta')}\right|$$

and consider the functions

$$u = \left(\left(\frac{d\mu(x)}{dx}\right)^{1/2} \left(\frac{d\mu(y)}{dy}\right)^{-1/2} q_0 d_\Psi\right) \circ \iota,$$

$$u' = \left(\left(\frac{d\mu(x)}{dx}\right)^{1/2} \left(\frac{d\mu(y)}{dy}\right)^{-1/2} q'_0 d_{\Psi'}\right) \circ \iota'$$

on Λ. These functions are connected with each other in such a way that they can be regarded as the same section of a certain linear bundle over Λ, which is called the *Keller-Maslov bundle* in (Tréves 1982). The choice of the phase functions determines a local trivialization of the bundle. The functions u and u' are, in general, expressed by the above formulae in different trivializations. One more invariant object, the section of the Keller-Maslov bundle, can therefore be associated with the integral (13.18). It is natural to adopt this section as the principal symbol of the corresponding Fourier integral operator. If $I(\Psi, q)$ and $I(\Psi', q')$ differ by a smooth function, then the principal symbols of the corresponding Fourier integral operators are the same.

Remark 13.1. The principal symbol constructed in this way depends on the choice of coordinates on the Lagrange manifold. It transforms like a *half-density*, i.e., it is multiplied by the modulus of the Jacobian to the power 1/2 when the coordinates on Λ^t are changed. However, in cases that are of interest to us Λ^t can be parametrized by the canonical 'coordinates' $(y, \eta) \in T'X$, the Jacobian of any transformation of the latter being equal to one.

Thus, on a manifold without boundary we have

$$U(t) = \sum_j U^{(j)}(t)$$

to within an operator with a smooth kernel, $U^{(j)}(t)$ being local Fourier integral operators with kernels of the form (13.18). It turns out that the Lagrange manifolds of the operators $U^{(j)}(t)$ are open subsets of a single global Lagrange manifold $\Lambda^t \subset T'X \times T'X$. In such cases $U(t)$ is said to be the *global Fourier integral operator* with Lagrange manifold Λ^t. The sum of the principal symbols of $U^{(j)}(t)$ is called the principal symbol of $U(t)$.

The global Lagrange manifold Λ^t corresponding to the fundamental solution of (13.4) can be described with the aid of the *bicharacteristic flow* $F^t : T'X \to T'X$, i.e., the family of displacements along the trajectories of the Hamilton system (13.13) (*bicharacteristics*) such that $F^t(x^0, \xi^0) = (x(t; x^0, \xi^0), \xi(t; x^0, \xi^0))$. The transformation F^t is *canonical*, i.e., it preserves the symplectic 2-form $dx \wedge d\xi$ on $T'X$. Besides, F^t preserves the Hamiltonian $B_1(x, \xi)$, and $F^t(x_0, \lambda\xi^0) = (x, \lambda\xi)$, where $(x, \xi) = F^t(x_0, \xi^0)$ and $\lambda > 0$. In this connection, it is sometimes more convenient to consider the flow F^t on

$$S^*X = \{(x, \xi) \in T'X : B_1(x, \xi) = 1\},$$

rather than on $T'X$. The graph

$$\Lambda^t = \{(x, \xi, y, \eta) \in T'X \times T'X : (x, \xi) = F^t(y, -\eta)\}$$

of the mapping $(y, \eta) \mapsto F^t(y, -\eta)$ is the Lagrange manifold of $U(t)$. Thus the bicharacteristics and the Lagrange manifold are the global geometric objects to be associated with the differential operator A.

The knowledge of Λ^t makes it possible to describe the singularities of the kernel $U(t; x, y)$, or, more precisely, the *singular support* (sing supp) and *wave front* (WF) of $U(t; x, y)$. We recall (see, for example, (Taylor 1981; Tréves 1982; Hörmander 1983–1985; Egorov and Shubin 1988b)) that sing supp v, where v is a distribution in a domain $X \subset \mathbb{R}^n$, is the smallest closed set outside of which $v \in C^\infty$. Furthermore, a point $(x^0, \xi^0) \in T'X = X \times (\mathbb{R}^n \setminus \{0\})$ does not belong to WF(v) if and only if there is a function $\varphi \in C_0^\infty(X)$ such that $\varphi(x^0) \neq 0$ and the Fourier transform $(\widetilde{\varphi v})(\xi)$ decreases rapidly as ξ tends to infinity while remaining inside an open cone that contains ξ_0. The definitions of the singular support and the wave front can both be carried over in a natural way to the case of manifolds. We remark that sing supp v coincides with the image of WF(v) under the natural projection of $T'X$ onto X.

The spectral properties of an operator A are to a large extent determined by the geometry of the bicharacteristics of A. The *periodic bicharacteristics*, i.e., trajectories along which $F^T(x,\xi) = (x,\xi)$, play a special role. For a fixed $x \in X$ one can also consider the *bicharacteristic loops*, i.e., trajectories $F^t(x,\xi)$ that return to $T'_x X$ after time $T \in \mathbb{R}^1$ (i.e., $F^T(x,\eta) = (x,\xi)$, where $\xi, \eta \in T'_x X$ and $T \in \mathbb{R}^1$). Both the set $\mathcal{L} = \mathcal{L}(A)$ of periods of closed bicharacteristics and the set

$$\mathcal{L}_x = \mathcal{L}_x(A) = \{T : (x,\xi) = F^T(x,\eta),\ \xi,\eta \in T'_x X\}$$

of 'lengths' of bicharacteristic loops always contain $T = 0$, since $F^0(x,\xi) = (x,\xi)$ for any $(x,\xi) \in T'X$.

It can be demonstrated (see Tréves 1982) that the wave front of the distribution (13.18) is contained in the image of the Lagrange manifold Λ^t_Ψ under the transformation $(x,\xi,y,\eta) \mapsto (x,\xi,y,-\eta)$. It follows that $\operatorname{sing\,supp} U(t;\cdot,\cdot)$ is contained in the projection of Λ^t onto $X \times X$ for every t. Hence one can derive the following result.

Theorem 13.3 (Colin de Verdière 1973; Chazarain 1974; Duistermaat and Guillemin 1975; Safarov 1988a). *The following inclusions are true:*

$$\mathrm{WF}(U(t;\cdot,\cdot)) \subset \{(x,\xi), F^t(x,\xi)\}, \tag{13.20}$$

$$\operatorname{sing\,supp} U(\cdot\,;x,x) \subset \mathcal{L}_x \tag{13.21}$$

for every fixed $x \in X$,

$$\operatorname{sing\,supp} \sigma \subset \mathcal{L}. \tag{13.22}$$

From (13.20) it follows, in particular, that if $u(t) = U(t)u(0)$ is a solution of equation (13.4), then

$$\mathrm{WF}(u(t)) \subset F^t(\mathrm{WF}(u(0))).$$

Thus the singularities of the solutions of equation (13.4) propagate along the bicharacteristics.

The singularities of the distributions $\sigma(t)$ and $U(t;x,x)$ at zero correspond to the zero period of every bicharacteristic. The character of the singularity of $\sigma(t)$ at any other point $T \in \mathcal{L}$ depends on the size of the set of T-periodic bicharacteristics, while the singularity of $U(t;x,x)$ at $T \in \mathcal{L}_x$ depends on the size of the set of bicharacteristic loops of 'length' T. Let Π be the set of points in S^*X that lie on periodic trajectories of 'length' zero, and let $\Pi_x = \{\xi \in S^*_x X : F^T(x,\xi) = (x,\eta) \text{ for some } T \neq 0,\ \eta \in S^*_x X\}$. If Π is a set of measure zero in S^*X, then every non-zero singularity of $\sigma(t)$ is weaker than the singularity of $\sigma(t)$ at zero. Similarly, if the measure of Π_x in the fibre $S^*_x X$ is equal to zero, then every non-zero singularity of $U(t;x,x)$ is weaker than the one at zero. In these cases the non-zero singularities introduce a contribution of order $o(\lambda^{n-1})$ into (13.2), and the following results can be established with the aid of Theorem 13.1.

We set

$$B_s(x,\xi) = B_0(x,\xi) - \frac{1}{2i}\sum_j \frac{\partial^2 B_1(x,\xi)}{\partial x_j \partial \xi_j} - \frac{1}{2i}\sum_j \frac{\partial B_1(x,\xi)}{\partial \xi_j}\frac{\partial}{\partial x_j}\ln\frac{d\mu}{dx}.$$

This is clearly a homogeneous function of degree zero in ξ.

Theorem 13.4 (Duistermaat and Guillemin 1975). *Let the measure of Π in S^*X be equal to zero. Then the asymptotic formula (9.21) with*

$$a_1 = -n(2\pi)^{-n} \int\limits_{B_1(x,\xi)<1} B_s(x,\xi)\, dx d\xi \tag{13.23}$$

is satisfied.

Theorem 13.5 (Safarov 1988a). *Let the measure of Π_x in the fibre $S^*_x X$ be equal to zero for any $x \in K$, where K is a compact subset of X. Then $e^A(\lambda^m; x, x) \sim \omega_0(x)\lambda^n + \omega_1(x)\lambda^{n-1} + o(\lambda^{n-1})$ uniformly with respect to $x \in K$, where $\omega_0(x)$ is defined by (9.15) and*

$$\omega_1(x) = -n(2\pi)^{-n} \frac{dx}{d\mu} \int\limits_{B_1(x,\xi)<1} B_s(x,\xi)\, d\xi. \tag{13.24}$$

In particular, if A is an elliptic differential operator in the space of functions, then B_s is an odd homogeneous function in ξ (see Duistermaat and Guillemin 1975). The expressions (13.23) and (13.24) are therefore equal to zero.

Remark 13.2. In (Duistermaat and Guillemin 1975; Hörmander 1983–1985; Safarov 1988a) two-term spectral asymptotic formulae were obtained for pseudodifferential operators acting in spaces of half-densities. For such pseudodifferential operators the *subprincipal symbol*

$$\operatorname{sub} B(x,\xi) = B_0(x,\xi) - \frac{1}{2i} \sum_j \frac{\partial^2 B_1(x,\xi)}{\partial x_j \partial \xi_j}$$

can be defined in an invariant way, and B_s is replaced by $\operatorname{sub} B$ in (13.23) and (13.24). Theorems 13.4 and 13.5 can be reduced to the results of (Duistermaat and Guillemin 1975; Hörmander 1983–1985; Safarov 1988a) if the operator $\tilde{B} = (d\mu)^{1/2} B (d\mu)^{-1/2}$ in the space of half-densities is considered instead of B. If this is so, then $B_s = \operatorname{sub} \tilde{B}$.

In some cases it is also possible to obtain an estimate of the remainder in (9.21), provided information on the behaviour of the flow F^t as $t \to \pm\infty$ is available (see Volovoj 1987). For example, if A is the Laplace-Beltrami operator ($m = 2$) on a manifold of negative curvature, then $o(\lambda^{n-1})$ can be replaced by $O(\lambda^{n-1}/\log \lambda)$ in (9.21) (Bérard 1977). On the other hand, if all the conditions for the flow F^t are removed from Theorems 13.4 and 13.5, then the results will no longer be true. For the Laplace-Beltrami operator on the sphere S^n (see Example 4.7) the function $N(\lambda; A)$ has a jump of order $\lambda_k^{(n-1)/2}$ at $\lambda = \lambda_k$. It follows that a formula of the type (9.21) is no longer possible in this case. Operators which, as the latter one, have a rich set of periodic bicharacteristics are considered in §14.

13.4. Remarks on Other Problems. Reflection and Branching of Bicharacteristics

The constructions described above constitute the basis of the realization of the hyperbolic equation method for other classes of operators too. Unfortunately, in many cases, sometimes even for small t, it is difficult to construct $U(t)$ and results on the propagation of singularities similar to Theorem 13.3 are no longer true, in general.

These obstacles can often be overcome as follows. To start with, let Γ be a pseudodifferential operator of order zero on X with symbol $\Gamma(x,\xi)$ under the conditions of Sect. 13.3. The kernel $U_\Gamma(t;x,y)$ can be represented locally in the form (13.18) in exactly the same way as $U(t;x,y)$. One must only replace the initial conditions for the transport equations for small t by setting $q(0,x,y,\theta) = \Gamma(y,\theta)$. The relations (13.20), (13.21), and (13.22) must be replaced by

$$\operatorname{WF}\big(U_\Gamma(t;x,y)\big) \subset \big\{\big((x,\xi),F^t(x,\xi)\big) : (x,\xi) \in \operatorname{supp}\Gamma\big\}, \qquad (13.25)$$

$$\operatorname{sing\,supp} U_\Gamma(\cdot\,;x,x) \subset \mathcal{L}_{x,\Gamma}, \qquad (13.26)$$

$$\operatorname{sing\,supp} \sigma_\Gamma \subset \mathcal{L}_\Gamma, \qquad (13.27)$$

where $\mathcal{L}_{x,\Gamma}$ is the set of 'lengths' of bicharacteristic loops starting from $(x,\xi) \in T'_xX \cap \operatorname{supp}\Gamma$, and where $\mathcal{L}_\Gamma$ is the set of periods of closed bicharacteristics intersecting the support of $\Gamma(x,\xi)$.

As we shall see below, for many classes of operators one can specify closed conic subsets $\Xi_T \subset T'X$ (as a rule, of measure zero), each of which contains all the 'troubles' of the problem on the time interval $(-T,T)$. If $\Gamma(x,\xi)$ vanishes in a conic neighbourhood of Ξ_T, then it turns out to be possible to construct local representations similar to (13.18) for $U_\Gamma(t;x,y)$, where $|t| < T$, and, on generalizing the notion of the 'bicharacteristic flow,' to prove the analogues of (13.25)–(13.27). Then, with the aid of Theorem 13.1, one can obtain an asymptotic formula for $N_\Gamma(\lambda;B) = \operatorname{Tr}(E^B(\lambda)\Gamma)$, which becomes even more accurate as T increases, since the discrepancy in this formula can be estimated by $c_T\lambda^{n-1} + o(\lambda^{n-1})$, where $c_T \to 0$ as $T \to \infty$, and where $o(\lambda^{n-1})$ depends on T. For $N_{I-\Gamma}(\lambda;B)$ a similar formula can be obtained by other methods. Moreover, if the measure of Ξ_T is equal to zero, then the coefficient multiplying λ^{n-1} in the discrepancy can be made as small as required for a fixed T by a suitable choice of the pseudodifferential operator Γ. This makes it possible to determine the asymptotic behaviour of $N(\lambda;B)$ with accuracy up to $o(\lambda^{n-1})$. An expression of the form (9.21) is true for $N(\lambda;B)$ in 'regular' cases.

By a similar argument, using the analogue of (13.26), it is, in principle, also possible to obtain the two-term asymptotics of the spectral function (see, for example, Theorem 13.10).

In the present section we shall explain what complications may arise in the construction of $U(t)$ in various problems and how to overcome these complications. We shall also present results on the propagation of singularities for $U_\Gamma(t)$.

Let X be a manifold with boundary $Y = \partial X$. The boundary can be straightened locally by assuming that $X = \{x : x_n > 0\}$, where $x = (x', x_n)$. We consider a second-order elliptic differential operator A in X with an arbitrary regular boundary condition on Y. Let Γ be a pseudodifferential operator on X. We assume that its symbol $\Gamma(x,\xi)$ vanishes in a neighbourhood of $\Xi^{(0)} = T'X\big|_Y$ and has a sufficiently small conic support. Because the propagation speed is finite, $U_\Gamma(t;x,y)$ can be constructed in the form (13.18) in the interior of X up to the moment when the bicharacteristics starting from $\operatorname{supp}\Gamma(x,\xi)$ reach the boundary. Then it becomes necessary to ensure that the boundary condition is also fulfilled.

A bicharacteristic $\gamma^t(y,\eta)$ starting from (y,η) and hitting the boundary at $(x', 0, \xi', \xi_n)$ at time t_0 is said to be *transversal* to the boundary if

$$\dot{x} = \frac{\partial A_2}{\partial \xi_n}(x', 0, \xi', \xi_n) < 0$$

on the boundary. Such a bicharacteristic can be extended according to the laws of geometrical optics. By the extension we mean the *reflected bicharacteristic* $\tilde{\gamma}^t(y,\eta)$ starting from $(x', 0, \xi', \tilde{\xi}_n)$ at the moment t_0, where $\tilde{\xi}_n$ is a solution of the equations

$$\begin{aligned} A_2(x', 0, \xi', \xi_n) &= A_2(x', 0, \xi', \zeta), \\ \operatorname{sign} \frac{\partial A_2}{\partial \xi_n}(x', 0, \xi', \xi_n) &= -\operatorname{sign} \frac{\partial A_2}{\partial \xi_n}(x', 0, \xi', \zeta) \end{aligned} \tag{13.28}$$

with respect to ζ.

We assume that all the bicharacteristics starting in $\operatorname{supp}\Gamma$ are transversal to the boundary. It turns out that in this case $U_\Gamma(t;x,y)$ can be represented as the sum of two oscillatory integrals (13.18) near the boundary:

$$U_\Gamma(t;x,y) = \int e^{i\Psi_1} q^{(1)}\, d\theta + \int e^{i\Psi_2} q^{(2)}\, d\theta. \tag{13.29}$$

As before, the Lagrange manifold $\Lambda^t_{\Psi_1}$ is generated by the bicharacteristic flow, while $\Lambda^t_{\Psi_2}$ is generated by the *reflected bicharacteristic flow*:

$$\Lambda^t_{\Psi_2} = \big\{(x,\xi),(y,\eta) : (x,\xi) = \tilde{\gamma}^t(y,-\eta),\ (y,-\eta) \in \operatorname{supp}\Gamma\big\}.$$

For small t the functions Ψ_j and $q^{(j)}$ can be found from the eikonal and transport equations. The equations and the boundary conditions for Ψ_1 and $q^{(1)}$ will both be the same as on a manifold without boundary. The reflected trajectories correspond to the equations for Ψ_2 and $q^{(2)}$. The initial conditions for these functions are set on the surface $\{x_n = 0\}$ and can be obtained (upon finding Ψ_1 and $q^{(1)}$) by substituting (13.29) into the boundary condition.

As soon as all the bicharacteristics starting in $\operatorname{supp}\Gamma$ arrive at the boundary the first term in (13.29) becomes a smooth function. Therefore only the second term needs to be left in (13.29) for such t. Furthermore, $U_\Gamma(t;x,y)$ can, as before, be constructed as an oscillatory integral of the form (13.18) up to the moment when the reflected trajectories arrive at the boundary. If all these trajectories are transversal, then $U_\Gamma(t;x,y)$ can be represented in the form (13.29) again, and so on.

This discussion indicates that on a manifold with boundary the role of bicharacteristics should be played by the *billiard trajectories*, i.e., trajectories obtained by consecutive reflections of bicharacteristics. As F^t it is natural to take the *billiard flow*, i.e., the family of displacements along billiard trajectories. Since the Hamiltonian $B_1(x,\xi) = (A_2(x,\xi))^{1/2}$ is preserved under the reflection of bicharacteristics, it follows that, as before, $B_1(F^t(y,\eta)) = B_1(y,\eta)$ and $F^t(y,\lambda\eta) = (x,\lambda\xi)$ if $F^t(y,\eta) = (x,\xi)$ and $\lambda > 0$.

As opposed to bicharacteristics, not every billiard trajectory can be extended to all $t \in \mathbb{R}^1$. Firstly, the transversality condition may be violated. Secondly, the trajectory may undergo an infinite number of reflections during a finite time interval (in which case it is called a *dead-end trajectory*). However, the 'bad' trajectories are rare: for any T the set $\Xi^{(1)} \subset T'X$ of points belonging to any trajectory that cannot be extended to $[-T,T]$ is closed and has measure zero in $T'X$ (Cornfel'd et al. 1980). The flow F^t is therefore defined for all t on a set of full measure in $T'X$.

Thus, if $\Xi_T^{(1)} \cap \operatorname{supp}\Gamma = \emptyset$, then $U_\Gamma(t;x,y)$ can be represented locally in the form (13.18) or (13.29) for $|t| < T$. (In this case the support of $\Gamma(x,\xi)$ does not have to be small any longer.) This makes it possible to think of $U_\Gamma(t)$ as the global Fourier integral operator with the Lagrange 'manifold'

$$\Lambda_\Gamma^t = \{(x,\xi),(y,\eta) : (x,\xi) = F^t(y,-\eta),\ (y,-\eta) \in \operatorname{supp}\Gamma\},$$

where F^t is the billiard flow. As in Sect. 13.3, this implies the following result on the propagation of singularities.

Theorem 13.6 (Chazarain 1973; Guillemin and Melrose 1979). *Let A be a second-order elliptic operator with regular boundary conditions and let Γ be a pseudodifferential operator whose symbol vanishes both in a neighbourhood of $\Xi^{(0)}$ and in a neighbourhood of $\Xi_T^{(1)}$. Then, for $|t| < T$ the wave front of $U_\Gamma(t;x,y)$ satisfies (13.25), where F^t is the billiard flow, and $\sigma_\Gamma(t)$ satisfies the inclusion*

$$\operatorname{sing\,supp}\sigma_\Gamma \cap (-T,T) \subset \mathcal{L}_T \cap (-T,T),$$

where $\mathcal{L}_\Gamma$ is the set of periods of closed trajectories of the flow F^t passing through the points from the support of the symbol of Γ.

For elliptic boundary value problems of order $m > 2$ one should consider the equation

$$\left(i\frac{d}{dt}\right)^m U_\Gamma - AU_\Gamma = 0 \tag{13.30}$$

in order the determine $U_\Gamma(t)$ (see Vasil'ev 1984, 1986). Although the Cauchy problem for equation (13.30) is ill posed, the operators $\exp(\pm itA^{1/m})\Gamma$ can be distinguished from the other solutions of this equation by the common initial condition $U_\Gamma(0) = \Gamma$ and the requirement that they should be uniformly bounded with respect to t. This makes it possible to find effectively a representation of $U(t; x, y)$ inside X in terms of oscillatory integrals.

Since several boundary conditions must be satisfied on ∂X for an elliptic boundary value problem of order $m > 2$, a representation of the type (13.29) for $\exp(-itA^{1/m})$, which can be obtained with the aid of equations of the form (13.28), must contain more terms. Because the symbol $A_m(x, \xi)$ of order $m > 2$ is now present in these equations, they have, in general, more than one solution $\tilde{\xi}_n$. In the representation (13.29) real solutions $\tilde{\xi}_n$ generate ordinary oscillatory integrals, while complex solutions generate oscillatory integrals with complex phase (which correspond to exponentially decaying reflected waves).

To every real $\tilde{\xi}_n$ there corresponds its own reflected bicharacteristic. Several different reflection laws will therefore arise if equations (13.28) have more than one real solution. In such cases, instead of the billiard flow, one can consider the *semiflow* $F^t : T'X \to T'X$, that is, the family of many-valued functions obtained by displacements along different billiard trajectories. Subject to this modification, Theorem 13.6 can be carried over to higher-order operators – one must only replace $(-T, T)$ by $[0, T)$ and take as $\Xi_T^{(1)}$ the complement of the set of points $(x, \xi) \in T'X$ for which all the billiard trajectories $F^t(x, \xi)$ are defined and transversal to the boundary for all $0 \leq t \leq T$ (this result is essentially contained in (Vasil'ev 1984)).

In the situation at hand the set $\Xi_T^{(1)}$ may not be small any longer. As before, the set of points from which there starts at least one trajectory that fails to be transversal to the boundary is of measure zero. The measure of the set of initial points of dead-end trajectories may, however, turn out to be different from zero (see Safarov and Vasil'ev 1988). This is connected with the fact that the set of distinct trajectories starting from any fixed point is uncountable, in general. If equations of the form (13.28) always have only one real solution (this is the case, for example, for the biharmonic operator Δ^2), then the trajectories do not undergo branching and the measure of $\Xi_T^{(1)}$ is equal to zero, as for a second-order operator.

Furthermore, let A be an elliptic operator (of order one, for the sake of simplicity) on a k-dimensional Hermitian bundle $\mathcal{E}$ over a compact manifold X without boundary. The principal symbol $A_1(x, \xi)$, which is a Hermitian automorphism of the fibre $\mathcal{E}_x$, has k continuous branches $\tau_j(x, \xi)$ of eigenvalues, which are called *terms*. The analysis of the hyperbolic equation (13.4) depends to a large extent on the behaviour of the multiplicities of the terms.

If all the terms are equal to one another, i.e., $A_1(x, \xi) = \tau(x, \xi) \cdot \mathbf{1}_x$, where $\mathbf{1}_x$ is the identity automorphism of $\mathcal{E}_x$, then the fundamental solution $U(t)$ can be constructed in the form (13.9) in the same manner as in the scalar case, except that the amplitude $q(t, x, y, \theta)$ will now be a homomorphism from $\mathcal{E}_y$

to $\mathcal{E}_x$. In the case at hand Theorems 13.2 and 13.3 remain valid without any modifications.

Next, we consider the case when not all of the terms $\tau_j(x,\xi)$ are the same, but their multiplicity is independent of (x,ξ). Then A can be microlocally reduced by a unitary pseudodifferential transformation to the form

$$A^{(1)} \oplus A^{(2)} \oplus \ldots \oplus A^{(s)}, \quad s \leq k, \tag{13.31}$$

where each $A^{(l)}$ is an operator with principal symbol $\tau_l(x,\xi)\cdot \mathbf{1}_x$ acting in a bundle $\mathcal{E}^{(l)}$ over X, so that $\mathcal{E} = \oplus \mathcal{E}^{(l)}$. In this case equation (13.4) can be decomposed into the direct sum of hyperbolic equations of the form

$$\frac{1}{i}\frac{dU^{(l)}}{dt} = A^{(l)}U^{(l)}, \quad U(t) = \oplus U^{(l)}(t),$$

each of which can be studied by means of the previous methods. Each term generates a bicharacteristic flow F_l^t with its own set of periods $\mathcal{L}^{(l)}$.

Theorem 13.7 (Ivrii 1982). *Subject to the above-mentioned conditions,*

$$\mathrm{WF}\,(U(t;x,y)) \subset \bigcup_l \left((x,\xi), F_l^t(x,\xi)\right),$$

$$\operatorname{sing\,supp} \sigma(t) \subset \bigcup_l \mathcal{L}^{(l)}.$$

If the multiplicity of the terms varies on $T'X$, then the situation becomes more complicated, since a decomposition of the form (13.31) can no longer be obtained. The terms can behave in various ways near a point at which the multiplicity changes. As was shown in (Arnol'd 1972), in a typical situation (i.e., for symbols from an everywhere dense open set in the C^∞ topology) the terms differ from one another everywhere, except on a subset $\Sigma \subset T'X$ of codimension two. On Σ at least two terms coincide with each other, the coinciding terms failing to be smooth. On some subsets of positive codimension in Σ the multiplicity of the terms may be even higher. The behaviour of the singularities of the kernel $U(t;x,y)$ near the set in which the multiplicity of the terms varies is very complex. All the same, an analogue of Theorem 13.7 is valid outside this set.

Theorem 13.8 (Ivrii 1982). *Let $\Xi_T^{(2)} \subset T'X$ be a subset such that the terms have constant multiplicity on each component of the complement of $\Xi_T^{(2)}$ and for $|t| \leq T$ the multiplicity does not change on the bicharacteristics of any term starting from $T'X \setminus \Xi_T^{(2)}$. Let $\Gamma(x,\xi) = 0$ in a neighbourhood of $\Xi_T^{(2)}$. Then*

$$\mathrm{WF}\,(U_\Gamma(t;x,y)) = \bigcup_l \left\{\left((x,\xi), F_l^t(x,\xi)\right) : \Gamma(x,\xi) \neq 0\right\}, \quad |t| < T,$$

$$\operatorname{sing\,supp} \sigma_\Gamma \cap (-T,T) \subset \bigcup_l \mathcal{L}_\Gamma^{(l)} \cap (-T,T).$$

Finally, for operators acting on a bundle over a manifold with boundary one can encounter difficulties connected both with the fact that the multiplicity of the terms may vary and with the multitude of reflection laws. Moreover, whenever a reflection occurs the singularities of a solution can pass from the bicharacteristic of one term to that of another term (the following well-known property of elastic waves is a manifestation of this effect: when a transversal wave is reflected, it can produce both a transversal wave and a longitudinal one). In this situation one can also introduce a many-valued semiflow F^t corresponding to the bicharacteristic flows of the terms undergoing all possible branchings whenever a reflection occurs at the boundary (see Safarov and Vasil'ev 1988). If neighbourhoods of $\Xi^{(0)}$, $\Xi_T^{(1)}$, and $\Xi_T^{(2)}$ are removed from $T'X$ by introducing a suitable pseudodifferential operator Γ as in Theorems 13.6 and 13.8, then similar theorems can be obtained for $U_\Gamma(t)$ and $\sigma(t)$ (see Vasil'ev 1986).

13.5. Normal Singularity. Two-Term Asymptotic Formulae

The approach to obtaining sharp asymptotic formulae for the spectrum developed by Ivrii rests on the results of the type of Theorems 13.6 and 13.8 concerned with the propagation of singularities, as well as on the energy estimates for hyperbolic equations. With the aid of such estimates, it can be proved that $\sigma(t)$ has a *normal singularity* at zero in the problems under consideration, i.e., there exists $s \in \mathbb{R}^1$ and a small t_0 such that $(tD_t)^N\sigma(t) \in H^s(-t_0, t_0)$ for all N. This property makes it possible to prove that the natural successive approximation procedure for constructing $U(t)$ (similar to that described in §12 for a parabolic equation), which involves freezing the coefficients, straightening the boundary, etc., yields the correct expansion of $\sigma(t)$ with respect to smoothness in a small neighbourhood of zero (although it fails to yield such an expansion for the kernel $U(t; x, y)$ itself). It turns out that $\sigma(t)$ has a power singularity at zero:

$$(\widetilde{\rho\sigma})(\lambda) \sim na_0\lambda^n + (n-1)a_1\lambda^{n-1} + \dots, \quad \lambda \to +\infty \tag{13.32}$$

for any function ρ with sufficiently small support.

The expansion (13.32) (with Theorem 13.1 taken into account) suffices to prove an asymptotic formula for the spectrum with a sharp estimate of the remainder of the form (9.20), which is valid, in particular, for all the classes of problems discussed in the present section. We remark that no examples of regular differential operators violating this estimate are known to date.

In a number of cases a similar argument applied to $U(t; x, x)$, rather than $\sigma(t)$, makes it also possible to obtain a one-term asymptotic formula for the spectral function with a sharp estimate of the remainder, which is uniform up to the boundary (see Ivrii 1982, 1984).

To prove the two-term asymptotic formula (13.5) for $N(\lambda; B)$ one must impose an additional restriction (which, in the end, turns out to have a geometric character) on the operator A being studied. It is required that for any T there be a closed conic set $\Xi'_T \subset T'X$ of measure zero such that

$$\operatorname{sing\,supp} \sigma_\Gamma \cap (-T, T) = \{0\} \tag{13.33}$$

for any pseudodifferential operator Γ whose symbol vanishes in a neighbourhood of Ξ'_T. Then the application of the results of Sect. 13.4 and Theorem 13.4 yields the estimate

$$\left| N_\Gamma(\lambda; B) - a_0^{(1)} \lambda^n - a_1^{(1)} \lambda^{n-1} \right| \le c_T a_0^{(1)} \lambda^{n-1} + o(\lambda^{n-1}), \tag{13.34}$$

where $c_T \to 0$ as $T \to \infty$.

With the aid of the same energy estimates, it can now be proved that $\sigma_{I-\Gamma}(t)$ has a normal singularity at zero too, and satisfies (13.32) with a_0 and a_1 replaced by $a_0^{(2)} = a_0 - a_0^{(1)}$ and $a_1^{(2)} = a_1 - a_1^{(1)}$. Theorem 13.1 with $T = T_0$ yields an estimate of the form (13.34) for $N_{I-\Gamma}(\lambda; B)$ with right-hand side $c_{T_0} a_0^{(2)} \lambda^{n-1} + o(\lambda^{n-1})$. The coefficient $a_0^{(2)}$ can be computed explicitly. It turns out that it can be made as small as required by reducing the support of the symbol of $I - \Gamma$. Consequently, the two-term asymptotic formula (9.21) follows from the estimates for N_Γ and $N_{I-\Gamma}$.

In particular, let $A = A_\pm$ be the Laplace-Beltrami operator on a compact Riemannian manifold X with boundary Y. The minus sign corresponds to the Dirichlet problem, while the plus sign corresponds to the Neumann or third boundary value problem (2.15). As has already been mentioned, formula (9.20) is always true. We denote by Π_T the set of points in $T'X$ that lie on those periodic trajectories of the billiard flow whose periods are distinct from zero and not greater than T. The union of all Π_T, where $T \neq 0$, will be denoted by Π. In the described scheme $\Xi_T \cup \Pi_T$, where $\Xi_T = \Xi^{(0)} \cup \Xi_T^{(1)}$, can be taken as Ξ'_T (see Sect. 13.4). By Theorem 13.6, if $\Gamma(x, \xi) = 0$ in a neighbourhood of Ξ'_T, then (13.33) is satisfied. We have therefore obtained the following result.

Theorem 13.9 (Ivrii 1980). *If Π is a set of measure zero in $T'X$, then the two-term asymptotic formula (9.21), (9.22) is valid.*

As was demonstrated by Petkov and Stojanov (1988), $\operatorname{mes} \Pi = 0$ for any domain in $\mathbb{R}^n$ typical in the sense of Sect. 8.2.

A similar result concerned with the two-term asymptotics of the spectral function of the Laplace-Beltrami operator on a manifold with boundary is contained in (Safarov 1988a). In this case the role of the set $\Xi_{x,T}$ of 'bad' points is played by the conic subset of the fibre $T'_x X$ consisting of all $\xi \in T'_x X$ for which the billiard trajectory $F^t(x, \xi)$ cannot be extended to the time interval $[-T, T]$ (only transversal reflections are admitted). The measure of $\Xi_x = \bigcup_{T \neq 0} \Xi_{x,T}$ in $T'_x X$ is not necessarily equal to zero. The points x for which Ξ_x has measure zero are called *regular*. Exactly as in Sect. 13.3, we define the set $\Pi_x \subset T'_x X$ of initial directions of those billiard trajectories that return to $T'_x X$ after some length of time.

Theorem 13.10 (Safarov 1988a). *Let $x \notin Y$ be a regular point. If the measure of Π_x in $T'_x X$ is equal to zero, then the asymptotic formula from Theorem 13.5 with $\omega_1(x) = 0$ is valid for $e^A(\lambda; x, x)$. The asymptotics is uniform in x on strictly internal compact sets that do not contain any irregular points or any points for which the measure of Π_x does not vanish.*

Now let A be an elliptic operator on a bundle $\mathcal{E}$ over a manifold X without boundary. The property that $\sigma(t)$ has normal singularities at zero implies again that the estimate (9.20) is always satisfied. As Ξ'_T one can take $\bigcup_l \Pi_T^{(l)} \cup \Xi_T^{(2)}$, where $\Pi_T^{(l)}$ is the set of periodic points of the flows F_l^t whose periods are distinct from zero and do not exceed T. If the measure of Ξ'_T is equal to zero for any T (i.e., if the set of initial points of all periodic trajectories and all trajectories that fall into the zone in which the multiplicity changes has measure zero), then the two-term asymptotics (9.21) is valid for $N(\lambda)$.

The method developed by Ivrii made it also possible to obtain sharp estimates of the remainder and two-term asymptotic formulae in many other cases: for very general boundary value problems for first- and second-order elliptic operators acting in sections of vector bundles over a manifold with boundary; for certain boundary value problems of the form $Au = \lambda Bu$, where A is an operator of order one or two and B is an operator of order one or zero; for the Maxwell operator, etc. Sharp spectral asymptotic formulae have also been obtained in problems concerned with the discrete spectrum of the Schrödinger and Dirac operators (Ivrii 1985), the discrete spectrum of the multiparticle Schrödinger operator, and the like. The most detailed presentation of Ivrii's method is contained in his book (Ivrii 1984), where further references can also be found.

By combining the hyperbolic equation method with the estimates from §5, Ivrii has recently obtained very accurate estimates of the number of negative eigenvalues of the Schrödinger operator with a singular potential and the number of eigenvalues of the Dirac operator in a fixed open interval (Ivrii 1986b,c,d). In these estimates the dependence of the operator on the parameters is taken into account effectively, which makes it possible to derive various asymptotic formulae for the spectrum.

13.6. Other Results

Sharp spectral asymptotic formulae for boundary value problems with branching billiard trajectories were obtained by Vasil'ev (the conditions in (Ivrii 1982, 1984) exclude branching). In such problems the realization of the scheme described in Sect. 13.4 presents difficulties, because the measure of $\Xi_T^{(1)}$ may be positive (see Sect. 13.4). In this connection Vasil'ev introduced the notion of a *non-dead-end* many-valued billiard flow F^t. It is a flow such that the measure of the set of points $(x, \xi) \in T'X$ from which there starts at least one dead-end trajectory $F^t(x, \xi)$ is equal to zero. Sufficient conditions for a non–dead-end flow are given in (Vasil'ev 1984).

In (Vasil'ev 1986) the one-term asymptotic formula (9.20) was established for elliptic operators of order $2m$ on a compact manifold X with boundary Y and with regular boundary conditions on Y, without any restrictions on the billiard semiflow. The two-term asymptotic formula (9.21) was proved in (Vasil'ev 1986) under the condition that the semiflow F^t is a non–dead-end one and the set of points (x, ξ) from which there starts at least one periodic trajectory $F^t(x, \xi)$ has measure zero. In this asymptotic formula

$$a_1 = (2\pi)^{1-n} \int\limits_{T'Y} \left(n_B(1; x', \xi') + (2\pi)^{-1}\varphi_B(1; x', \xi')\right) dx'd\xi', \qquad (13.34')$$

where, for any fixed x' and ξ', n_B and φ_B are the distribution function of the eigenvalues and the scattering phase of the following auxiliary problem on the half-axis $x_n > 0$:

$$\begin{aligned} A'(x', 0, \xi', D_{x_n})u(x_n) &= \lambda u(x_n), \\ B'_j(x', \xi', D_{x_n})\big|_{x_n=0} &= 0. \end{aligned} \qquad (13.35)$$

The operators $A'(x', 0, \xi', D_{x_n})$ and $B'_j(x', \xi', D_{x_n})$ can be obtained by replacing ξ_n by D_{x_n} in the principal symbols of A and the boundary operators B_j.

The *scattering phase* φ_B is defined as follows. We consider the continuous spectrum of the problem (13.35). It occupies the infinite half-interval $[\lambda_*^{(1)}, +\infty)$ with $\lambda_*^{(1)} = \min \sigma_{2m}(\xi_n)$, where $-\infty < \xi_n < +\infty$, and where σ_{2m} is the complete symbol of $A'(x', 0, \xi', D_{x_n})$. We say that λ_* is a *singular point* of the continuous spectrum of the problem (13.35) if the equation $\sigma_{2m}(\xi_n) = \lambda_*$ has a multiple real ξ-root. We rearrange all singular points in such a way that $\lambda_*^{(1)} < \lambda_*^{(2)} < \ldots < \lambda_*^{(\nu)}$. It is clear that $1 \le \nu \le 2m - 1$.

Let λ be a point belonging to the continuous spectrum that is neither a singular point nor an eigenvalue of the problem (13.35). We denote by

$$\xi_1^- < \xi_1^+ < \xi_2^- < \xi_2^+ < \ldots < \xi_q^- < \xi_q^+$$

the real ξ_n-roots of the equation $\sigma_{2m}(\xi_n) = \lambda$. We note that $\sigma'_{2m}(\xi_l^-) < 0$ and $\sigma'_{2m}(\xi_l^+) > 0$. Every eigenfunction of the continuous spectrum that corresponds to λ has the form

$$\sum_{l=1}^{q} \left(\frac{c_l^- \exp(ix_n\xi_l^-)}{(-2\pi\sigma'_{2m}(\xi_l^-))^{1/2}} + \frac{c_l^+ \exp(ix_n\xi_l^+)}{(2\pi\sigma'_{2m}(\xi_l^+))^{1/2}} \right) + \sum_{l=1}^{m-q} \tilde{c}_l v_l(x_n),$$

where v_l are decreasing functions as $x_n \to \infty$, and where c_l^+, c_l^-, and $\tilde{c}_l$ are constants. The space of such eigenfunctions is q-dimensional. It is said that q

is the *multiplicity of the continuous spectrum at* λ. The columns c^+ and c^- formed by c_l^+ and c_l^- are related to one another by

$$c^+ = S(\lambda)c^-,$$

where $S(\lambda)$ is a unitary matrix (called the *scattering matrix*). The matrix-valued function $S(\lambda)$ admits a regular extension to those eigenvalues within the continuous spectrum that are not singular points at the same time.

By definition, $\varphi_B(\lambda) = 0$ in $(-\infty, \lambda_*^{(1)})$. In the remaining intervals of the form $(\lambda_*^{(k)}, \lambda_*^{(k+1)})$,

$$\varphi_B(\lambda) = \operatorname{Arg}(\det S(\lambda)) + c_k, \quad c_k = \text{const}.$$

The numbers c_k can be determined from the following normalization relations:

$$\varphi_B(\lambda_*^{(\varkappa)} + 0) - \varphi_B(\lambda_*^{(\varkappa)} - 0) = 2\pi q_*^{(\varkappa)},$$

$$q_*^{(\varkappa)} = -p_*^{(\varkappa)} - \lim_{\lambda \to \lambda_*^{(\varkappa)}} \left((\lambda - \lambda_*^{(\varkappa)}) \int_0^\infty (r_B(\lambda; x_n, x_n) - r(\lambda; x_n, x_n))\, dx_n \right), \quad (13.36)$$

where $p_*^{(\varkappa)}$ is the multiplicity of $\lambda_*^{(\varkappa)}$ as an eigenvalue of the problem (13.35) (if $\lambda_*^{(\varkappa)}$ is not an eigenvalue, then $p_*^{(\varkappa)} = 0$), and where $r_B(\lambda; x_n, y_n)$ and $r(\lambda; x_n, y_n)$ are the integral kernels of the resolvent of the problem (13.35) on the half-axis and of the problem on the whole axis with no boundary condition at zero. The limit in (13.36) is taken with respect to those λ that do not lie on the real axis. The limit exists and is finite: $|p_*^{(\varkappa)} + q_*^{(\varkappa)}| \le 2m$. If the equation $\sigma_{2m}(\xi_n) = \lambda_*^{(\varkappa)}$ has only one multiple real root ξ_* being is a double root (this is the case in the general situation), then formula (13.36) can be simplified and reduced to

$$q_*^{(\varkappa)} = \pm 1/4. \quad (13.37)$$

In (13.37) the plus sign is taken if for $\lambda = \lambda_*^{(\varkappa)}$ the problem (13.35) has a solution of the form $u(x_n) = \exp(ix_n\xi_n) + v(x_n)$, where $v(x_n) \to 0$ as $x_n \to +\infty$. The minus sign is taken if there is no such solution.

Similar results have also been obtained by Vasil'ev for even-order elliptic systems on a manifold with boundary. Vasil'ev's method of proof differs from that of Ivrii, even though the general idea (see Sect 13.4) remains the same. When studying the asymptotic behaviour of $N_{I-\Gamma}(\lambda)$ by this method one can use either the ordinary representation of $U(t)$ inside the domain for small t, or (near the boundary) an approximation of the spectral function of A by that of an operator with constant coefficients, which can be obtained by local straightening of the boundary and freezing the coefficients of A. This method turns out to be simpler from the technical point of view, but it has a smaller domain of applicability (in particular, for systems it is necessary to require that the eigenvalues of the principal symbol have constant multiplicity).

The *transmission problem* is a specific modification of the boundary value problems for elliptic systems. Let $X_\pm$ be n-dimensional manifolds with common boundary Y and let $A_\pm$ be elliptic operators of order $2m$ on $X_\pm$. An operator A in $L_2(X)$, where $X = X_+ \cup X_-$, can be defined in the natural way on $D(A) \subset H^{2m}(X_+) \oplus H^{2m}(X_-)$, the domain $D(A)$ being defined by the appropriate consistency conditions for the boundary values of $u_\pm \in H^{2m}(X_\pm)$ on Y. A semiflow F^t appears again in this case: any bicharacteristic arriving at the boundary splits into a reflected trajectory, which goes back into X_+, and a refracted trajectory, which escapes into X_-. The asymptotic formulae for the spectrum of a problem of this kind can be derived by practically the same methods. In the case when $A_\pm$ are the Laplace-Beltrami operators on $X_\pm$ the asymptotic formulae were obtained by Safarov (1987).

§14
Bicharacteristics and Spectrum

There is a deep connection between the spectrum of a differential operator and the properties of the bicharacteristic flow. The existence of such a connection is suggested, in the first place, by physical concepts: in classical and quantum systems the periodic trajectories and, respectively, the eigenfunctions of the operator describe objects whose evolution is periodic in time. This connection has already been mentioned in §13 (Theorems 13.3 and 13.6), where it has been used to outline the proof of the theorems on the two-term asymptotics of $N(\lambda)$. In Sects. 14.1 and 14.2 we shall generalize these results by discarding the condition that the set of periodic trajectories must be small. When more is known about the behaviour of the trajectories of the flow, it is possible to obtain additional information on other asymptotic characteristics of the spectrum, which cannot be reduced to the two-term asymptotics of $N(\lambda)$. One possible way of achieving this end is to study 'weak' non-zero singularities of the distribution $\sigma(t)$ (Sect. 14.3). In this way one can obtain, in particular, a generalization of the classical Poisson formula. Another method of studying the spectrum involves a direct construction of approximate eigenfunctions (Sects. 14.4 and 14.5). As a rule, the latter leads to a description of a part of the spectrum only.

In the present section A denotes either a self-adjoint elliptic (pseudo)differential operator on a manifold X without boundary, or a self-adjoint elliptic differential operator on a manifold X with boundary Y and an arbitrary regular boundary condition. Moreover, $A > 0$, $m = \operatorname{ord} A > 0$, and $N(\lambda^m) = N(\lambda^m, A)$. We denote by F^t the bicharacteristic flow in the former case and the billiard flow in the latter (throughout this section it is assumed that no branching of trajectories occurs on the boundary). The flows were introduced in Sects. 13.3 and 13.4. As a rule, X is assumed to be a compact manifold. Any exceptions are clear from the context. We set

$$A_s(x,\xi) = m^{-1}\big(A_m(x,\xi)\big)^{-(m-1)/m}\Bigg(A_{m-1}(x,\xi) - \frac{1}{2i}\sum_j \frac{\partial^2}{\partial x_j \partial \xi_j} A_m(x,\xi) - \frac{1}{2i}\sum_j \frac{\partial A_m}{\partial \xi_j}\frac{\partial}{\partial x_j}\ln \mu(x)\Bigg).$$

On a manifold without boundary $A_s(x,\xi)$ is the subprincipal symbol of the operator

$$\left(\frac{\partial \mu}{\partial x}\right)^{1/2} A^{1/m} \left(\frac{\partial \mu}{\partial x}\right)^{-1/2}$$

(cf. Remark 13.2).

14.1. The General Two-Term Asymptotic Formula

The period T and phase shift $b(T,x,\xi)$ (defined to within $2\pi k$, where $k \in \mathbb{Z}$) constitute the most important characteristics of a closed trajectory $\gamma_T = F^t(x,\xi)$, where $0 \le t < T$, such that $F^T(x,\xi) = (x,\xi)$. For operators on a manifold without boundary

$$\beta(T,x,\xi) = \beta_c(T,x,\xi) + \beta_s(T,x,\xi)$$

with

$$\beta_s(T,x,\xi) = -\int_0^T A_s\big(F^t(x,\xi)\big)\,dt, \quad \beta_c(T,x,\xi) = \alpha\pi/2,$$

where α is the *Maslov index* of γ_T (see Hörmander 1983–1985, Vol. 3; Maslov 1987, 1988), and where β_c can be interpreted as the *phase shift* due to the passage of the trajectory through the caustic (see Sect. 14.5), and β_s as the phase shift generated by the subprincipal symbol. The Maslov index α can take one of the values $0,1,2,3$ only, while β_s can assume any value from $\mathbb{R}^1$. On a manifold with boundary a phase shift will also take place when the trajectory is reflected by the boundary. If no branching occurs, then the scattering 'matrix' $S(\lambda)$ from Sect. 13.6 has dimensions 1×1, i.e., it is a complex number with modulus one. The number $\operatorname{Arg} S(A_m(x,\xi))$ is called the phase shift associated with reflection. It depends on the principal symbols of A and the operators appearing in the boundary conditions. On a manifold with boundary

$$\beta(T,x,\xi) = \beta_c(T,x,\xi) + \beta_s(T,x,\xi) + \beta_r(T,x,\xi),$$

where β_r is the total phase shift caused by the reflections of γ_T at the boundary. It is clear that $\beta(T, x, \lambda\xi) = \beta(T, x, \xi)$ for any $\lambda > 0$.

Example 14.1. For the Laplace-Beltrami operator on a Riemannian manifold $\beta_s \equiv 0$. On a manifold with boundary and with the Dirichlet conditions every reflection changes the phase by π. In the case of the Neumann or third boundary conditions the phase shift is equal to 0 (i.e., there is no phase shift).

A closed trajectory γ_T is said to be *primitive* if it is not a multiple iteration of a trajectory whose period is less than T. For $(x, \xi) \in \Pi$ (where Π is the set of periodic points of the flow F^t) we denote by $T^0(x, \xi)$ the positive period of the primitive trajectory $F^t(x, \xi)$. It is clear that $T^0(x, \lambda\xi) = T^0(x, \xi)$ for any $\lambda > 0$. Let $\beta^0(x, \xi) = \beta(T^0(x, \xi), x, \xi)$. The set $\Pi \subset T'X$ and the functions T^0 and β^0 defined on Π are measurable (Safarov 1986). The leading non-zero singularities of $\sigma(t)$ can be described with the aid of these functions in the following way.

Theorem 14.1 (Safarov 1986, 1988b). *Let X be a manifold without boundary and let $\chi \in S(\mathbb{R}^1)$, $\tilde{\chi} \in C_0^\infty(\mathbb{R}^1)$, and $0 \in \operatorname{supp} \chi$. Then*

$$\int \chi(\lambda - \mu)\, dN(\mu) = n(2\pi)^{-n} \int\limits_{\Pi \cap \{A_m < 1\}} \sum_{k \in \mathbb{Z}} \tilde{\chi}(kT^0)$$

$$\exp\left(ik\lambda T^0 + ik\beta^0\right) dx d\xi\, \lambda^{n-1} + o(\lambda^{n-1}). \quad (14.1)$$

On a manifold with boundary a similar theorem is true for $\sigma_\Gamma(t)$, where Γ is a pseudodifferential operator, under the condition that $\operatorname{supp} \chi \subset (-T, T)$ and $\Gamma(x, \xi) = 0$ in a neighbourhood of the set Ξ_T (for definitions see §13).

The proof of Theorem 14.1 is based on the representation of the kernel of the operator $\exp(-itA^{1/m})$ as the sum of oscillatory integrals (13.18). To within terms of higher smoothness, these integrals are determined by the Lagrange manifold and the principal symbol of the Fourier integral operator $\exp(-itA^{1/m})$. Thus the 'leading' singularities of $\sigma(t)$ also depend only on the Lagrange manifold and the principal symbol. The principal symbol can, in turn, be expressed in terms of the phase shift, which yields (14.1) as a result.

With the aid of Theorem 13.1 (the decomposition of $N(\lambda)$ into $N_\Gamma(\lambda) + N_{I-\Gamma}(\lambda)$ is used for an operator on a manifold with boundary; see Sect. 13.4), one can derive the following result from (14.1) and (13.16).

Theorem 14.2 (Safarov 1986, 1988b). *We set*

$$Q(\lambda) = n(2\pi)^{-n} \int\limits_{\Pi \cap \{A_m < 1\}} \frac{\left[\pi - \lambda T^0(x, \xi) - \beta^0(x, \xi)\right]_{2\pi}}{T^0(x, \xi)}\, dx d\xi,$$

where $[\,\cdot\,]_{2\pi}$ designates the remainder modulo 2π, i.e., $[t]_{2\pi} = t - 2k\pi$, where $k \in \mathbb{Z}$ and $-\pi < [t]_{2\pi} \leq \pi$. Then, for any $\varepsilon > 0$,

$$a_0\lambda^n + a_1\lambda^{n-1} + Q(\lambda - \varepsilon)\lambda^{n-1} - \varepsilon n a_0\lambda^{n-1} - o(\lambda^{n-1}) \le N(\lambda)$$
$$\le a_0\lambda^n + a_1\lambda^{n-1} + Q(\lambda + \varepsilon)\lambda^{n-1} + \varepsilon n a_0\lambda^{n-1} + o(\lambda^{n-1}), \qquad (14.2)$$

where $o(\lambda^{n-1})$ may depend on ε. If $Y = \emptyset$, then

$$a_1 = -n(2\pi)^{-n} \int\limits_{\{A_m<1\}} A_s(x,\xi)\, dx d\xi,$$

and if $Y \neq \emptyset$, then a_1 is the constant defined by (13.34').

The advantage of writing the asymptotics in this form is that no additional restrictions on the flow F^t are required. By imposing conditions on Q (and so, also on F^t), it is possible to obtain other less general formulae from (14.2).

For example, if Q is a uniformly continuous function on $\mathbb{R}^1$, then, by the arbitrariness of ε, (14.2) implies the formula

$$N(\lambda) = a_0\lambda^n + a_1\lambda^{n-1} + Q(\lambda)\lambda^{n-1} + o(\lambda^{n-1}). \qquad (14.3)$$

It is called the *quasi-Weyl asymptotics*. In all examples known to us $T^0(x,\xi)$ assumes no more than countably many values in a set of full measure in Π. In this situation Q is an almost periodic function (in the sense of H. Bohr) and is uniformly continuous if and only if it is continuous at every point. In particular, if $T^0(x,\xi) \equiv T^0 = \text{const}$ on a set of full measure, then Q is a periodic function with period $2\pi T_0^{-1}$. It is seen from the definition that Q is *oscillatory*, that is,

$$\int\limits_0^\lambda Q(\mu)\, d\mu \le \text{const}.$$

If it is a periodic function, then its integral over the whole period is equal to zero.

If $Q \equiv 0$, then (14.3) implies the ordinary Weyl formula (9.21). In order that the equality $Q \equiv 0$ be true it is sufficient that the measure of Π be equal to zero (see Theorems 13.4 and 13.9). However, Q may also vanish identically when Π has non-zero measure (see Safarov 1986).

We shall now assume that Q has discontinuities at μ_k, where $\mu_k \to +\infty$. Then, by (14.2), one can construct a positive sequence $\{\varepsilon_k\}$ converging to zero such that

$$N(\mu_k + \varepsilon_k) - N(\mu_k - \varepsilon_k)$$
$$= \big(Q(\mu_k + 0) - Q(\mu_k - 0)\big)\mu_k^{n-1} + o\big(\mu_k^{n-1}\big). \qquad (14.4)$$

Moreover, (14.4) holds for any positive sequence $\{\varepsilon_k'\}$ that converges to zero at a slower rate than $\{\varepsilon_k\}$. If $Q(\mu_k+0)-Q(\mu_k-0) \ge c > 0$, then (14.4) means that in the vicinity of the points μ_k there are groups of eigenvalues of $A^{1/m}$ converging to μ_k as $k \to \infty$ and having total multiplicity of order μ_k^{n-1}. Such groups of eigenvalues are called *clusters*. A more detailed discussion of cluster asymptotics (under certain additional restrictions on A) will be presented in Sect. 14.2. For the time being, we just remark that

$$Q(\mu+0)-Q(\mu-0)=(2\pi)^{1-n}\int_{\Omega_\mu\cap\{A_m<1\}}\left(T^0(x,\xi)\right)^{-1}dxd\xi,$$

where $\Omega_\mu=\left\{(x,\xi)\in\Pi:\mu T^0(x,\xi)+\beta^0(x,\xi)=0\ (\mathrm{mod}\,2\pi)\right\}$. Therefore Q has discontinuities at those points μ_k for which the measure of Ω_μ is different from zero. It follows, in particular, that Q is continuous everywhere, except for a set containing at most countably many points.

Remark 14.1. In (Safarov 1988a) a two-sided asymptotic estimate of the form (14.2) was also obtained for the spectral function $e^A(\lambda^m;x,x)$ in the case when the measure of Π_x (see §13) is different from zero. On a manifold with boundary it is assumed that x is a regular point (see Sect. 13.5) and $x\notin Y$.

Remark 14.2. For operators with branching billiard trajectories a two-sided estimate of the form (14.2) was established in (Safarov 1988b). Examples indicate that for such operators, clusters may also arise in the vicinity of points $\mu_k\to+\infty$ at which Q is continuous. Of course, uniform continuity is then violated at the points of the sequence μ_k.

14.2. Operators with Periodic Bicharacteristic Flow

Let A be an operator on a connected manifold X without boundary. We assume that all the bicharacteristics of A are periodic. If this is so, then they have a common period T (see Besse 1978). We also assume that $\beta(T,x,\xi)\equiv\beta_0=$ const. Since $\beta_c(T,x,\xi)$ is a continuous function of (x,ξ) on each connected subset of $T'X$, the assumption imposes a restriction on $A_s(x,\xi)$ only. For example, the Laplace operator on a manifold with a periodic geodesic flow has the above properties (in addition to the obvious examples of symmetric spaces of rank one, namely, S^n, $\mathbb{RP}^n$, $\mathbb{C}^n\mathbb{P}^n$, $\mathbb{HP}^n$, and $\mathrm{Ca}\mathbb{P}^n$, there is quite a large family of such manifolds; see (Besse 1978)).

By (14.2), under these conditions, all the eigenvalues of $A^{1/m}$ are contained, with accuracy up to $o(\lambda^{n-1})$, in the union of intervals of the form $(\mu_k+\varepsilon_k,\mu_k-\varepsilon_k)$, where $\mu_k=(2\pi k-\beta_0)T^{-1}$ and $\varepsilon_k\to0$. This result can be made more precise.

Theorem 14.3 (Colin de Verdière 1973; Duistermaat and Guillemin 1975). *Subject to the above assumptions, all the eigenvalues of $B=A^{1/m}$ are contained in the union of intervals of the form*

$$(\mu_k-k^{-1}r,\mu_k+k^{-1}r)\tag{14.5}$$

for some $r>0$. Conversely, if for any $\varepsilon>0$ all sufficiently large eigenvalues of an elliptic pseudodifferential operator B of order one are contained in the *intervals* $(\nu_k - \varepsilon, \nu_k + \varepsilon)$, *where* ν_k *is an arithmetic progression, then all the bicharacteristics of* B *are periodic and have a common period* T, *the phase shift* $\beta(T, x, \xi) = \beta_0$ *is independent of* (x, ξ), *and* $\nu_k = (2\pi k - \beta_0)T^{-1}$.

We shall present an outline of the proof of the former assertion. Consider the Fourier integral operator $U(T) = \exp(-iTB)$. Since the bicharacteristic flow is T-periodic, $U(T)$ is a pseudodifferential operator (see Example 13.1). Computations indicate that the principal symbol of $U(T)$ is equal to $\exp(-i\beta_0)$ (see, for example, (Hörmander 1983–1985, Vol. 4)). It follows that $U(T) = \exp(-i\beta_0)I + C$, where C is a pseudodifferential operator of order -1 commuting with B. The relation $\exp(-iT\lambda_j) = (1 + O(\lambda_j^{-1}))\exp(i\beta_0)$ or $i(T\lambda_j + \beta_0) = \mathcal{L}n(1 + O(\lambda_j^{-1}))$, where $\mathcal{L}n$ is one of the values of the logarithm, is therefore satisfied for the eigenvalues λ_j of B. Hence those eigenvalues λ_j that are close to $2\pi kT^{-1}$ are seen to have the form $2\pi kT^{-1} - \beta_0 T^{-1} + O(k^{-1})$, as has been claimed.

We denote by δ_k the total multiplicity of the eigenvalues λ_{jk} in the interval (14.5).

Theorem 14.4 (Colin de Verdière 1979). *Let us assume that all bicharacteristics of 'length'* T *are primitive. Then* $\delta_k = R(k)$ *for any sufficiently large* k, *where* $R(k)$ *is a polynomial of degree* $n-1$ *such that*

$$R(k) = n\big(k - \beta_c(T)/2\pi\big)^{n-1} T^{-n} \int\limits_{\{A_m < 1\}} dx d\xi + O(k^{n-3}),$$

where $\beta_c(T) = \beta_c(T, x, \xi)$ *(in this case* $\beta_c(T, x, \xi)$ *is independent of* x, ξ*).*

The distribution of eigenvalues in a cluster can be described with the aid of the sequence of measures

$$\mathcal{M}_k(t) = \sum_{j_k} \delta\big(t - (\lambda_{j_k}^2 - \mu_{j_k}^2)\big),$$

which have supports in the intervals

$$\big((\mu_k - k^{-1}r)^2 - \mu_k^2, (\mu_k + k^{-1}r)^2 - \mu_k^2\big).$$

Each of the measures $\mathcal{M}_k$ admits the asymptotic expansion (in the sense of distribution theory)

$$\mathcal{M}_k(t) \sim \sum_{j=0}^{\infty} k^{n-1-j} m_j(t),$$

where $m_j(t)$ are measures on the interval $(-4\pi rT^{-1} - \varepsilon, 4\pi rT^{-1} + \varepsilon)$, which are referred to as the *cluster invariants* (Colin de Verdière 1979).

Example 14.2. The Schrödinger operator on a compact manifold. Let $A = -\Delta + q$, where $-\Delta$ is the Laplace operator on a Riemannian manifold X with a T-periodic geodesic flow, and where $q \in C^\infty(X)$. Let $\tilde{q}(x, \xi)$ be the mean value of $q(x)$ (as a function on $T'X$) along the geodesic starting from x in the direction $\xi/|\xi|$. (On a sphere the mapping $q \mapsto \tilde{q}$ coincides with the Radon transform.) Then the measure $m_0(t)$ has the form

$$\langle m_0, f\rangle = n \int_{\{A_2<1\}} f(\tilde{q}(x,\xi))\, dx d\xi. \tag{14.6}$$

Formula (14.6) can be regarded as a far reaching generalization of the classical Szegö theorem on the asymptotic behaviour of the eigenvalues of a Toeplitz matrix (this theorem corresponds to the case when $X = S^1$). For more details on this aspect of the theory see (Boutet de Monvel and Guillemin 1981).

14.3. 'Weak' Non-Zero Singularities of $\sigma(t)$

In the one-dimensional case the classical Poisson formula reads

$$\sum_k \exp(ikt) = 2\pi \sum_k \delta(t - 2\pi k).$$

The left-hand side can be regarded as the trace of the operator $2\cos(-itA^{1/2}) - 1$, where $(-A)$ is the Laplace operator on the circle S^1. The numbers $2\pi k$ on the right-hand side can be regarded as the periods of closed geodesics on S^1. Thus the Poisson formula provides a complete description of the singularities of the distribution $\operatorname{Tr}\cos(-itA^{1/2})$ for the Laplace operator on S^1. A similar (but considerably less accurate) result can also be obtained for more general operators.

To begin with, let X be a manifold without boundary, let $\vartheta \in T'X$, and let $\gamma = F^t\vartheta$, where $0 \leq t < T$, be a closed bicharacteristic of an operator A with period T. We consider the tangent mapping

$$\left(dF^T\right)_\vartheta = T_\vartheta T'X \to T_\vartheta T'X$$

at ϑ. It has an eigenvector ω (the tangent vector to γ at ϑ) with eigenvalue 1. If all the remaining eigenvalues of $(dF^T)_\vartheta$ are distinct from one, then γ is said to be a *non-degenerate* bicharacteristic and the restriction of $(F^T)_\vartheta$ to the invariant subspace complementary to ω is called the *Poincaré map* and denoted by P_γ. The Poincaré maps for various points $\vartheta \in \gamma$ are conjugate to one another (i.e., $P_\gamma(\vartheta_1) = C^{-1}P_\gamma(\vartheta_2)C$, where C is a non-singular linear mapping).

Theorem 14.5 (Duistermaat and Guillemin 1975). *Under the conditions of Theorem 13.3, let $T > 0$ be an arbitrary isolated element of the set of periods $\mathcal{L}(X)$. We assume that there exist only finitely many bicharacteristics γ with period T, all of which are non-degenerate. Then*

$$\sigma(t) - (2\pi i)^{-1} \sum_{\gamma} \left| \det(I - P_\gamma) \right|^{-1/2} \exp(i\beta_\gamma) T_\gamma^0 (t - T - i0)^{-1} \in L_{1,\mathrm{loc}} \quad (14.7)$$

in a neighbourhood of T. Here β_γ is the phase shift on γ and T_γ^0 is the period of the primitive bicharacteristic of which γ is a multiple.

Remark 14.3. In the analogous formula in (Duistermaat and Guillemin 1975) and (Guillemin and Melrose 1979) the coefficient preceding the sum over γ is incorrect.

The non-degeneracy condition in Theorem 14.5 can be relaxed. It suffices to assume that the set of bicharacteristics with period T consists of several d_j-dimensional smooth submanifolds Z_j that are *clean*, i.e., for any $\vartheta \in Z_j$ the set of fixed points of $(dF^T)_\vartheta$ coincides with the tangent space to Z_j at ϑ (the case $d_j = 1$ corresponds to non-degeneracy). The character of the singularity of $\sigma(t)$ at $t = T$ depends on d_j: if $\rho \in C_0^\infty(\mathbb{R}^1)$ and $\operatorname{supp} \rho \cap \mathcal{L}(X) = T$, then to each Z_j there corresponds a series of the form

$$\sum_{k=0}^{\infty} c_{jk} \lambda^{-k+(d_j-1)/2}, \qquad c_{jk} = \mathrm{const}$$

in the asymptotics of $(\widetilde{\rho\sigma})(\lambda)$ as $\lambda \to +\infty$ (see Duistermaat and Guillemin 1975).

Next, let X be a Riemannian manifold with boundary Y and let $-A$ be the Laplace operator with the Dirichlet or third boundary conditions. We assume that X is *geodesically strictly convex* (which means that any interval of a geodesic γ connecting two sufficiently close points x_1 and x_2 on the boundary is contained in $X \setminus Y$ and $\max_{x \in \gamma} \operatorname{dist}(x, Y) \geq c(\operatorname{dist}(x_1, x_2))^2$). Then the billiard flow F^t is defined on the whole space $T'X$, since any billiard trajectory can be infinitely extended in both directions and consists locally of finitely many segments. We denote by G^t the geodesic flow on $T'Y$ corresponding to the Riemannian metric induced from X. The trajectories of G^t are referred to as the *gliding rays*.

Theorem 14.6 (Guillemin and Melrose 1979). *Subject to the above assumptions, the singular support of the distribution $\sigma(t) = \operatorname{Tr} \exp(-itA^{1/2})$ is contained in the union of the set $\mathcal{L}(X)$ of periods of closed billiard trajectories and the set $\mathcal{L}(Y)$ of periods of closed gliding rays.*

This result refines Theorem 13.6, which was proved by an explicit construction of $U_\Gamma(t)$ by the methods of the theory of Fourier integral operators. For the time being, no construction of $U(t)$ is available on a manifold with boundary, and the proof of Theorem 14.6 is based on a deep analysis of the propagation of singularities of hyperbolic boundary value problems.

Exactly as on a manifold without boundary, we can define non-degenerate closed billiard trajectories and the Poincaré maps P_γ. In the situation at hand the singularities of $\sigma(t)$ corresponding to such trajectories can be described by the same formula.

Theorem 14.7 (Guillemin and Melrose 1979). *Let $T > 0$ be an isolated point of the set of periods $\mathcal{L}(X) \cup \mathcal{L}(Y)$ and let $T \notin \mathcal{L}(Y)$. We assume that there exist only finitely many billiard trajectories γ with period T, all of which are non-degenerate. Then (14.7) is satisfied in a neighbourhood of T (the phase shift due to reflections is of course included in β_γ).*

For degenerate closed trajectories (14.7) makes no sense any longer. It can, nevertheless, be demonstrated that such trajectories also generate singularities of $\sigma(t)$.

The results of the present section can be carried over in a certain sense to operators whose essential spectrum is non-empty. For example, let A be the operator $-\Delta_D$ in a domain $X \subset \mathbb{R}^n$ being the complement of a compact set. In analogy with the solutions of the wave equation in a bounded domain, which have the form

$$\sum_k \Big(w_k^+ \exp\big(i\sqrt{\lambda_k} t\big) + w_k^- \exp\big(-i\sqrt{\lambda_k} t\big) \Big),$$

where λ_k are the eigenvalues of the Laplace operator, in the case under consideration the solutions of the wave equation can be represented as

$$u(x,t) = \sum_k w_k(x) \exp(i\mu_k t).$$

Here μ_k are complex numbers such that $\operatorname{Im} \mu_k > 0$ and $w_k(x)$ are the solutions of the Helmholtz equation $\mu_k^2 w_k + \Delta w_k = 0$ in X with the Dirichlet conditions on ∂X and the radiation conditions $w_k \sim c_k |x|^{-(n-2)} \exp(\operatorname{Im} \mu_k |x|)$ at infinity. The distribution $\sum \exp(i\mu_k t)$ plays the role of $\sigma(t)$. Under certain assumptions, one can prove that the singularities of $\sigma(t)$ are contained in the set of periods of periodic billiard trajectories in X. In particular, if the complement of X is strictly convex, then only the zero period is possible and $\operatorname{sing\,supp} \sigma = \{0\}$. Theorem 14.7 can be generalized to the case under consideration (Majda and Ralston 1978a, 1978b, 1979; Chazarain 1980; Bardos et al. 1982; Melrose 1982). On the connection of these results with scattering theory see (Melrose 1982).

14.4. Quasimodes

By the spectral theorem (Theorem 1.2), if A is a self-adjoint operator with discrete spectrum, $u \in D(A)$, and $\|Au - \lambda u\| < \varepsilon \|u\|$, then the ε-neighbourhood of λ contains an eigenvalue of A. Besides, as follows from the estimate

$$\|E^A(\lambda - \delta, \lambda + \delta)u - u\| \leq \delta^{-1} \|Au - \lambda u\|,$$

u is then close to its projection onto the invariant subspace of A corresponding to the spectrum in the neighbourhood of λ. It is customary to refer to such 'approximate eigenfunctions' u as *quasimodes*.

If there is only one eigenvalue (which is simple) in the δ-neighbourhood of λ, then the quasimode u does not differ from the real eigenfunction by more than $\delta^{-1}\varepsilon\|u\|$. But if the total multiplicity of the spectrum in this neighbourhood is greater than one, then, in general, the quasimode u approximates a certain linear combination of the eigenfunctions, rather than any single eigenfunction. If this fact is overlooked, then one may be led to false results on the multiplicity of the eigenvalues. In particular, according to Theorem 8.3, in a flat domain X admitting the isometry group $\mathbb{Z}_p$ the eigenvalues have multiplicity one or two. At the same time, for such a domain the ordinary methods of constructing quasimodes (see Sect. 14.5) lead to p quasimodes, which undergo cyclic permutations when the domain is rotated. As was proved by Arnol'd (1972), these quasimodes are certainly not close to the real eigenfunctions.

To obtain correct results on the multiplicity of the eigenvalues it is required that the constructed quasimodes be 'almost orthogonal.' The corresponding abstract result is due to Lazutkin (1981, 1987).

Theorem 14.8. *Let A be a self-adjoint operator with discrete spectrum and let the estimate $\|Au_k - \lambda u_k\| \leq c\lambda_k^{-\nu}\|u_k\|$ be satisfied for a sequence of quasimodes u_k and numbers $\lambda_k \to \infty$. If, in addition, $|(u_k, u_l)| \leq c(\lambda_k + \lambda_l)^{-1}\|u_k\|\|u_l\|$, then there exists a sequence of eigenvalues μ_k of A such that $|\mu_k - \lambda_k| \leq c^2\lambda_k^{-\nu}$.*

14.5. Construction of Quasimodes

The formalism first developed in the mathematical theory of diffraction and short wave propagation can be used to construct quasimodes. As in the hyperbolic equation method, in this case there arise the transport equations, the Hamilton flow, and the Lagrange manifolds. This is not an accidental coincidence. The methods of constructing the asymptotics of the solutions of evolution equations with respect to smoothness outlined in §13 and the methods of constructing the asymptotics of solutions with respect to a large parameter are, in fact, realizations of the same mathematical theory, namely, the *Maslov canonical operator method*. On this subject, see (Fedoryuk and Maslov 1976; Maslov 1976, 1987, 1988).

For simplicity, let A be the Laplace-Beltrami operator on a n-dimensional Riemannian manifold X. To begin with, let X be a manifold without boundary. We shall seek a quasimode, i.e., an approximate solution of the equation $-\Delta u = \lambda u$ with $\lambda = \mu^2$, in the form

$$u = u(x;\mu) = e^{i\mu\tau(x)}\varphi(x;\mu), \tag{14.8}$$

where $\varphi(x;\mu)$ can be expanded into an asymptotic power series in μ. The substitution of (14.8) into the original equation leads to the eikonal equation $|\nabla\tau| = 1$, where $|\xi|$ denotes the length of a covector ξ in the Riemannian metric. The latter equation can be solved by the Hamilton-Jacobi method (cf. Sect. 13.3). To this end one can consider the Hamilton system

$$\frac{dx}{dt} = \frac{\partial g}{\partial \xi}, \quad \frac{d\xi}{dt} = -\frac{\partial g}{\partial x} \tag{14.9}$$

on $T'X$, where g is the principal symbol of $-\Delta$. The initial values $\tau = \tau_0$ and $\nabla\tau_0$ of the eikonal and its gradient such that $|\nabla\tau_0| = 1$ are specified on a hypersurface $S \subset X$. We consider the trajectories of the system (14.9) starting from $(x, \nabla\tau_0(x)) \in T'X|_S$. They span an n-dimensional Lagrange manifold $\Lambda \subset T'X$. For any sufficiently small t the corresponding part of the Lagrange manifold can be projected onto X without singularities. On this projection τ is defined by the formula

$$\tau(x_1) = \tau(x_0) + \int\limits_{\gamma(x_0,x_1)} \xi\, dx,$$

where $\gamma(x_0, x_1)$ is the trajectory connecting $x_0 \in S$ with x_1. Next, as in Sect. 13.3, for the individual terms of the expansion of the amplitude $\varphi(x,\mu)$ one can obtain a recurrent system of linear transport equations along the trajectories (14.9). It follows that the quasimode can be constructed in the form (14.8) up to any order of accuracy on a portion of X close to S.

If the trajectories are extended sufficiently far in t, then Λ can no longer be projected diffeomorphically onto X, in general. In a neighbourhood of the set of singular points $x \in X$ of the projection of Λ (the caustic) the representation (14.8) is unsuitable. In the vicinity of the caustic a quasimode can be represented either in terms of special functions (in particular, the *Airy functions* (see Babich 1987)), or with the aid of the canonical operator (Maslov 1987, 1988). These representations are also connected with the manifold Λ; they are analogues of (13.18). Moving further down the trajectories we find ourselves again in a part of Λ that can be projected onto X without singularities. Here the representation (14.8) is valid again, but now with a different eikonal τ', the difference between τ' and τ on the caustic being equal to $k\pi/2$ for some $k \in \mathbb{Z}$. In other words, a phase shift by $k\pi/2$ occurs when the caustic is crossed.

In order that the described process be concluded by the construction of a quasimode a number of conditions must be satisfied. It is necessary that Λ be a compact manifold without boundary. Then, on the projection of Λ onto X outside the caustic, a quasimode can be constructed as the sum of several terms of the type (14.8) over all the inverse images of the projection. In the complement of the projection the quasimode will decay exponentially with respect to μ. Besides, for the quasimode to be unique it is required that the eikonal should change by a multiple of 2π as one moves around any closed contour on Λ. If $\Gamma_1, \dots, \Gamma_N$ are cycles on Λ that generate the one-dimensional homology group of Λ, then the requirement corresponds to N quantization conditions of the form

$$\mu \int_{\Gamma_j} \xi \, dx - \frac{\pi}{2} \operatorname{Ind} \Gamma_j = 0 \pmod{2\pi}, \tag{14.10}$$

where $\operatorname{Ind} \Gamma_j$ is the Maslov index of Γ_j (see Hörmander 1983–1985; Maslov 1988), which accounts for the jumps of the eikonal as the caustic is crossed.

The Lagrange manifold Λ from which to construct the quasimode can be chosen in several ways. It is defined by the initial conditions for the eikonal equation. Of course, Λ may turn out not to be compact. Moreover, the quantization conditions may fail to be satisfied for some μ. In this connection, one needs, as a rule, a sufficiently rich supply of suitable Lagrange manifolds to realize the presented scheme of constructing quasimodes. For example, suppose that we have succeeded in constructing an $(N-1)$-parameter continuous family of compact Lagrange manifolds invariant under displacements along the trajectories (14.9). Then the system of N equations (14.10) can be used to determine the 'admissible' values of the parameters and 'quasi-eigenvalues' $\lambda = \mu^2$. However, in the many-dimensional situation it is difficult to find a sufficiently rich family of compact Lagrange manifolds. As is known from the Kolmogorov-Arnol'd-Moser theory (Arnol'd et al. 1985), no such continuous families exist in a typical situation. Because of this, the construction of quasimodes almost always requires that the presented method be improved.

In (Babich 1987) one starts from a single closed trajectory $\gamma \in \Lambda$. If the Hamilton flow is stable on γ in the first approximation, then in the vicinity of γ it is possible to construct a Hamilton system being close to the original system and having the required family of invariant Lagrange manifolds near γ. The quasimodes constructed from those manifolds of this family that are sufficiently close to γ can also serve as quasimodes for the original operator.

This method can be extended to the case of manifolds with boundary (see Babich 1987). In the case of the Laplace operator in a convex domain two classes of quasimodes can be constructed in this way, namely, quasimodes of the *bouncing ball* type, which are concentrated near a closed stable trajectory of the billiard flow, and *whispering gallery* quasimodes, which correspond to a sequence of such closed trajectories converging to a closed geodesic on the boundary of the domain. Quasimodes of the latter kind turn out to be concentrated in a small neighbourhood of the boundary.

In another version (see, for example, Maslov 1976) a complex-valued eikonal $\tau(x)$ can be constructed, which is real on the projection $\gamma_0 = \pi\gamma \in X$ of a closed trajectory γ and has positive imaginary part in the complement of γ_0. In this case the complex eikonal equation (as well as the transport equation) can be solved in the complement of γ_0 with accuracy up to $O((\operatorname{Im} \tau)^k)$, where $k > 0$. In this way a germ of complex Lagrange manifolds is defined on γ. Since the quasimodes constructed from this germ decay exponentially with respect to λ in the complement of γ_0, it is possible to discard the requirement that Λ must be closed. Of all the quantization conditions (14.10), only the single condition corresponding to γ will remain. The constructed quasimodes turn out to be concentrated near γ_0.

One must note that, as a rule, all these methods provide only an approximation of a small portion of the spectrum, since the set of closed trajectories is usually rather sparse (for example, on a manifold of negative curvature the set of periodic trajectories is countable). This obstacle was overcome to a large extent by Colin de Verdière (1977) and Lazutkin (1981), who succeeded in eliminating the requirement that the family be continuous from the original method of constructing quasimodes from a family of Lagrange manifolds. Instead, they confined themselves to discontinuous families parametrized by Cantor sets of positive measure. The presence of such families turns out to be a typical property (see Lazutkin 1981). In this way, on constructing the quasimodes, it was possible to approximate a sequence of eigenvalues, the relative density of which is equal to the relative measure of the part of the phase space $T'X$ occupied by the Lagrange manifolds under consideration. An analysis of the billiard flow proved that for a flat convex domain being close to a circle, the measure can be arbitrarily close to one (Lazutkin 1981). The conjecture that a detailed analysis of the Hamilton system can make it possible to construct quasimodes approximating a portion of the spectrum of unit density was stated in (Lazutkin 1981).

In the conclusion we remark that for a (pseudo)differential operator $A(x, hD_x)$ with a small parameter multiplying the derivatives similar methods can also be employed to construct quasimodes, i.e., functions $u(x;h)$ yielding a small discrepancy in the equation $A(x, hD_x)u = \lambda(h)u$ as $h \to 0$ (see Maslov 1987, 1988).

§15
Approximate Spectral Projection Method

15.1. The Basic Concept

The approximate spectral projection method was proposed by Shubin and Tulovskij (1973) (see also the presentation of this work in (Shubin 1978a)). Afterwards it was used by Rojtburd (1976) (see also Shubin 1978a, Appendix 2), Fejgin (1976, 1978, 1979), Hörmander (1979b, 1979c), Bezyaev (1978, 1982), Bogorodskaya and Shubin (1986), and, especially, Levendorskij (1982, 1984, 1985, 1986a, 1986b, 1988a, 1988b, 1990, and references therein). The basic concept of the method is briefly presented in (Egorov and Shubin 1988b, §8) and in §9 of the present article. We shall present the idea in a more general context.

Suppose that we want to find the asymptotics as $t \to +\infty$ of the number $N(0; A_t)$ of negative eigenvalues of a self-adjoint operator A_t depending on a parameter $t > 0$ and defined in a functional Hilbert space $\mathcal{H}$. We shall assume, unless otherwise stipulated, that $\mathcal{H} = L_2(X)$, where X is a domain in $\mathbb{R}^n$. Instead of this it can, as a rule, be assumed that X is a smooth manifold (possibly with boundary). The latter case can usually be reduced to that of a domain in $\mathbb{R}^n$ with the aid of a local argument followed by the use of a partition of unity. Also, one can assume that $\mathcal{H} = (L_2(X))^p$, where $p \geq 1$ is an integer, i.e., the elements of $\mathcal{H}$ are vector-valued functions on X with values in $\mathbb{C}^p$, so that A_t is a $(p \times p)$-matrix–valued operator. As a rule, this assumption does not involve any major difficulties either.

Let A_t be a (pseudo)differential operator in X with Weyl symbol $a_t = a_t(x, \xi)$. Since A_t is assumed to be self-adjoint (for example, obtained as the Friedrichs extension from $C_0^\infty(X)$), it follows that a_t will be a real-valued function (Hermitian-valued in the matrix case).

Let $\chi_{(-\infty,0)}$ be the indicator function of the half-axis $(-\infty, 0)$. Then the spectral projection $\chi_{(-\infty,0)}(A_t)$ will, in a certain sense, be close to the operator with symbol $\chi_{(-\infty,0)}(a_t)$. It is the trace of the latter operator that determines the conjectured Weyl asymptotics

$$\begin{aligned} N(0; A_t) &\sim (2\pi)^{-n} \int\limits_{a_t(x,\xi)<0} dx d\xi \\ &= (2\pi)^{-n} \operatorname{mes}\{(x,\xi) : a_t(x,\xi) < 0\} \end{aligned} \tag{15.1}$$

in the scalar case, and

$$N(0; A_t) \sim (2\pi)^{-n} \int N(0; a_t(x,\xi))\, dx d\xi \tag{15.1'}$$

in the more general matrix case.

The operator with the discontinuous symbol $\chi_{(-\infty,0)}(a_t)$ cannot, however, be used directly, because no appropriate formalism has been developed. The basic idea of the approximate spectral projection method consists in considering a regularization of the symbol. The necessary estimates can be obtained with the aid of the technique of pseudodifferential operators. Any improvement in the technique usually results in some progress in using the approximate spectral projection method.

The approximate spectral projection method occupies an intermediate position between the variational and Tauberian methods. Being based on variational principles, it also uses certain arguments characteristic of Tauberian methods. It is applicable practically to all problems 'with smooth coefficients' and provides estimates of the remainder in asymptotic formulae of the type (15.1), even though less accurate than the hyperbolic equation method (in those cases in which both methods are applicable).

We shall demonstrate how to reduce other problems concerned with the asymptotics of the spectrum to that of the asymptotics of $N(0; A_t)$.

In the problem concerned with the asymptotics of the standard distribution function $N(\lambda; A)$ of the eigenvalues (see Sect. 1.11) one must set $A_t = A - tI$ for a lower semi-bounded operator A. Then the asymptotics (15.1) turns into the classical Weyl asymptotics (9.3) (with $\lambda = t$). Moreover, if A is an elliptic operator of order m, then $a_m(x,\xi) - t$, where a_m is the principal symbol of A, can be taken as a_t. Such a replacement of the complete symbol by the principal symbol leads to an equivalent asymptotics and can be applied in all cases in which it is possible to specify some principal terms of a_t.

When considering the semiclassical asymptotics (as $h \to +0$) of $N(\lambda; A_h)$, where $A_h = a(x, hD_x)$, one should take $t = h^{-1}$ while assuming that λ is fixed. In the eigenvalue problem $Au = \lambda Bu$ with a positive operator B and a lower semi-bounded operator A one should set $A_t = A - tB$. Then (15.1) takes the form

$$N(\lambda; A, B) \sim (2\pi)^{-n} \operatorname{mes}\{(x,\xi) : a(x,\xi) < \lambda b(x,\xi)\}, \tag{15.2}$$

where $N(\lambda; A, B)$ is the number of eigenvalues of the problem $Au = \lambda Bu$ that are not greater than λ, and where a and b are the (principal) symbols of A and B.

One often encounters problems in which it is necessary to find the asymptotics of the number $N((\alpha,\beta), A_t)$ of eigenvalues of A_t belonging to an interval (α, β), where $\alpha > -\infty$. In this case one can take a real function $f = f(\lambda)$ such that $N((\alpha,\beta), A_t) = N(0, f(A_t))$, and the problem is again reduced to the case already discussed. (As f one usually takes a rational function to make it easier to work with the operator $f(A_t)$.) In this way one can study, for example, the asymptotic behaviour of eigenvalues accumulating near the boundary of the essential spectrum of the operator under consideration.

It is sometimes necessary to consider *operator-valued symbols* when applying the approximate spectral projection method. This is so, for example, in the theory of degenerate operators and boundary value problems, as well as in the case of the Schrödinger operator with some special non-negative potentials that do not tend to $+\infty$. In this case formula (15.1′), in which a_t is an operator-valued symbol, is usually valid as before, but the corresponding scalar formula may turn out not to be satisfied any longer.

15.2. Operator Estimates

In order to realize the approximate spectral projection method one usually constructs self-adjoint operators $\mathcal{E}_t^{\pm}$, which, so to say, approximate the projection $\chi_{(-\infty,0)}(A_t)$ from above and below, and are such that if $L_t^{\pm}$ are the images of the spectral projections $\chi_{[1/2,+\infty)}(\mathcal{E}_t^{\pm})$, then

$$(A_t u, u) < 0 \quad \text{for } u \in L_t^- \setminus \{0\}, \tag{15.3}$$

$$(A_t u, u) \geq 0 \quad \text{for } u \in (L_t^+)^{\perp} \cap D(A_t), \tag{15.4}$$

where $(L_t^+)^\perp$ is the orthogonal complement of L_t^+ in $\mathcal{H}$, and where $D(A_t)$ is the domain of A_t. In view of Glazman's lemma (1.13), it follows from (15.3) and (15.4) that

$$\dim L_t^- \le N(0; A_t) \le \dim L_t^+. \tag{15.5}$$

One must, therefore, construct the operators $\mathcal{E}_t^\pm$ and estimate the dimension of each of the corresponding subspaces $L_t^\pm$. An estimate of the dimension was first obtained with the aid of an estimate of the trace class norm $\|(\mathcal{E}_t^\pm)^2 - \mathcal{E}_t^\pm\|_1$, where $\|K\|_1 = \operatorname{Tr}\sqrt{K^*K}$ (see Shubin and Tulovskij 1973; Shubin 1978a; Hörmander 1979b, 1979c). Namely, if $\mathcal{E}$ is a trace class operator in $\mathcal{H}$ and $L = \chi_{[1/2,+\infty)}(\mathcal{E})\mathcal{H}$ is the image of the corresponding spectral projection, then

$$|\dim L - \operatorname{Tr}\mathcal{E}| \le 2\|\mathcal{E}^2 - \mathcal{E}\|_1, \tag{15.6}$$

which yields a two-sided estimate for $\dim L$. The inequalities (15.3) and (15.4) can be verified starting from certain inequalities, which do not contain $L_t^\pm$ explicitly. Namely, if a weak inequality is allowed in place of the strict inequality (15.3), then it suffices, for example, to prove the inequalities

$$\big(A_t\mathcal{E}_t^- u, \mathcal{E}_t^- u\big) \le 0, \quad u \in \mathcal{H}, \tag{15.7}$$

$$\big(A_t(u - \mathcal{E}_t^+ u), (u - \mathcal{E}_t^+ u)\big) \ge 0, \quad u \in D(A_t), \tag{15.8}$$

which, essentially, are alternative versions of the Gårding inequality. Here the lower-order terms, which are, as a rule, present in Gårding type inequalities, can usually be eliminated by choosing a sufficiently large t or by means of a suitable correction of the 'almost-projections' $\mathcal{E}_t$. For example, suppose that we are concerned with the asymptotic behaviour of the ordinary distribution function $N(\lambda)$ for an operator A, that is, $A_t = A - tI$. This being the case, if self-adjoint operators $\mathcal{F}_t$ such that

$$\big((A - tI)\mathcal{F}_t u, \mathcal{F}_t u\big) \le M(u, u), \quad u \in \mathcal{H}$$

are constructed, then one can easily prove the inequality

$$\big((A - tI - 4MI)\mathcal{F}_t u, \mathcal{F}_t u\big) \le 0, \quad u \in \mathcal{H}.$$

Hence, on replacing t by $t - 4M$, we get

$$\big((A - tI)\mathcal{F}_{t-4M} u, \mathcal{F}_{t-4M} u\big) \le 0, \quad u \in \mathcal{H},$$

which means that $\mathcal{F}_{t-4M}$ can be taken as $\mathcal{E}_t^-$. By analogy, if operators $\mathcal{F}_t$ such that

$$\big((A - tI)(u - \mathcal{F}_t u), u - \mathcal{F}_t u\big) \ge -M(u, u), \quad u \in D(A)$$

are constructed, then

$$\big((A - (t - 4M)I)(I - \mathcal{F}_t)u, (I - \mathcal{F}_t)u\big) \ge 0, \quad u \in D(A),$$

and we can set $\mathcal{E}_t^+ = \mathcal{F}_{t+4M}$.

The trace class norm of $\mathcal{E}^2 - \mathcal{E}$ appearing in (15.6) is awkward to estimate, because it has no analytic expression. Fejgin (1976) proposed a method which makes it possible to dispense with this estimate, although it yields a weaker estimate of the remainder. It consists in verifying the inequality $-1/2I \le \mathcal{E}_t \le 3/2I$ and then considering the operator $\mathcal{E}_{1t} = \mathcal{E}_t^2(3 - 2\mathcal{E}_t)$. We note that the function $f(x) = x^2(3-2x)$ has the following properties: $f(0) = 0$, $f(1) = 1$, and $0 \le f(x) \le 1$ for $x \in [-1/2, 3/2]$. Since $\mathcal{E}_{1t} = f(\mathcal{E}_t)$, it follows that $0 \le \mathcal{E}_{1t} \le I$, and so $0 \le \mathcal{E}_{1t}^2 - \mathcal{E}_{1t} \le I$. Moreover, if $\mathcal{E}_t$ is a projection, then $\mathcal{E}_{1t}$ is also a projection, and the now obvious equality $\|\mathcal{E}_{1t}^2 - \mathcal{E}_{1t}\|_1 = \operatorname{Tr}(\mathcal{E}_{1t}^2 - \mathcal{E}_{1t})$ indicates that the estimate of the trace class norm can be replaced by an estimate of the trace. As an example, let us present a precise result, which was proved by Fejgin (1976) and, in an abstract form, by Bezyaev (1982). It is concerned with the case $A_t = A - tI$.

Proposition 15.1. *Let A be a lower semi-bounded operator in a Hilbert space $\mathcal{H}$. Suppose that there exist two families $\{\mathcal{E}_t^{\pm}, t \ge 1\}$ of trace class operators in $\mathcal{H}$ such that the operator $A\mathcal{E}_t^-$ is bounded, $\mathcal{E}_t^+\mathcal{H} \subset D(A)$, and the following conditions, in which $\mathcal{E}_{1t}^{\pm} = (\mathcal{E}_t^{\pm})^2(3 - 2\mathcal{E}_t^{\pm})$, are satisfied:*

1) $-\frac{1}{2}I \le \mathcal{E}_t^{\pm} \le \frac{3}{2}I$;

2) $\mathcal{E}_{1t}^-(A - tI)\mathcal{E}_{1t}^- \le ct^{1-\nu}I$, *where* $0 < \nu < 1$, *and where c is a positive constant;*

3) $(I - \mathcal{E}_{1t}^+)(A - tI)(I - \mathcal{E}_{1t}^+) \ge -ct^{1-\nu}I$;

4) $\operatorname{Tr}\mathcal{E}_{1t}^{\pm} = V(t) + O(W(t, ct^{1-\nu}))$ *as $t \to +\infty$, where $V(t)$ is a non-decreasing positive function such that $V(t) \to +\infty$ as $t \to +\infty$, and where*

$$W(t,\tau) = V(t+\tau) - V(t-\tau), \quad t, \tau \ge 0;$$

5) $\operatorname{Tr}\left(\mathcal{E}_{1t}^{\pm}(I - \mathcal{E}_{1t}^{\pm})\right) = O(W(t, ct^{1-\nu}))$ *as $t \to +\infty$.*

Then A has discrete spectrum and the asymptotic formula

$$N(t) = V(t) + O(W(t, ct^{1-\nu})) \tag{15.9}$$

is valid for the corresponding distribution function $N(t)$.

15.3. Construction of an Approximate Spectral Projection

Following Levendorskij (1986b), we shall present a rough outline of the construction of an approximate spectral projection for the Friedrichs extension A_t in $L_2(X, H)$ of the pseudodifferential operator

$$r_+ a_{t,w} e_+ : C_0^\infty(X, H) \to L_2(X, H),$$

where H is an auxiliary Hilbert space, X is a domain in $\mathbb{R}^n$. e_+ is the operator that extends any function from X by setting it equal to zero on $\mathbb{R}^n \setminus X$, r_+ is the restriction from $\mathbb{R}^n$ to X, and a_w denotes a pseudodifferential operator in $\mathbb{R}^n$ with an operator-valued Weyl symbol $a(x,\xi)$ (the values of a are self-adjoint operators in H), i.e., an operator $a_w : C_0^\infty(X, H_1) \to C_0^\infty(X, H)$ defined by the formula

$$a_w u(x) = (2\pi)^{-n} \int e^{i(x-y)\cdot\xi} a\left(\frac{x+y}{2}, \xi\right) u(y)\, dy d\xi,$$

in which the integral is computed over $\mathbb{R}_y^n \times \mathbb{R}_\xi^n$, H_1 being a Hilbert space densely embedded in H. It is assumed that a satisfies the standard estimates of the theory of pseudodifferential operators, which ensure, in particular, that a_w is well defined.

Now, we introduce a smoothed indicator function $\chi_t \in C_0^\infty(\mathbb{R})$ of the half-axis $(-\infty, 0)$ and a function $\psi_{t,X} \in C_0^\infty(X)$ approximating the indicator function of X. It is assumed that both χ_t and ψ_t converge to the corresponding indicator functions as $t \to +\infty$, however, the convergence is not too fast in the sense that their derivatives do not grow too rapidly.

The integrand in (15.1′) may fail to be finite (this situation can be encountered, for example, when considering degenerate equations). We therefore also introduce an auxiliary symbol $d_t = d_t(x,\xi)$ (which is also Hermitian-valued) such that $d_t > 0$ and

$$N\big(0; (a_t - d_t)(x,\xi)\big) < \infty, \quad (x,\xi) \in X \times \mathbb{R}^n.$$

Next, we set

$$\mathring{e}_t^{\pm} = \chi_t\left(d_t^{-\frac{1}{2}} a_t d_t^{-\frac{1}{2}} \mp 1\right), \quad e_t^{\pm} = \psi_{t,X}\, \mathring{e}_t^{\pm}\, \psi_{t,X},$$

and, finally, we take $\mathcal{E}_t^{\pm} = e_{t,w}^{\pm}$.

It is necessary to verify various estimates of the type of Proposition 15.1 for the approximate spectral projections $\mathcal{E}_t^{\pm}$ constructed in this way. Whether or not this can be accomplished successfully depends on the presence of a good enough algebra of pseudodifferential operators in which all the operators involved could be included, and in which a good theorem on composition would be valid and Gårding type inequalities for operators with positive symbols could be obtained. All this is possible under the following conditions:

1) the right-hand side of (15.1′) is finite;

2) a_t is a hypoelliptic symbol in an algebra of pseudodifferential operators, the algebra depending on t, in general.

The latter condition usually means that there exist scalar weight functions $\Phi_t(x,\xi)$ and $\varphi_t(x,\xi)$ and an operator-valued function $q_t(x,\xi)$ such that

$$\left\|(q_t^{-1})^*(\partial_\xi^\alpha \partial_x^\beta a_t) q_t^{-1}\right\| \le c_{\alpha\beta} \Phi_t^{-|\alpha|} \varphi_t^{-|\beta|}, \quad (x,\xi) \in X \times \mathbb{R}^n, \tag{15.10}$$

$$a_t(x,\xi) \ge c q_t(x,\xi)^* q_t(x,\xi) \quad \text{for } |x| + |\xi| \ge C(t). \tag{15.11}$$

The weight functions Φ_t, φ_t, and q_t must satisfy certain conditions such as those of Beals and Fefferman (1974), among which the following *sharpened uncertainty principle* is the basic one: if $h_t = \Phi_t^{-1}\varphi_t^{-1}$, then

$$h_t(x,\xi) \leq C(1+|x|+|\xi|)^{-\varepsilon}, \tag{15.12}$$

where $\varepsilon > 0$, or

$$\sup_{x,\xi} h_t(x,\xi) \to 0 \quad \text{as } t \to +\infty. \tag{15.12$'$}$$

In this situation one can, as a rule, set $d_t = q_t^* q_t h_t^\delta$, where $\delta > 0$.

Under these conditions it is possible to prove the asymptotic formula (15.1′) with an estimate of the remainder involving the measures of sets connected with the boundary $\partial X \times \mathbb{R}^n$ and with the level surfaces $\lambda_{t,j}(x,\xi) = 0$ of the eigenvalues $\lambda_{j,t}$ of $a_t(x,\xi)$.

Algebras constructed on the basis of the classes of Weyl symbols proposed by Hörmander (1979a and 1983–1985, §18.5) can often serve as the required algebras of pseudodifferential operators. In particular, the role of the uncertainty principle is clarified in Hörmander's calculus.

We remark that the role of the uncertainty principle in spectral theory has recently been clarified explicitly by Fefferman (1983). In particular, for a scalar operator A a violation of the principle of the form (15.12), (15.12′) may lead to a violation of the classical Weyl formula (15.1) (even though it has a well-defined meaning), in which case it becomes necessary to introduce an operator symbol.

15.4. Some Precise Formulations

A. We begin with the formulation of Hörmander's result (Hörmander 1979b) on the asymptotic behaviour of the ordinary distribution function $N(\lambda)$ of the spectrum for an operator a_w in $\mathbb{R}^n$ defined by a positive Weyl symbol $a = a(z)$, where $z = (x,\xi)$. In this example we shall use the terminology and notation from (Hörmander 1983–1985, §18.5) (see also Hörmander 1979a). Let $a \in S(a,g)$, where g is a Riemannian metric on $\mathbb{R}_z^{2n} = \mathbb{R}_x^n \times \mathbb{R}_\xi^n$ that is σ temperate with respect to the canonical symplectic structure σ. Setting $h(z) = [\sup(g_z/g_z^\sigma)]^{1/2}$, we assume that

$$h(z) \leq Ca(z)^{-\gamma}, \quad 1+|z| \leq a(z)^N. \tag{15.13}$$

This being the case, if $0 < \nu < 2\gamma/3$, then

$$N(\lambda) = V(\lambda) + O\big(V(\lambda+\lambda^{1-\nu}) - V(\lambda-\lambda^{1-\nu})\big), \tag{15.14}$$

as $\lambda \to +\infty$, where

$$V(\lambda) = (2\pi)^{-n} \text{mes}\,\{(x,\xi) : a(x,\xi) < \lambda\}. \tag{15.15}$$

The remainder in (15.14) can be estimated more explicitly under additional assumptions about a. Namely (see Shubin 1978a, §28), if

$$V'(\lambda)V(\lambda)^{-1} = O(\lambda^{\nu-\varepsilon-1}), \tag{15.16}$$

then

$$V(\lambda+\lambda^{1-\nu}) - V(\lambda-\lambda^{1-\nu}) = O(\lambda^{-\varepsilon}V(\lambda)).$$

In turn, the estimate (15.16) will be fulfilled if it is required, for example, that the condition

$$|z\cdot\nabla a(z)| \geq ca(z)^{1+\varepsilon-\nu}, \quad |z| \geq R_0, \quad c>0, \quad z=(x,\xi) \tag{15.17}$$

be satisfied. Thus, if (15.17) is satisfied, then (15.14) takes the simpler form

$$N(\lambda) = V(\lambda)(1+O(\lambda^{-\varepsilon})). \tag{15.18}$$

Let us consider, for example, the case when the symbol is *polyhomogeneous* in z, i.e., admits the asymptotic expansion

$$a \sim a_m(z) + a_{m-1}(z) + \dots ,$$

where $a_{m-j}(z)$ is positively homogeneous of degree $m-j$ in z and $a_m(z)>0$ for $z\neq 0$. (On the precise meaning of such an expansion see, for example, (Shubin 1978a, §23) or (Egorov and Shubin 1988b, §1).) For such symbols conditions (15.17) and, consequently, (15.16) are satisfied if $\varepsilon \leq \nu$, since

$$z\cdot\nabla a(z) \sim ma_m(z) + (m-1)a_{m-1}(z) + \dots$$

by the Euler formula. It is clearly convenient to take $\varepsilon=\nu$, because this choice yields the best estimate (15.18). Furthermore, all the conditions will be satisfied for the metric $g=(1+|z|^2)^{-1}dz^2$, in which case $h(z)=(1+|z|^2)^{-1}$ and one can take $\gamma = 2/m$, so that (15.18) can be obtained for any $\varepsilon < 4/(3m)$. By computing $V(\lambda)$, it is easy to show that

$$\begin{aligned} V(\lambda) &= (2\pi)^{-n}\lambda^{2n/m}\operatorname{mes}\{z: a_m(z)<1\} \\ &\quad - (2\pi)^{-n}\lambda^{(2n-1)/m} \int\limits_{a_m(z)=1} \frac{a_{m-1}(z)}{|\nabla a_m(z)|}\,dS_z + O(\lambda^{(2n-2)/m}), \end{aligned}$$

where dS_z is the Euclidean area element of the surface $\{z: a_m(z)=1\}$. Thus in our case the asymptotics (15.18) takes the form

$$V(\lambda) = c_0\lambda^{2n/m} + c_1\lambda^{(2n-1)/m} + O(\lambda^{(2n-1-\varkappa)/m}), \tag{15.19}$$

where $\varkappa < 1/3$. Considering, for example, the harmonic oscillator, one can expect that $\varkappa = 1$ is the optimal estimate. This result can be obtained either by the hyperbolic equation method, which was applied in this situation for the first time by Helffer and Robert (1982a) (see also Helffer 1984), or by studying the Fourier transform of the measure $dN(\lambda^{1/2m})$ with the aid of the heat kernel (Aramaki 1987).

The described situation is quite typical for the approximate spectral projection method. For very general operators it gives a weaker estimate of the remainder in the spectral asymptotic formulae than that obtained by the hyperbolic equation method. However, the latter has a narrower domain of applicability (the hyperbolic equation method requires that the symbols should have a certain structure, which is unnecessary for the approximate spectral projection method).

We have described a situation in which the approximate spectral projection method is applied to operators in $\mathbb{R}^n$ whose symbols are subject to conditions in which x and ξ have the same status. This, of course, does not have to be so, and the approximate spectral projection method can be easily adapted to a situation that lacks symmetry under the exchange of x and ξ. For example, Fejgin (1978) and Levendorskij (1982) used the approximate spectral projection method to consider differential operators in a domain $X \subset \mathbb{R}^n$ with coefficients that can grow rapidly at infinity. For example, for the Schrödinger operator $A = -\Delta + q(x)$ the conditions for the potential have the form $1 \le q \in C^1$ and

$$|\nabla q(x)| \le Cq(x)^{1+\delta_0}, \quad 0 \le \delta_0 < 1, \quad q(x) \to +\infty \tag{15.20}$$

as $|x| \to \infty$ or $x \to \partial X$.

Generally, let $a(x,\xi)$ ba a Hermitian $(N \times N)$-matrix–valued symbol that meets the following conditions, which presuppose the knowledge of a function q satisfying (15.20):

$$\left\| \left[\partial_\xi^\alpha \partial_x^\beta a(x,\xi)\right] a(x,\xi)^{-1} \right\| \le C_{\alpha\beta}(|\xi| + q(x))^{-\rho|\alpha|+\delta|\beta|}, \tag{15.21}$$

where $\delta_0 \le \delta < \rho \le 1$, and

$$\|a(x,\xi)\| \le C(|x| + q(x))^k,$$
$$\|a(x,\xi)^{-1}\| \to 0 \quad \text{as } |x| + |\xi| \to \infty \text{ or } x \to \partial X.$$

Then the closure of a_w in $(L_2(X))^N$ (from the domain $(C_0^\infty(X))^N$) is self-adjoint and has discrete spectrum. Moreover, if

$$V_\pm(\lambda) = (2\pi)^{-n} \sum_{j=1}^{N} \operatorname{mes}\left\{(x,\xi) \in X \times \mathbb{R}^n : 0 < \pm\lambda_j < \lambda\right\}, \tag{15.22}$$

where $\lambda_j = \lambda_j(x,\xi)$ are the eigenvalues of the matrix a, and $W_\pm(t;\tau) = V_\pm(t+\tau) - V_\pm(t-\tau)$, then the asymptotic formulae

$$N_\pm(\lambda; A) = V_\pm(\lambda) + O\big(\lambda^{-\varkappa} V_\pm(\lambda) + W_\pm(\lambda, \lambda^{1-\varkappa})\big), \tag{15.23}$$

where $\varkappa \in (0, (\rho-\delta)/(3k))$ if $N > 1$, and $\varkappa \in (0, (\rho-\delta)/(2k))$ if $N = 1$ (i.e. in the scalar case), are satisfied for the distribution functions $N_\pm(\lambda; A)$.

To prove these asymptotic formulae it is necessary to construct the approximate spectral projections as a family of operators defined with the aid of amplitudes $a(x, y, \xi)$ which are equal to zero for $|x-y| \geq c(q(x)+q(y))^{-\gamma}$, where $\gamma > \delta_0$ and $c > 0$ (see Fejgin 1978).

The condition (15.20) permits not only rapidly growing potentials (for example, $\exp(\exp(\exp\ldots(\exp|x^2|))\ldots)$, but also potentials that, on the contrary, grow very slowly. For example, if $n = 1$, then one can take

$$q(x) = \underbrace{\ln\ln\ldots\ln}_{k}|x|$$

for large $|x|$ and consider the Schrödinger operator $A = -d^2/dx^2+q$ in $L_2(\mathbb{R}^1)$. Formula (15.23) is inapplicable in this case, because $W(\lambda, \lambda^{1-\varkappa})$ grows faster than $V(\lambda)$. However, a somewhat more general argument presented in (Levendorskij 1986b, §2, Theorem 1) makes it possible to obtain the asymptotics

$$N(\lambda; A) = V(\lambda)\big[1 + O\big((\underbrace{\exp\ldots\exp}_{k}\lambda)^{-r}\big)\big],$$

where $r \in (0, 2/3)$ (see Levendorskij 1986b, §5). We also remark that in the case of the Schrödinger operator the smoothness conditions contained in (15.21) are unessential and can be fulfilled by means of smoothing (see Fejgin 1978).

B. We shall describe Fejgin's results on the semiclassical asymptotics (Fejgin 1979). We consider an elliptic self-adjoint h-differential operator A_h in $\mathbb{R}^n$, i.e., an operator of the form

$$A_h = \sum_{|\alpha|\leq m,\, k\leq m_1} a_{\alpha k}(x)h^{|\alpha|+k}D^\alpha \tag{15.24}$$

with a small parameter h. Let

$$a_0(x, h\xi) = \sum_{|\alpha|\leq m} a_{\alpha 0}(x)h^{|\alpha|}\xi^\alpha$$

be a function playing the role of the principal symbol of A in the problem of the *semiclassical spectral asymptotics*, i.e., the problem concerned with the asymptotic behaviour as $h \to +0$. We assume that $a_{\alpha k} \in C^\infty(\mathbb{R}^n)$ and

$$\begin{gathered}\left|\partial_x^\gamma a_{\alpha k}(x)\right| \leq C_{\alpha\gamma k}a_{00}(x)^{\max\left(1-\frac{|\alpha|+k}{m}-\rho(|\gamma|+k),0\right)},\\ x \in \mathbb{R}^n, \quad |\alpha| \leq m, \quad k \leq m_1\end{gathered} \tag{15.25}$$

for some $\rho > 0$ and any multi-index γ. Moreover, let the 'ellipticity' condition

$$|a_0(x, h\xi)| \geq c_0\left(h^{2m}|\xi|^{2m} + a_{00}^2(x)\right)^{1/2},$$

where $c_0 > 0$, be satisfied, and let the finiteness condition

$$\operatorname{mes}\{(x,\xi) : a(x,\xi,h) < 2\} < \infty$$

for the phase-space volume be fulfilled, $a(x,\xi,h)$ being the complete Weyl symbol of A_h. Hence one can deduce that the spectrum of A is discrete in $(-\infty, 1)$ for small $h > 0$. Then it turns out that for any $R \leq R_0 = (\rho + 1/m)^{-1}$, $h \leq 1$, $H \in \mathbb{R}$, and $M \in [M_0, h^{-R}]$, where M_0 is large enough, the following estimate is true for the distribution function of the eigenvalues of A_h:

$$\begin{aligned} &\left|N(Mh^R; A_h) - (2\pi)^{-n}\operatorname{mes}\{(x,\xi) : a_0(x,h\xi) < Mh^R\}\right| \\ &\quad \leq C \operatorname{mes}\{(x,\xi) : \left|a_0(x,h\xi) - Mh^R\right| \leq CM^s h^{R+\Delta}\} \\ &\quad\quad + C_H \operatorname{mes}\{(x,\xi) : a_0(x,h\xi) \leq Mh^R + CM^s h^{R+\Delta}\} M^{-H} h^{\varepsilon H}, \end{aligned} \tag{15.26}$$

where

$$1 - (\rho + 1/m)/2 < s < 1, \quad \varepsilon = (1 - R(\rho + 1/m))/2 - \Delta,$$

and where $\Delta \geq 0$ is chosen in such a way that $\varepsilon > 0$ for $R < R_0$ and $\varepsilon = 0$ for $R = R_0$. The constants C and C_H depend only on the constants $C_{\alpha\gamma k}$ and c_0 in the estimates of the symbols, as well as on m and H.

Hence, by choosing the parameters M and R, it is possible to obtain various kinds of information on the eigenvalues. For example, if $R = 0$, then (15.26) yields an asymptotic formula for $N(\lambda; A_h)$ as $h \to +0$ for any fixed λ. But if $R = R_0$, then, subject to certain additional assumptions, (15.26) yields the Bohr-Sommerfeld asymptotics for an individual eigenvalue $\lambda_N = \lambda_N(h)$ as $h \to +0$ and $N \to \infty$ (see Fejgin 1979).

The Schrödinger operator

$$A_h = -h^2\Delta + q(x) \tag{15.27}$$

is the most important special case of (15.24). The asymptotics (15.26) is applicable in this case, provided that $q > 0$ and q satisfies the estimates (15.25) (with $\alpha = k = 0$ and $a_{00} = q$). Levendorskij (1986b) obtained the asymptotics of $N(\lambda; A_h)$ as $h \to +0$ for a fixed λ under weaker restrictions on the potential q, which can be a Hermitian $(p \times p)$-matrix–valued function in a (possibly unbounded) domain $X \subset \mathbb{R}^n$ with Lipschitz boundary (in which case A_h is to be understood as the Friedrichs extension). Namely, let

$$q \geq cI, \quad \left\|q^{(\alpha)}(x)q^{-1}(x)\right\| \leq C_\alpha, \quad |\alpha| = 1, 2, \dots .$$

This being the case, if λ is such that $q(x) \geq \lambda I$ outside a compact set in X, then, as $h \to +0$,

$$N(\lambda; A_h) = h^{-n}V(\lambda; q; X) + O(h^{-n+\varkappa}) \tag{15.28}$$

with

$$V(\lambda; q; X) = (2\pi)^{-n} v_n \sum_{j=1}^{p} \int_X \left(\lambda - \lambda_j^0(x)\right)_+^{n/2} dx,$$

where v_n is the volume of the unit ball in $\mathbb{R}^n$, $\lambda_j^0(x)$ are the eigenvalues of the matrix $q(x)$, $a_+ = \max(a,0)$, and $\varkappa \in (0,1/3)$. The quality of the estimate of the remainder in (15.28) is determined by $\varkappa$ and can be improved in the following cases:

a) if q is locally diagonalizable in a smooth way, then one can take $\varkappa \in (0,1/2)$;

b) if, in addition, the boundary ∂X is piecewise smooth, then one can take $\varkappa \in (0,2/3)$.

C. We shall now consider the problem concerned with the asymptotic behaviour of the spectrum of the linear *operator pencil*

$$Au = \lambda Bu \tag{15.29}$$

in a bounded domain $X \subset \mathbb{R}^n$. Let A and B be classical (polyhomogeneous in ξ) symmetric pseudodifferential operators on $\bar{X}$ of order m_1 and m_2, respectively, and let $m_1 > m_2$. Let the principal symbol $a(x,\xi)$ of A be positive, and let

$$(Au,u) \geq c\|u\|^2_{(m_1)}, \quad u \in C_0^\infty(X), \tag{15.30}$$

where $\|\cdot\|_{(m_1)}$ is the ordinary Sobolev norm of order m_1. Let $N_\pm(\lambda)$ be the distribution function of the positive and negative eigenvalues of the problem (15.29) with the Dirichlet conditions in the ordinary variational setting, so that, by the variational principle, $N_\pm(\lambda)$ is equal to the number of negative eigenvalues of the Friedrichs extension of $A \mp \lambda B$. We set

$$(\partial X)_\varepsilon = \{x : \operatorname{dist}(x, \partial X) < \varepsilon\}.$$

In this situation the approximate spectral projection method yields the following results (Levendorskij 1984, 1985, 1986b):

a) If B is an elliptic operator with principal symbol $b(x,\xi)$ and X is a domain such that

$$\operatorname{mes}(\partial X)_\varepsilon \leq C\varepsilon, \quad \varepsilon > 0, \tag{15.31}$$

then

$$N_\pm(\lambda) = c_\pm \lambda^{n/m} + O\big(\lambda^{(n-r)/m}\big), \tag{15.32}$$

where $m = m_1 - m_2$, $r \in (0,1/3)$, and the constants $c_\pm$ can be written in terms of the principal symbols a and b in the usual way. One of the constants $c_\pm$ vanishes, depending on the sign of b.

Subject to certain additional restrictions, this result can be generalized to matrix-valued operators (in which case both constants $c_\pm$ can be distinct from zero).

If X has Lipschitz boundary, then one can take $r \in (0,1/2)$ in the asymptotic formula (15.32), and, for a domain with piecewise smooth boundary, even $r \in (0,2/3)$.

b) Now, suppose that B is not elliptic and $\varkappa \in [0,1)$ is a number such that

$$\int_{X \times S^{n-1}} |b(x,\xi)|^{-\varkappa} dx dS_\xi < \infty,$$

where $S^{n-1} = \{\xi : |\xi| = 1\}$ and dS_ξ is the Euclidean surface element of the sphere S^{n-1}. Then

$$N_\pm(\lambda) = c_\pm \lambda^{n/m} + O(\lambda^{\rho(r)}), \tag{15.33}$$

where $\rho(r) = \max\{(n-r)/m, n(1-\varkappa)/(m+r-\varkappa m)\}$ with $r \in (0, 1/3)$, provided that X satisfies (15.31). If the boundary ∂X is Lipschitzian, then one can take $r \in (0, 1/2)$, as above, or even $r \in (0, 2/3)$ for a domain with piecewise smooth boundary. This result can also be carried over to matrix-valued operators, subject to certain additional restrictions on the symbols. The case of general boundary conditions can also be considered in the framework of this scheme (Levendorskij 1986b).

D. Degenerate operators provide the basic examples indicating that even though the 'scalar' Weyl asymptotics (15.1) or (15.14) may cease to make sense or become invalid, the application of operator-valued symbols can restore its validity.

Following Levendorskij (1988a), we describe the basic idea of the necessary construction. Let A be a lower semi-bounded formally self-adjoint differential operator in a domain $X \subset \mathbb{R}^n$ with boundary conditions on X degenerate on a manifold $\Gamma \subset \bar{X}$, the degeneracy being defined by functions of the type $\rho(x) = \operatorname{dist}(x, \Gamma)$ depending only on the normal variables to Γ. Then in many cases A can be realized in a narrow strip X'_t adjoining Γ as a non-degenerate operator on Γ with an operator-valued symbol $\tilde{a}_t$, while remaining non-degenerate in $X_t = X \setminus \bar{X}'_t$. Let $A_{1,t}$ and $A_{2,t}$ be the operators of the boundary value problems for A in X_t and X'_t with the Dirichlet boundary conditions on $\bar{X}_t \cap \bar{X}'_t$ and the original boundary conditions on $\partial X_t \cap \partial X$ and $\partial X'_t \cap \partial X$, respectively. If the domain X'_t contracts not too rapidly as $t \to +\infty$, then it can be found with the aid of the standard variational argument that

$$N(t; A) \sim N(t; A_{1,t}) + N(t; A_{2,t}). \tag{15.34}$$

The classical Weyl asymptotics ((15.1) with $A_t = A_{1,t} - tI$) is valid for $N(t; A_{1,t})$, while the operator asymptotics

$$N(t; A_{2,t}) \sim (2\pi)^{n_1 - n} \iint_{T^*\Gamma} N\big(t; \tilde{a}_t(\tilde{x}, \tilde{\xi})\big)\, d\tilde{x} d\tilde{\xi}, \tag{15.35}$$

where $n_1 = \operatorname{codim} \Gamma$, is true for $A_{2,t}$. Unfortunately, it may turn out to be a hard problem to compute explicitly the asymptotics of the right-hand side of (15.35). The problem is the more easily solved the narrower the domain X'_t. But formula (15.34) and the Weyl asymptotics for $N(t; A_{1,t})$ can be proved only if X'_t is not too narrow. The choice of X'_t is dictated by the feasibility of the subsequent application of the approximate spectral projection method.

In this way it is possible to consider a very general anisotropic degeneracy, including, in particular, the cases of weak, strong, and moderate degeneracy at the boundary ∂X described in §10 (for more details see Levendorskij 1988a).

Another example of degeneracy is provided by the Schrödinger operator with a homogeneous in x (of degree $a > 0$) non-negative potential (see Example 10.5) that can have a zero of finite order $a_1 < a$ on an n_1-dimensional cone $\hat{K}$ such that the intersection $K = \hat{K} \cap S^{n-1}$ with the unit sphere is a smooth ($(n_1 - 1)$-dimensional) manifold (here $n_1 < n$). Solomyak's results (Solomyak 1985) described in §10 can be obtained by the approximate spectral projection method. Levendorskij (1988a) extended these results to the Schrödinger operator with a matrix-valued potential.

We shall describe the generalization of the Weyl formula put forward by Levendorskij (1988a), which makes it possible to predict the asymptotics of the spectrum in cases of weak or moderate degeneracy, and also the order of the asymptotics in the case of strong degeneracy.

Let $d = d(x)$ be a function on X such that $d \geq 0$ and

$$d \leq c_1 A + c_2 I, \tag{15.36}$$

where d is identified with the multiplication operator by d. Then, for any $c > 0$ there exist $c', c'' > 0$ such that

$$N(t; A + cd) \leq N(t; A) \leq N(c't + c''; A + cd).$$

It follows that if $N(t; A + cd)$ has a power or logarithmic power asymptotics as $t \to +\infty$, then

$$c'N(t; A) \leq N(t; A + cd) \leq c''N(t; A),$$

which often makes it possible to obtain sharp-order estimates for $N(t; A)$. The reason for replacing A by $A + cd$ is that for an operator with discrete spectrum, d can be chosen in such a way that $d(x) \to +\infty$ as $x \to \Gamma$ if $\Gamma \subset \partial X$. Moreover, it turns out that in the case of weak or moderate degeneracy the Weyl formula (15.1) with $a_t(x, \xi) = a(x, \xi) + cd(x) - t$ (and with $A_t = A - tI$) predicts correctly not only the order, but also the leading term of the asymptotics if d is chosen properly.

Let us present an example involving the choice of d. Outside a fixed neighbourhood of Γ one should take $d(x) = 1$. Suppose that in a coordinate system $y = (y', y'')$ the operator A has the form

$$A = \sum_{\alpha,\beta} D_y^\beta a'_{\alpha\beta}(y) \rho_\alpha(y'') \rho_\beta(y'') D_y^\alpha$$

in a neighbourhood U of $x_0 \in \Gamma$. The functions ρ_α define the degeneracy (for $y'' = 0$). Then one should set

$$d(x) = \max_{\alpha \neq 0} \rho_\alpha(y''(x))^2 |y''(x)|^{-2|\alpha''|},$$

where $\alpha = (\alpha', \alpha'')$ is the splitting of the multi-index α corresponding to $y = (y', y'')$ (see Levendorskij 1988a).

Finally, we remark that the approximate spectral projection method makes it also possible to consider a number of other problems not mentioned in this section (for example, the spectral asymptotics for problems with constraints (Levendorskij 1986a) and problems in shell theory (Levendorskij 1985), or the asymptotics of the integrated density of states of almost periodic and random operators (Bezyaev 1978; Bogorodskaya and Shubin 1986)). For more details we refer the reader to the above-mentioned articles.

§16
The Laplace Operator on Homogeneous Spaces and on Fundamental Domains of Discrete Groups of Motions

16.1. Preliminary Remarks

If G is an *isometry group* acting on a Riemannian manifold X, then the Laplace-Beltrami operator $-\Delta$ commutes with the action of that group. It follows that G has an induced representation in each eigenspace of $-\Delta$. As a consequence, it becomes possible to study the spectrum by means of representation theory, using the algebraic structure of X. The richest structure is encountered in the case when X is a *homogeneous space*, i.e., the quotient space of a Lie group G relative to a closed subgroup H. The algebraic approach is particularly successful for manifolds of constant curvature, i.e., spaces that can be obtained by the factorization of $\mathbb{R}^n$, S^n, or $\mathbb{H}^n$ (the *Lobachevski space*) with respect to a discrete group of motions. For the Laplace operator on a manifold of this kind one can often succeed in carrying out the complete spectral analysis. The operators of the basic boundary value problems on fundamental domains can also be included in such an analysis. All this supplements the contents of §4 substantially.

Furthermore, a rich algebraic structure often leads to important relations (such as the Poisson or Selberg formulae) connecting the spectrum of the Laplace operator with the geometric characteristics of the manifold – in the first place with the lengths of closed geodesics on that manifold. In the general situation there is also such a connection, but it turns out to be concealed at a deeper level (see §14).

Within the range of questions under consideration it is possible to obtain a number of results for the problem of whether or not 'one can hear the shape of a drum,' concerned with the possibility of reconstructing a differential operator from its spectrum. Our attention will be devoted mainly to this and similar problems. We shall not touch upon the spectral theory of automorphic functions on the Lobachevski plane; on the latter subject see (Venkov 1979).

16.2. The Automorphic Laplace Operator

Let X be a Riemannian space and let G be a discrete group of isometries of X. A *fundamental domain* of G on X is a closed connected set $\Omega_0 \subset X$ such that every orbit of G intersects Ω_0, and if γ is an element of G other than the unity and $\Omega = \operatorname{Int} \Omega_0$, then $\Omega \cap \gamma\Omega = \emptyset$. The operators $-\Delta_D = -\Delta_{D,\Omega}$ and $-\Delta_N = -\Delta_{N,\Omega}$ of the Dirichlet and Neumann problems for the Laplace-Beltrami operator on Ω can be introduced in the usual way. Furthermore, let us fix a *character* χ of G, i.e., a homomorphism from G into the group $S^1 = \{z \in \mathbb{C} : |z| = 1\}$. We consider the space $C^\infty_\chi(X)$ of smooth functions u on X such that $u(\gamma x) = \chi(\gamma)u(x)$ and the space $C^\infty_\chi(\Omega)$ of restrictions of functions from $C^\infty_\chi(X)$ to Ω. If $\operatorname{mes} \Omega < \infty$, then the Laplace-Beltrami operator is symmetric and essentially self-adjoint on $C^\infty_\chi(\Omega)$. Its closure is called the *automorphic Laplace operator* and denoted by $-\Delta_\chi = -\Delta_{\chi,\Omega}$. If another fundamental domain is chosen, then $-\Delta_{\chi,\Omega}$ changes into a unitarily equivalent operator. In the special case when the transformations $\gamma \in \Omega$ have no fixed points and χ is the *trivial* character (i.e., $\chi(\gamma) \equiv 1$), $-\Delta_\chi$ can be regarded as the Laplace operator on the quotient manifold $M = G/X$. An important case is also that of G generated by a finite set G_0 of *reflections*, i.e., isometries γ such that $\gamma^2 = 1$ that leave invariant the points of a submanifold X_γ of codimension one. The system of generators G_0 can be chosen in such a way that the submanifolds X_γ, where $\gamma \in G_0$, constitute the boundary of a fundamental domain Ω_0 of G. For such a domain Ω_0 the automorphic Laplace operator with the trivial character χ coincides with $-\Delta_{N,\Omega}$, and if $\chi(\gamma) = -1$ for any $\gamma \in G_0$, then it coincides with $-\Delta_{D,\Omega}$. Other real-valued characters define combinations of the Dirichlet and Neumann conditions.

16.3. The Laplace Operator on a Flat Torus. The Poisson Formula

In the case of zero curvature one can most easily describe the spectrum of the operator Δ on a *flat torus*, i.e., $M_G = \mathbb{R}^n/G$, where G is a group of motions isomorphic to $\mathbb{Z}^n$. There is a natural embedding of G as a *lattice* in $\mathbb{R}^n$. The *conjugate lattice* (and *conjugate group*) $G^\#$ can be defined as $\{\alpha \in \mathbb{R}^n : \alpha \cdot \gamma \in \mathbb{Z} \text{ for all } \gamma \in G\}$. The equality $M^\# = \mathbb{R}^n/G^\#$ defines the dual torus. The spectrum of $-\Delta_M$ consists of the eigenvalues $4\pi^2|\alpha|^2$, where $\alpha \in G^\#$, with eigenfunctions $\varphi_\alpha(x) = \exp(2\pi i\alpha \cdot x)$. Each $\alpha \in G^\#$ defines an element of the fundamental group $\pi_1(M^\#)$, the number $2\pi|\alpha|$ being equal to the length of a closed geodesic on $M^\#$ from the corresponding homotopy class.

Given a compact Riemannian manifold X, let us agree to denote by $\mathcal{L}(X)$ the set of lengths of all possible closed geodesics on X. We can see that in the case when $X = M = \mathbb{R}^n/G$ is a flat torus, the spectrum of $-\Delta_M$ and the set $\mathcal{L}(M^\#)$ are uniquely determined by one another. Moreover, the spectrum of $-\Delta_M$ is uniquely connected with $\mathcal{L}(M)$. The connection is given by the classical *Poisson summation formula*, according to which

$$\sum_{\gamma' \in G^\#} \tilde{f}(\gamma') = \operatorname{vol} M \sum_{\gamma \in G} f(2\pi\gamma) \tag{16.1}$$

for any function $f \in S(\mathbb{R}^n)$ and any lattice $G \subset \mathbb{R}^n$ ($\tilde{f}$ is the Fourier transform of f).

The *Jacobi formula* for the θ-function $\theta_M(t) = \theta_{-\Delta_M}(t)$ of $-\Delta_M$ (see Sect. 12.4) is a special case of (16.1). By (16.1),

$$\theta_M(t) = \sum_{\gamma' \in G^\#} e^{-4\pi^2 t|\gamma'|^2} = \frac{\operatorname{vol} M}{(2\pi t)^{n/2}} \sum_{\gamma \in G} e^{-|\gamma|^2/t},$$

or

$$\theta_M(t) = \frac{\operatorname{vol} M}{(2\pi t)^{n/2}} \theta_{M^\#}(t)((4\pi^2 t)^{-1}).$$

An example of non-isometric isospectral manifolds was first constructed in the class of flat tori. The construction of lattices was found by Witt in 1941 (Witt 1941), but it was only in 1964 that Milnor noticed its connection with the spectral properties of flat tori (Milnor 1964) (this is why it is customary to talk of the *'Milnor example'*).

Let $G_0 = \left\{\gamma \in \mathbb{Z}^n : \sum_j \gamma_j \in 2\mathbb{Z}\right\}$. We consider the lattice $G(n)$ in $\mathbb{R}^n$ generated by G_0 and the element $w_n = (1/2, \ldots, 1/2)$. The lattice G(n) is self-conjugate and $\operatorname{vol} M_{G(n)} = 1$. Next, let $G = G(8) \oplus G(8)$ and $H = G(16)$ be lattices in $\mathbb{R}^{16}$. They are not isomorphic, i.e., cannot be mapped onto one another by an orthogonal transformation of $\mathbb{R}^{16}$, since G has a basis consisting of elements whose length is not greater than $\sqrt{2}$, while the length of any element of H of the form $\gamma + w_{16}$, where $\gamma \in G_0$, is greater than $\sqrt{2}$, so that H does not have such a basis. It follows that the tori M_G and M_H are not isometric. On the other hand, to prove isospectrality it suffices to verify that the θ-functions of M_G and M_H coincide. Since the lattices are self-conjugate, it follows from (16.1) that the functional equation $\Phi(t) = t^{-n/2}\Phi(t^{-1})$ is satisfied by either of the functions $\Phi(t) = \theta(it/2\pi)$. The function Φ is periodic with period one, since the square of the length of every element of G and H is an even number. It follows that $\Phi(t)$ is a modular function with weight $t^{-n/4}$. For $n = 16$ the space of such functions is one-dimensional (see Serre 1970). The θ-functions of G and H can therefore differ only by a multiplicative constant. Finally, since $\operatorname{Vol} M_G = \operatorname{Vol} M_H$, it follows by (16.1) that the leading terms of the asymptotic formulae for the θ-functions as $t \to 0$ are equal to one another, and so $\theta_G \equiv \theta_H$.

Afterwards, Kneser (1967) constructed a similar example in dimension 12 and demonstrated that no such examples can exist for $n = 2$. Next, Wolpert (1978) established that, nevertheless, isospectral tori are isometric in the general situation. If $G = A\mathbb{Z}^n$ and $H = B\mathbb{Z}^n$, where A and B are square matrices, and the tori M_G and M_H are isometric, then the elements of the matrix A^*A are linear combinations of the elements of B^*B with rational coefficients, and both matrices belong to a subset of codimension 1 in the space of positive matrices. It follows, in particular, that any continuous family of isospectral tori is isometric. The total number of isospectral pairwise non-isometric tori is finite in any dimension (Wolpert 1978). In particular, for $n = 3$ the number is not greater than 15 (Berry 1981).

Furthermore, let G be an arbitrary discrete group of isometries of $\mathbb{R}^n$ with a compact fundamental domain. Any such group contains a subgroup G_0 of finite index that is isomorphic to $\mathbb{Z}^n$ (see Shvartsman and Vinberg 1988). It follows that the description of the spectrum of the automorphic Laplace operator on M_G corresponding to the trivial character on G_0 can be reduced to finding those λ for which a linear combination of $\varphi_\alpha(x)$, where $|\alpha|^2 = \lambda/4\pi^2$, satisfies the automorphicity conditions for $\gamma \in G/G_0$. To date this problem has been studied only for groups G generated by reflections (*affine Weyl groups*); see (Shvartsman and Vinberg 1988). For polyhedral domains of a special form (*Weyl chambers*) a description of the spectra of the Dirichlet and Neumann problems has been given in terms of the solutions of certain Diophantic equations connected with G.

16.4. The Case of Spaces of Constant Negative Curvature

To begin with, we direct our attention to the two-dimensional case. As opposed to the isometry group of $\mathbb{R}^n$, the group $PGL(2, \mathbb{R})$ of matrices with determinant one acting as isometries on $\mathbb{H}^2$ (linear-fractional transformations) has an extraordinarily rich set of discrete subgroups. Here we confine ourselves to subgroups $G \subset PGL(2, \mathbb{R})$ that posses a compact fundamental domain (*Fuchsian groups of the first kind*). Such subgroups are characterized by the fact that all their elements (except, of course, the unity) are *hyperbolic*, i.e., are matrices whose trace is greater than two. Hyperbolic transformations have no finite fixed points, and so we shall consider compact manifolds $M = M_G = G \setminus \mathbb{H}^2$ without boundary.

Selberg's trace formula (see Venkov 1979) constitutes the basis of the spectral theory of the manifolds under consideration. The formula can be regarded as a far reaching generalization of the Jacobi formula from flat tori, which are Riemannian surfaces of genus one, to Riemannian surfaces of genus $g \geq 2$. Selberg's formula gives an explicit expression for the trace of any function of the operator Δ_M that decays fast enough (in particular, $\theta_M(t) = \operatorname{Tr}\exp(t\Delta_M)$ for $t > 0$) in terms of the lengths of closed geodesics on M.

An element γ of G is said to be *primitive* if it cannot be represented as $\gamma = (\gamma')^k$, where $k > 1$. Any non-primitive element can be represented as a positive integral power of a primitive element. To every non-unit $\gamma \in G$ there corresponds a family of closed geodesics on M of the same length $l(\gamma)$. For $\theta(t)$ the Selberg formula reads

$$\theta(t) = \operatorname{vol} M \frac{e^{-t/4}}{(4\pi t)^{3/2}} \int_0^\infty \frac{\tau e^{-\tau/4t}}{\sinh(\tau/2)} d\tau + \frac{1}{2} \sum_{k=1}^\infty \sum_\gamma \frac{l(\gamma)}{\sinh(l(\gamma^k)/2)} \frac{e^{-t/4-(l(\gamma^k))^2/4t}}{4\pi t^{1/2}}, \tag{16.2}$$

where the sum with respect to γ extends over all primitive elements $\gamma \in G$.

In particular, it follows from (16.2) that the spectrum of $-\Delta_M$ and the spectrum of lengths of geodesics determine each other uniquely. The lengths of geodesics can, in turn, be expressed explicitly in terms of the transformation matrices: $l(\gamma) = 2\cosh^{-1}(\operatorname{tr}\gamma/2)$. This argument made it possible for Gel'fand (1963) to establish that in the class under consideration there are no more than countably many non-isometric manifolds with a given spectrum. According to (Ishii 1973), this number is even finite. Buser (1980) obtained an estimate, according to which the number of isospectral pairwise non-isometric surfaces of genus g does not exceed $\exp(507g^3)$.

On the other hand, examples of non-isometric isospectral surfaces were constructed in (Vignéras 1980). From the general point of view such examples are, however, untypical. According to (Wolpert 1979), in the Teichmüller space of classes of conformally equivalent metrics the spectrum determines the metric to within conformal equivalence in the complement of an analytic submanifold.

In the many-dimensional case even more striking examples can be constructed: in dimension $n \geq 3$ there are isospectral manifolds that fail to be diffeomorphic (Vignéras 1980).

On other aspects of spectral theory on manifolds of constant negative curvature, for instance in connection with boundary value problems, and also on non-compact fundamental domains see (Berger et al. 1971; McKean 1972; Bérard 1986) and, in particular, (Venkov 1979).

16.5. The Case of Spaces of Constant Positive Curvature

In this case we have the most complete description of the spectrum. This is connected with the simplicity of the classification of discrete groups of isometries and with the multitude of analytic and algebraic structures on S^n, which complement one another.

For the sphere S^n itself the eigenvalues have been computed in Example 4.7.

Now let G be a group of isometries of S^n, no one of which (except for $\gamma = 1$) has fixed points, and let $M = G \setminus S^n$. For any even number n the only non-trivial group consists just of the identity and antipodal mappings. It corresponds to $M = \mathbb{R}P^n$.

We consider the case of an odd number $n = 2m - 1$, where $m > 1$. If φ is an eigenfunction of the Laplace operator on S^n, then the function

$$(P\varphi)(x) = (\operatorname{card} G)^{-1} \sum_{\gamma \in G} \varphi(\gamma x)$$

is invariant under the action of G. It can therefore be regarded as an eigenfunction on M. It follows that the spectrum of the Laplace operator on M is contained in the spectrum on S^n, and the problem is reduced to the task of determining the multiplicities d_k of the eigenvalues $k(n+k+1)$. The multiplicity generating function, i.e., the so-called *Poincaré series*

$$F_G(z) = \sum_k z^k d_k, \tag{16.3}$$

serves as a convenient tool for describing the multiplicities. The following identity is satisfied for the groups of motions of S^n under consideration (see Ikeda 1980a):

$$F_G(z) = (\operatorname{card} G)^{-1} {\sum_{g \in G}}' \frac{1 - z^2}{\det(1 - gz)}, \quad |z| < 1, \tag{16.4}$$

where g is regarded as an element of the group $O(n+1)$. This describes the spectrum of $-\Delta_M$ in terms of G. Conversely, suppose that the spectrum of $-\Delta_M$ is known. Then, by (16.4), it is possible to find the union $\sigma(G)$ of the spectra of the matrices $g \in G$ by means of the Poincaré series (16.3). The spectra of g can be obtained from $\sigma(G)$ only in a finite number of ways. As a result, it turns out that the problem of determining the group G (to within conjugacy with an element from $O(n+1)$) and the manifold M itself from the spectrum of $-\Delta_M$ can have only finitely many solutions. It is known that the solution is unique in three dimensions, and, for lens spaces, also in five dimensions (Ikeda 1980b). At the same time, in seven dimensions there are examples of lens spaces that are isospectral, but not even homeomorphic (Ikeda 1983). We recall that a *lens space* $\Lambda(q; q_1, \ldots, q_m)$ corresponds to the cyclic group $G = \mathbb{Z}_q$ generated by a transformation being the direct sum of rotations of m two-dimensional planes by $2\pi q_j/q$.

For another class of groups of isometries of S^n, namely, the groups G generated by reflections, we are in a position to study the spectra of the Dirichlet and Neumann boundary value problems on fundamental domains by considering the Poincaré series. Suppose that G is the group generated by reflections with respect to a system R of hypersurfaces in $\mathbb{R}^{n+1}$ passing through the centre of S^n. Each connected component $C = C(R)$ of the complement of the planes from R is called a *Weyl chamber*. We take $M = \bar{C} \cap S^n$ as the fundamental domain.

The irreducible groups generated by reflections admit a complete classification. If for every plane H_i being a wall of $C(R)$ we consider the unit vector e_i orthogonal to H_i and directed towards the half-space that contains C, then the system of vectors e_i has the following property: $e_i \cdot e_k = -\cos(\pi/m_{ik})$, where m_{ik} are integral numbers. The graph Γ that consists of vertices corresponding to the edges of C such that vertices i, k are connected $m_{ik} - 2$ times is called the *Coxeter graph*. All Coxeter graphs of irreducible groups have been enumerated. Their list reads: A_l for $l \geq 1$, B_l for $l \geq 2$, D_l for $l \geq 4$, E_6, E_7, E_8, F_4, G_2, H_3, H_4, and $I(p)$ for $p = 5$ or $p \geq 7$ (see Shvartsman and Vinberg 1988).

If G is a reducible group, then its Coxeter graph splits into connected components corresponding to the decomposition of G into irreducible groups. The fundamental domains Ω and Ω' of two groups G and G' on S^n are isometric if and only if the corresponding Coxeter graphs are isomorphic. Let $\mathcal{H}_k$ be the space of homogeneous harmonic polynomials of degree k in $\mathbb{R}^{n+1}$, and let $\mathcal{H}_k^N$ be the subspace of G-invariant polynomials with $h_k^N = \dim \mathcal{H}_k^N$. Each number h_k^N is equal to the multiplicity of the eigenvalue $\lambda_k = k(n+k+1)$ of the operator $-\Delta_{N,\Omega}$ of the Neumann problem. The Poincaré series $F_G^N(z) = \sum_k h_k^N z^k$ can be represented explicitly in terms of the numerical characteristics of G as follows. We order the walls of C in an arbitrary way and consider a transformation T composed of consecutive reflections relative to all the walls of C. The eigenvalues of T are independent of the order of reflections and have the form $\exp(2\pi i m_j/m)$, where m is an integer (the *Coxeter number*), and where the integers m_j are called the *indices* of G.

Theorem 16.1 (Bérard 1980a, 1980b; Urakawa 1982). *The identity*

$$F_G^N(z) = (1 - z^2) \prod_j \left(1 - z^{m_j+1}\right)^{-1}$$

is satisfied, $\{m_j\}$ being the set of all indices of G. It follows that in the case of the Neumann problem two fundamental domains Ω and Ω' are isospectral if and only if the corresponding groups G and G' have the same set of indices.

A similar result is also true for the Dirichlet problem in Ω. The eigenfunctions of this problem are the *anti-invariant* (i.e., such that $\varphi(\gamma x) = \det \gamma \varphi(x)$ for any $\gamma \in G$) eigenfunctions of the Laplace operator on S^n. The corresponding Poincaré series $F_G^D(z)$ is equal to

$$F_G^D(z) = z^{n+1} F_G^N(z) = (1 - z^2) z^{n+1} \prod_j \left(1 - z^{m_j+1}\right)^{-1}.$$

Thus, in order to construct an example of isospectral non-isometric domains on S^n it suffices to construct groups G and G' that have different Coxeter graphs, i.e., different sets of numbers m_{ij}, but the same set of indices m_j. In

(Urakawa 1982) such examples were constructed for $n = 3$. In particular, one can take the group $A_3 \times A_1$ in $\mathbb{R}^4$ as G and $I_2(3) + I_2(4)$ as G'. The cone in $\mathbb{R}^4$ spanned by the vectors e_3, $e_1 - e_2 + e_3$, $e_1 + e_2 + e_3$, and e_4 serves as the Weyl chamber for G, while the direct product of two angles of size $\pi/3$ and $\pi/4$ in $\mathbb{R}^2$ is the Weyl chamber for G'.

The intersection of the above-mentioned angles with S^3 yields non-isometric domains on S^3, for which the Dirichlet as well as Neumann problems are isospectral. The intersection of these angles in $\mathbb{R}^4$ with a layer $|x| \in (\alpha, \beta)$ of a ball yields non-isometric domains in $\mathbb{R}^4$ for which the Dirichlet and Neumann problems are isospectral. In this way one can construct the, so far, most complete Euclidean counter-examples for the problem of whether or not 'one can hear the shape of a drum.' However, a common shortcoming of these examples is that the boundaries of the constructed domains in $\mathbb{R}^4$ are merely piecewise smooth. To date no counter-examples have been found for the original problem involving a domain in $\mathbb{R}^n$ with smooth boundary.

16.6. Isospectral Families of Nilmanifolds

In all examples described above the constructed non-isometric isospectral manifolds turn out to be, nevertheless, spectrally isolated. Here we shall present a construction due to Gordon and Wilson (1984), which makes it possible to build continuous families of non-isometric isospectral manifolds.

These manifolds are generalizations of the flat tori from Sect. 16.3. Let G be a simply connected *nilpotent* Lie group, for which the exponential mapping $\exp : g \to G$ is an epimorphism of the Lie algebra g onto G. Let Γ be a discrete subgroup of G that admits a compact quotient manifold $X = \Gamma \setminus G$ (the *nilmanifold*) equipped with the left-invariant Riemannian metric inherited from G.

To each automorphism $\Phi \in \mathrm{Aut}\,(G)$ there corresponds the manifold $X_\Phi = \Gamma_\Phi \setminus G$, where $\Gamma_\Phi = \Phi(\Gamma)$. It is possible to give a complete description of those automorphisms Φ for which the manifolds X and X_Φ are isometric. According to (Gordon and Wilson 1984), all these automorphisms are exhausted, to within isometries of G and automorphisms that leave the subgroup Γ invariant, by the inner automorphisms $\Phi \in \mathrm{Inn}\,(G)$ of the form $\Phi_y(G) = y^{-1}xy$, where $y \in G$.

Furthermore, we consider the class $AIA(G)$ of *almost inner automorphisms*: $\Phi \in AIA(G)$ if for any $x \in G$ there exists $\Psi \in \mathrm{Inn}\,(G)$ such that $\Phi(x) = \Psi(x)$. The set $AIA(G)$ is a Lie group, which is closed as a subset of $\mathrm{Aut}\,(G)$.

The examples constructed in (Gordon and Wilson 1984) demonstrate that $AIA(G)$ can be much richer than $\mathrm{Inn}\,(G)$. In order to describe these constructions, it proves more convenient to go over to the Lie algebras. The algebra $\mathrm{ad}\,(g) = g/z$ of inner differentiations of g (z denotes the centre of g) serves as the Lie algebra for $\mathrm{Inn}\,(G)$, while the algebra $AID(g)$ of *almost inner differentiations* of g, i.e., differentiations φ such that for any $P \in g$ the element $\varphi(P)$ finds itself in the orbit of P under the action of $\mathrm{ad}\,(g)$, serves as the Lie algebra for $AIA(G)$.

Example 16.1. We define a nilpotent Lie algebra g by means of generators $\{P_i, Q_i, R_i : i = 1, 2\}$ and the relations $[P_1, Q_1] = [P_2, Q_2] = R_1$ and $[P_1, Q_2] = R_2$ (all the remaining generators commute). Since $\dim z = 2$, it follows that $\dim \mathrm{ad}\,(g) = 4$. At the same time, in addition to $\mathrm{ad}\,(g)$, $AID(g)$ contains mappings φ_1 and φ_2 such that $\varphi_1(P_1) = \varphi_2(P_2) = R_2$, i.e.,

$$\dim\,(AIA(G)/\mathrm{Inn}\,(G)) = \dim\,(AID(g)/\mathrm{ad}\,(g)) = 2.$$

The introduced class $AIA(G)$ is important because of the following property, which can be proved with the aid of the method of orbits.

Lemma 16.1. *For $\Phi \in AIA(G)$ the Laplace operators on p-forms on X and X_Φ are isospectral.*

A detailed analysis of the isometry classes of nilmanifolds, which was carried out in (Gordon and Wilson 1984), makes it possible to give an explicit description of the set E of isometry classes of isospectral manifolds.

Theorem 16.2. *Let* $\dim\,(AIA(G)/\mathrm{Inn}\,(G)) = d > 0$. *Then E has the structure of a d-dimensional manifold.*

Since $d = 2$ in Example 16.1, we obtain a two-dimensional continuous family of isospectral non-isomeric deformations of the original manifold X.

16.7. Sunada's Technique and Solution of Kac's Problem

(*Added in the English edition.*) Since §16 was written in 1986 dramatic events have taken place in the problem of isospectral manifolds. Sunada's result (Sunada 1985) carried over the idea of almost inner automorphisms to finite groups acting on a manifold and made it possible to reduce the task of constructing non-isometric isospectral manifolds to certain well-studied problems of finite group theory.

Theorem 16.3. *Let X be a Riemannian manifold with a finite group G of isometries acting upon it. Let H and K be subgroups of G acting freely. Suppose that the groups H and K are almost conjugate, i.e., there exists a bijection $f : H \to K$ carrying each element $h \in H$ into an element $f(h) \in K$ that is conjugate to h in G. Then the quotient manifolds $X_H = H\backslash X$ and $X_K = K\backslash X$ are isospectral. In the case when G contains all the isometries of X and the subgroups H and K are not conjugate, these manifolds are not isometric.*

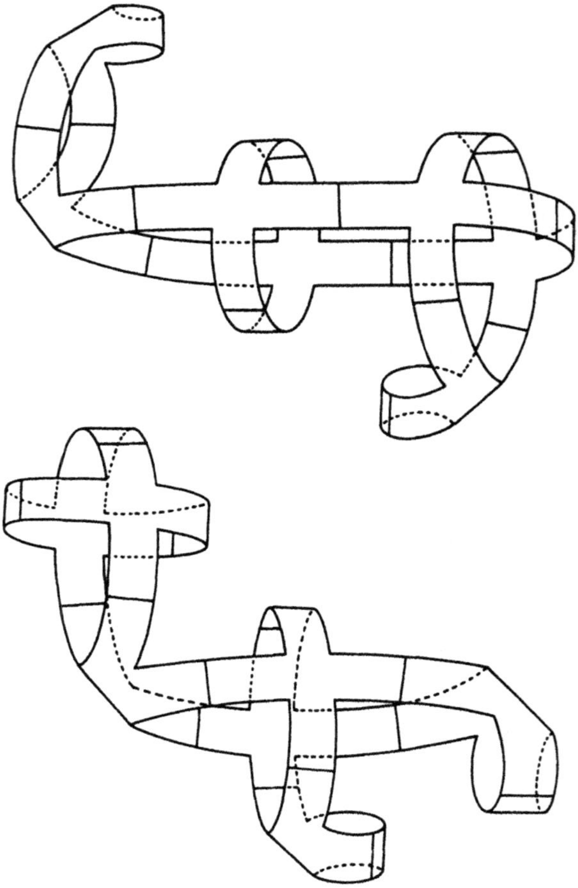

a

b

Fig. 1

Thus, in order to construct non-isometric isospectral manifolds, one must take a manifold X_0 and construct a covering manifold X such that the group G of the covering manifold contains two subgroups H and K meeting the requirements of Sunada's theorem. In this case both X_H and X_K will be covering manifolds for X. It is well known how to construct a covering with a group having the prescribed properties. Thus, Sunada's theorem was soon applied to find many examples of isospectral manifolds. In particular, Buser (1986) demonstrated that for every $g > 2$ there exists a pair of isospectral non-isometric Riemannian surfaces of genus g.

The next decisive step in the problem under consideration is connected with Berard's (1989) observation that it is, in fact, unnecessary for the subgroups H and K to act freely. This means that X_H and X_K regarded as quotient

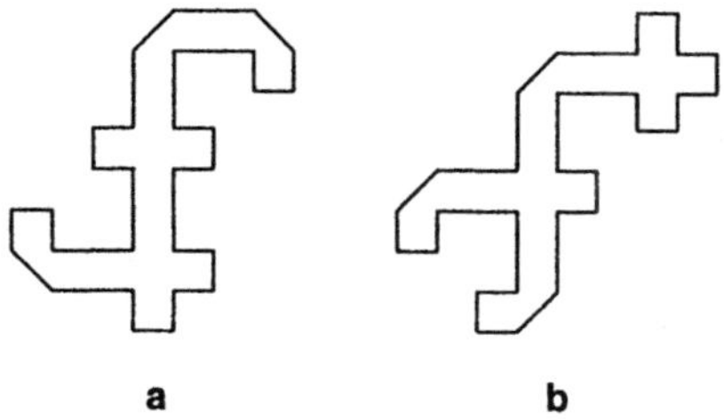

Fig. 2

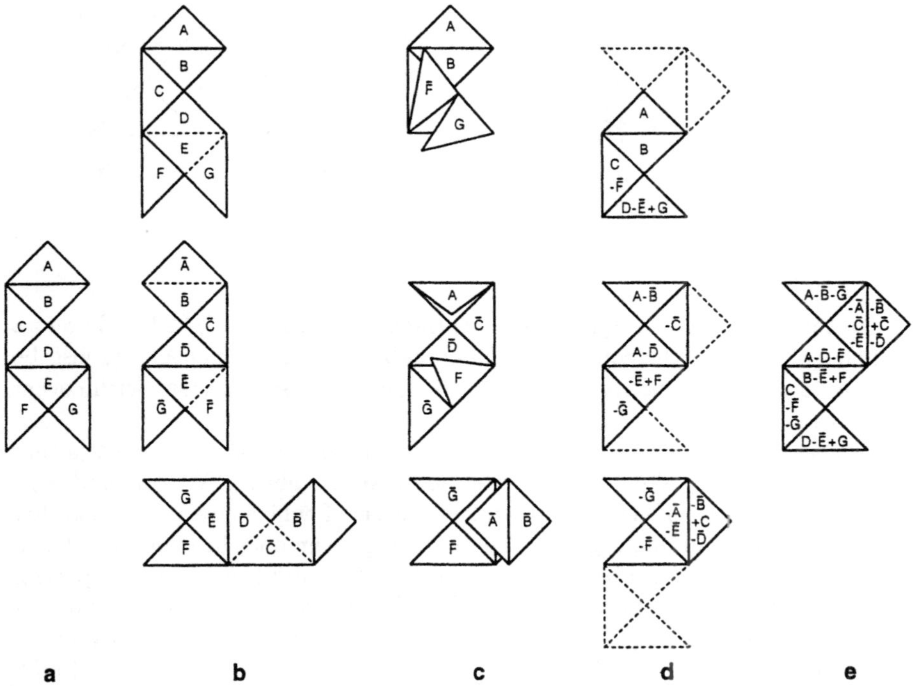

Fig. 3

sets must be described as the spaces of orbits of the actions of H and K on X. Such sets are called *Riemannian orbifolds*. Their topology is one of a manifold with boundary, the boundary being composed of the fixed points of the group action. On that boundary one can set either the Dirichlet or the Neumann boundary conditions. Provided H and K are almost conjugate, the Dirichlet and Neumann Laplacians are both isospectral. (In order to imagine an orbifold one can think of the sphere S^n with the generator of the group $\mathbb{Z}_2$ acting by reflection relative to the equatorial plane. This action is not free,

a b

Fig. 4

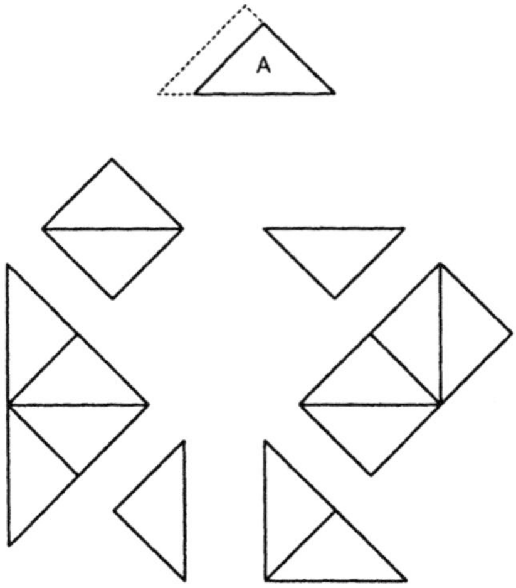

a b

Fig. 5

since the points on the equator are invariant. The orbifold $\mathbb{Z}_2/S^n$ is therefore a hemisphere with the equator as the boundary.) The original X may also be a manifold with boundary (and, from the metrical point of view, with corner points).

Thus, one can take a manifold X_0 with boundary and simply glue together some copies of X_0 along parts of the boundary to obtain X, X_H, and X_K. If fourteen copies of a cross are taken, seven of which are glued together first, followed by gluing together the remaining seven ones in a manner shown in Figs. 1a,b, then one obtains two (both Dirichlet and Neumann) isospectral manifolds with planar metrics. Finally, we observe that these manifolds posses the $\mathbb{Z}_2$ symmetry groups. After the last factorization we obtain two isospectral plane domains (Figs. 2a,b), thus solving Kac's problem (Gordon et al. 1992).

However, the answer turns out to be even simpler! Chapman (1992) has shown that isospectrality may be achieved merely by folding and cutting paper figures. Make three paper copies of the domain shown in Fig. 3a. Fold them along the dotted lines as shown in Figs. 3b–d. Here A, B, etc. designate triangles and $\bar{A}, \bar{B}, \ldots$ designate the same triangles in reversed positions. Also, each eigenfunction (e.g., Neumann) is transplanted to the folded domains. Thus, $D - \bar{A} + G$ in the upper part of Fig. 3d means that the new function on this triangle is the sum of the parts of the original eigenfunction on D and G minus the part of the same eigenfunction on E. The transplanted functions are not smooth enough to be eigenfunctions on the folded figures, since some parts of the original boundary go inside after folding, thus causing a discontinuity of the transplanted eigenfunction. If, however, we unite the three folded figures 3d, then the discontinuities will be successfully compensated. Thus we obtain a Neumann eigenfunction on the domain 3e. The letters indicate how the function is finally transplanted. A similar transplantation can be performed in the opposite direction, from 3e to 3a, thus proving Neumann isospectrality. Another way of folding guarantees Dirichlet isospectrality.

The original domain does not have to be composed of triangles. It is possible to fold figures 3a and 3e completely to only one triangle, cut a fancy design out of it, and, after unfolding, obtain new isospectral domains (Fig. 4). A special way of such cutting (Fig. 5) provides us with an example in which the eigenvalues can be found explicitly. On discarding one small triangle, the same for both domains, we have the domain X_1 being the disjoint union of a unit square and a triangle with sides $2\sqrt{2}, 2, 2$. The other domain X_2 consists of a rectangle with sides 1 and 2 and a triangle with sides $\sqrt{2}, \sqrt{2}, 2$. For the former domain the eigenvalues of the Dirichlet Laplacian (up to the common factor π^2) are

$$\Sigma_1 = \{\lambda = n^2 + m^2\} \cup \left\{\lambda = \left(\frac{i}{2}\right)^2 + \left(\frac{j}{2}\right)^2\right\}, \quad m, n > 0, \quad i > j > 0.$$

For the latter domain the eigenvalues are

$$\Sigma_2 = \left\{\lambda = \left(\frac{N}{2}\right)^2 + M^2\right\} \cup \left\{\lambda = \frac{1}{2}(I^2 + J^2)\right\}, \quad M, N > 0, \quad I > J > 0.$$

It is easily verified that these sets coincide if the multiplicity of each eigenvalue is taken into account.

As a result, Kac's problem is finally solved. There remains, however, the question of whether or not isospectral domains with smooth boundaries exist.

§17 Operators with Periodic Coefficients

17.1. Bloch Functions and the Zone Structure of the Spectrum of an Operator with Periodic Coefficients

Operators with periodic coefficients arise in the description of periodic structures of various kinds. This happens in the most natural way in the quantum theory of solids, for example, metals (see Ziman 1972). Namely, the ions of a metal forming a crystal lattice give rise to a periodic field, in which a free electron can be considered. Moreover, in a number of cases the interaction between the electrons can be neglected if compensating terms are added to the potential of each ion. Then, in accordance with the fundamental principles of quantum mechanics, the possible values of the energy of a free electron belong to the spectrum of the Schrödinger operator with a periodic potential and the corresponding eigenfunctions of the operator are the wave functions defining the complex amplitude, which characterizes the distribution of the coordinates of the electron (the square of the modulus of this amplitude can be interpreted as the probability density of finding the electron at a given point).

The periodicity of an n-dimensional structure can be characterized by a *lattice* in $\mathbb{R}^n$, that is, a discrete subgroup $\Gamma \subset \mathbb{R}^n$ of the form

$$\Gamma = \{z_1 e_1 + \ldots + z_n e_n,\ z_j \in \mathbb{Z}\},$$

where the vectors $e_1, \ldots, e_n$ form a fixed basis in $\mathbb{R}^n$. A function $a : \mathbb{R}^n \to \mathbb{C}$ is said to be *Γ-periodic* or *periodic with period lattice* Γ if $a(x+\gamma) = a(x)$ for any $\gamma \in \Gamma$. In this case $a(x)$ can clearly be considered as a function on the quotient group $\mathbb{R}^n/\Gamma$, which is an n-dimensional torus. The exponent $e_\xi(x) = e^{i\xi\cdot x}$ is Γ-periodic (in x) if and only if $\xi \cdot \gamma \in 2\pi\mathbb{Z}$ for any $\gamma \in \Gamma$. The points ξ that satisfy this condition also form a lattice Γ', which is called the *dual lattice* to Γ (in physical literature Γ' is often called the *inverse lattice*) and consists of all points of the form

$$\{z_1 e_1' + \ldots + z_n e_n', z_j \in \mathbb{Z}\},$$

where $\{e_1', \ldots, e_n'\}$ is the dual basis to $\{e_1, \ldots, e_n\}$, that is, $e_j' \cdot e_k = 2\pi\delta_{jk}$ (this definition differs by the factor 2π from the definition in Sect. 16.3). For example, if $\Gamma = 2\pi\mathbb{Z}^n$, which is the commonly used standard lattice, then $\Gamma' = \mathbb{Z}^n$.

A differential operator

$$A = a(x, D_x) = \sum_{|\alpha|\le m} a_\alpha(x) D_x^\alpha \tag{17.1}$$

is called *Γ-periodic* if all the coefficients a_α are Γ-periodic functions on $\mathbb{R}^n$. Introducing the shift (translation) operators by the formula $T_\gamma f(x) = f(x+\gamma)$, one can easily verify that A is a Γ-periodic operator if and only if it commutes with every operator from $\{T_\gamma,\ \gamma \in \Gamma\}$. In terms of the symbol

$$a(x,\xi) = \sum_{|\alpha|\le m} a_\alpha(x)\xi^\alpha,$$

the fact that A is Γ-periodic means that the function $a(x,\xi)$ is Γ-periodic in x for every fixed ξ. This definition can be extended in the obvious way to pseudodifferential operators. An important example of a Γ-periodic operator on $\mathbb{R}^n$ is the Schrödinger operator $A = -\Delta + q$ with a Γ-periodic potential $q = q(x)$.

Difference operators, that is, operators on a lattice, which we shall always take to be $\mathbb{Z}^n$ for simplicity, are often considered instead of operators on $\mathbb{R}^n$. If a sublattice $\Gamma \subset \mathbb{Z}^n$ is given, then an operator A in $l^2(\mathbb{Z}^n)$ is called Γ-periodic if it commutes with all the translation operators T_γ such that $\gamma \in \Gamma$, which can be defined in $l^2(\mathbb{Z}^n)$ in the same way as in the continuous case. An important example is provided by the *difference Schrödinger operator* $A = -\Delta + q$, where Δ is the *difference Laplace operator* on $\mathbb{Z}^n$:

$$\Delta u(x) = \sum_{y:|y-x|=1} (u(y) - u(x)),$$

and where q is a Γ-periodic function on $\mathbb{Z}^n$.

The translations T_γ, where $\gamma \in \Gamma$, transform the space of solutions of the equation $Au = \lambda u$ into itself. It is therefore natural to expect that the construction of the eigenfunction expansion of a Γ-periodic operator can be confined to those functions that are also eigenfunctions of each of the translations. Such functions are called *Bloch functions*. By definition, $\psi = \psi(x)$ is a Bloch function if it satisfies the condition

$$\psi(x + \gamma) = \chi(\gamma)\psi(x)$$

identically in x for all $\gamma \in \Gamma$. If $\psi \not\equiv 0$, then it is clear that $\chi(\gamma) \neq 0$ for all γ. It is also easily verified that $\chi(0) = 1$ and $\chi(\gamma_1 + \gamma_2) = \chi(\gamma_1)\chi(\gamma_2)$, that is, χ is a homomorphism from Γ into the multiplicative group $\mathbb{C}^* = \mathbb{C} \setminus \{0\}$. It is readily seen that if $|\chi(\gamma)| \neq 1$ for any $\gamma \in \Gamma$, then $\psi(x)$ grows exponentially in the direction $+\gamma$ or $-\gamma$ as a function on $\{x + n\gamma, n \in \mathbb{Z}\}$, where x is such that $\psi(x) \neq 0$. Therefore, if ψ grows no faster than a power function, or, more generally, if ψ is a tempered distribution (such distributions ψ are sufficient to construct the eigenfunction expansion of any self-adjoint operator), then it is necessary that $|\chi(\gamma)| = 1$, and we can write

$$\chi(\gamma) = e^{ip\cdot\gamma}, \quad \gamma \in \Gamma,$$

where the vector $p \in \mathbb{R}^n$ (which is independent of γ) is called the *quasimomentum*. In this case it is easily seen that

$$\psi(x) = e^{ip\cdot x}\varphi(x) \tag{17.2}$$

identically in x, the function (or distribution) $\varphi(x)$ being Γ-periodic. The quasimomentum p fails to be uniquely defined by the given Bloch function ψ, since any vector $\gamma' \in \Gamma'$, where Γ' is the dual lattice, can be added to p. One can therefore assume that $p \in B$, where B is a fundamental domain of the action of Γ' on $\mathbb{R}^n$ by translations, i.e., B is any (measurable) set containing one representation of each coset of $\mathbb{R}^n$ relative to Γ'. Such a set B is called an *elementary cell of* Γ' or a *Brillouin zone* corresponding to Γ.[6]

The parallelepiped

[6] In physical literature a Brillouin zone is meant to be a special elementary cell of the dual lattice.

$$B = \{\xi_1 e'_1 + \ldots + \xi_n e'_n,\ 0 \le \xi_j < 1,\ j = 1, \ldots, n\},$$

where $\{e'_1, \ldots, e'_n\}$ is the basis of the dual lattice Γ', is an example a Brillouin zone. In what follows we shall always assume for simplicity and definiteness that the Brillouin zone is chosen in this way, even though the majority of constructions are independent of the choice.

The space of all Γ-periodic functions is isomorphic to the space of all Bloch functions with fixed quasimomentum p, the isomorphism being defined by the multiplication operator I_p by $e_p(x) = e^{ip\cdot x}$. We observe that these spaces are finite-dimensional in the discrete case (the dimension of either of them being equal to the number of points of the quotient group $\mathbb{Z}^n/\Gamma$, or, equivalently, the number of points of a fundamental domain of Γ). In the continuous case, applying an operator $A = a(x, D_x)$ of the form (17.1) to a function of the form (17.2), we obtain the formula

$$a(x, D_x)\left[e^{ip\cdot x}\varphi(x)\right] = e^{ip\cdot x} a(x, p + D_x)\varphi(x) \tag{17.3}$$

(called the *translation formula*), where $a(x, p + D_x)$ denotes the operator A_p with symbol $a_p(x, \xi) = a(x, p + \xi)$. Hence

$$A_p = I_{-p} A I_p. \tag{17.4}$$

It follows that the operator $A = a(x, D_x)$ on the space of sufficiently smooth Bloch functions with quasimomentum p is similar to the operator $A_p = a(x, p + D_x)$ acting on the space of (also sufficiently smooth) Γ-periodic functions.

We remark that if A is the Schrödinger operator, i.e., $A = -\Delta + q$, where q is the multiplication operator by a Γ-periodic function $q(x)$, then $A_p = -\Delta + 2p \cdot D_x + p^2 + q$.

Now, denote by $\mathcal{H}_p$ the space of all Bloch functions with quasimomentum p that belong to $L_{2,\mathrm{loc}}(\mathbb{R}^n)$. It is clear that $\mathcal{H}_p$ is a Hilbert space equipped with the scalar product

$$(f, g)_{\mathcal{H}_p} = \frac{1}{\operatorname{mes} E} \int_E f(x)\overline{g(x)}\, dx,$$

where dx is the ordinary Lebesgue measure on $\mathbb{R}^n$, E is any elementary cell of the lattice Γ, and $\operatorname{mes} E$ is the Lebesgue measure of E. The operator I_p defines an isometric isomorphism $I_p : \mathcal{H}_0 \to \mathcal{H}_p$, where $\mathcal{H}_0$ is the space of Γ-periodic functions belonging to $L_{2,\mathrm{loc}}(\mathbb{R}^n)$. The multiplier $1/\operatorname{mes} E$ is introduced in order that the exponents $\left\{e^{i(p+\gamma)\cdot x}, \gamma' \in \Gamma'\right\}$ have unit norm. It follows that the exponents form an orthonormal basis in $\mathcal{H}_p$. One can now easily verify the natural direct integral decomposition

$$L_2(\mathbb{R})^n = \oplus \int_B \mathcal{H}_p\, dp, \tag{17.5}$$

where dp is the ordinary Lebesgue measure on the Brillouin zone B. This means that there is a one-to-one linear isometric correspondence between the elements $u \in L_2(\mathbb{R}^n)$ and measurable[7] vector-valued functions $p \mapsto s(p)$ mapping every point $p \in B$ into a vector $s(p) \in \mathcal{H}_p$ such that

$$\int_B \|s(p)\|^2\,dp < \infty.$$

(Here, by definition, the integral is equal to the square of the norm in the Hilbert space

$$\mathcal{H} = \oplus \int_B \mathcal{H}_p\,dp$$

consisting of the vector-valued functions described above.) The decomposition (17.5) can be obtained by expressing any function $f \in L_2(\mathbb{R}^n)$ as the Fourier integral involving all the exponents $\{e^{i\xi\cdot x},\ \xi \in \mathbb{R}^n\}$, followed by collecting the exponents belonging to the same coset of $\mathbb{R}^n$ relative to the subgroup Γ' (there is a one-to-one correspondence between the cosets and the points of the Brillouin zone B). This means that $f = \int f_p\,dp$, the Bloch function f_p being given by

$$f_p(x) = (2\pi)^{-n} \sum_{\gamma'\in\Gamma'} e^{i(p+\gamma')\cdot x}\tilde{f}(p+\gamma'),$$

where $\tilde{f}$ is the Fourier transform of f. Hence, by simple transformations, one can obtain the more explicit formula

$$f_p(x) = (\operatorname{mes} B)^{-1} \sum_{\gamma\in\Gamma} e^{-ip\cdot\gamma} f(x+\gamma).$$

It follows easily from the Parseval equality for the Fourier transform that the constructed decomposition is isometric.

Now let A be a formally self-adjoint elliptic differential operator. Then in (17.5) A can also be expressed as the direct integral of the self-adjoint operators $A\big|_{\mathcal{H}_p}$, which can be constructed, for example, as the closures of the restrictions of A to the set of all smooth Bloch functions with quasimomentum p. As follows from (17.4) such an operator is unitarily equivalent to the operator A_p on the torus $\mathbb{R}^n/\Gamma$, which means that it has discrete spectrum. Thus we obtain the complete decomposition of the given operator A in terms of Bloch eigenfunctions, which is due to Gel'fand (1950) (see also Eastham 1973; Reed and Simon 1978, Vol. 4). Assuming for definiteness that the principal symbol of A is positive (for $\xi \neq 0$), we can arrange the eigenvalues of $A\big|_{\mathcal{H}_p}$ (or A_p), taking their multiplicity into account, into a non-decreasing sequence

[7] In this case measurability can be understood, for example, as the measurability of the vector-valued function $p \mapsto I_p^{-1}s(p)$ on B with values in $\mathcal{H}_0$, which, in turn, means that the scalar function $p \mapsto (I_p^{-1}s(p), \varphi)$ is measurable for every φ in $\mathcal{H}_0$.

$$E_1(p) \leq E_2(p) \leq \ldots \leq E_l(p) \leq \ldots, \tag{17.6}$$

where $E_l(p) \to +\infty$ as $l \to +\infty$. An elementary argument involving perturbation theory (see §8) indicates that since only the lower-order terms of A_p depend on p (the dependence being polynomial), every $E_j = E_j(p)$ is a continuous Γ'-periodic function of p or a function on the torus $\mathbb{R}^n/\Gamma'$. Moreover, the eigenvalues of multiplicity one are even analytic in p, while every multiple eigenvalue can be locally represented as a system of branches of a many-valued analytic function of p, provided a suitable numbering is introduced in place of the increasing order. To be precise, the union of all the graphs of the functions E_j in the (p, E)-space has locally the form of the set of zeros of a polynomial $E^k + a_1(p)E^{k-1} + \ldots + a_k(p) = 0$ in a neighbourhood of any given point (p_0, E_0), where $a_1, \ldots, a_n$ are holomorphic functions of p and k is the multiplicity of the eigenvalue E_0 of A_{p_0}. The functions $E_j(p)$ or the appropriate analytic functions of p defining the same set of eigenvalues (17.6), possibly with a different numbering, are usually called the *band functions*, the *Bloch spectrum*, or the *Floquet spectrum*.

Because of the above-mentioned representation of A as a direct integral, the spectrum $\sigma(A)$ of A is equal to the union of the spectra of all operators A_p, where $p \in B$, i.e., the union of the sets of values of all functions E_j (this union is closed, since the functions E_j are continuous and $E_j(p) \to +\infty$ uniformly in p as $j \to \infty$). But since the torus $\mathbb{R}^n/\Gamma'$ is connected, the set of values of any of the continuous functions $E_j : \mathbb{R}^n/\Gamma' \to \mathbb{R}^n$ is an interval $[a_j, b_j]$, where $a_j = \min E_j(p)$ and $b_j = \max E_j(p)$. It follows that

$$\sigma(A) = [a_1, b_1] \cup [a_2, b_2] \cup [a_3, b_3] \cup \ldots, \tag{17.7}$$

where $a_1 \leq a_2 \leq a_3 \leq \ldots$, $b_1 \leq b_2 \leq b_3 \leq \ldots$, and $a_j \to +\infty$ as $j \to +\infty$. In general, the intervals $[a_j, b_j]$ can overlap, touch one another at a common end-point, be contained in one another (having a common end-point), or be equal to one another. It follows from (17.7) and the relation $\lim_{j\to\infty} a_j = +\infty$ that $\sigma(A)$ can also be represented in one of the following two forms:

$$1)\ \ \sigma(A) = [c_1, d_1] \cup [c_2, d_2] \cup \ldots \cup [c_l, d_l] \cup [c_{l+1}, +\infty), \tag{17.8}$$

where the intervals $[c_j, d_j]$ and the ray $[c_{l+1}, +\infty)$ are pairwise disjoint (i.e., $c_1 < d_1 < c_2 < d_2 < c_3 < \ldots < c_l < d_l < c_{l+1}$);

$$2)\ \ \sigma(A) = \bigcup_{l=1}^{\infty} [c_l, d_l],$$

where the intervals $[c_l, d_l]$ are pairwise disjoint (i.e., $c_1 < d_1 < c_2 < d_2 < c_3 < d_3 < \ldots$ in (17.8)) with $c_l \to +\infty$ as $l \to +\infty$. In both cases all the numbers c_j and d_j are uniquely determined by $\sigma(A)$. The intervals $[c_j, d_j]$ (and, in the former case, the half-interval $[c_{l+1}, +\infty)$) are called the *permitted zones*, since, in the quantum-mechanical interpretation, the values of the energy of the particle described by the quantum Hamiltonian A can lie only in these zones. This is also why the intervals $(-\infty, c_1), (d_1, c_2), (d_2, c_3), \ldots$, which have no common points with the spectrum, are called the *forbidden zones* (the bounded forbidden zones, i.e., all of them except for $(-\infty, c_1)$, are also called *gaps*). Skriganov (1985) and Veliev (1987) proved that if A is the Schrödinger operator and $n \geq 2$, then the first case is realized, that is, the spectrum contains a half-axis and there are only finitely many forbidden zones. In the one-dimensional case this is an exceptional rather than typical situation. Namely, for $n = 1$, i.e., for the one-dimensional Schrödinger operator with a periodic potential, which is also called the *Hill operator*, the intervals $[a_j, b_j]$ in (17.7) cannot even overlap (they can only have common end-points), because the multiplicity of any eigenvalue is not greater than two. The potentials $q(x)$ of those operators $A = -d^2/dx^2 + q(x)$ for which the first case (17.8) is realized are called *finite gap potentials* and can be described explicitly, namely, they can be expressed in terms of the θ-functions. (See, for example, the book (Manakov et al. 1980), which contains a description of an important relationship between finite gap potentials and certain nonlinear differential equations, namely, the Korteweg-de Vries equation and its higher-order analogues).

In particular, such potentials are certainly analytic functions. Potentials $q(x)$ with a fixed number of zones depend on finitely many parameters (for example, if there is only one forbidden zone $(-\infty, c_1)$, then $q = \text{const}$). Finite gap potentials are therefore rare in the obvious sense, even though they can approximate any smooth potential (Marchenko 1977). Moreover, Skriganov (1985) studied the growth of the overlapping multiplicity of the zones $[a_j, b_j]$ for the many-dimensional Schrödinger operator as $j \to \infty$ and obtained estimates of this growth.

The decomposition (17.5) is also valid in the discrete case, in which all the spaces $\mathcal{H}_p$ are finite-dimensional. Therefore both $A\big|_{\mathcal{H}_p}$ and A_p are operators in a finite-dimensional space and the number of functions $E_j(p)$ in (17.6) is finite, so that the spectrum of a self-adjoint Γ-periodic difference operator is also of the form (17.7), the number of intervals in (17.7) being always finite. Of interest are inverse problems concerned with the reconstruction of the coefficients of A from spectral data, for instance, form the band functions $E_j(p)$, from some of them, from their values at a fixed p (for example from the periodic eigenvalues), and the like. For the Schrödinger operator $A = -\Delta + q(x)$ these questions have been studied by Eskin, Ralston, and Trubowitz (see Eskin et al. 1986 and references therein), who used the hyperbolic equation method (cf. §14). They proved that if q is an analytic potential and Γ is a lattice such that the conditions $|\gamma| = |\gamma'|$ imply one of the equalities $\gamma = \pm\gamma'$ for any $\gamma, \gamma' \in \Gamma$, then all the functions $E_j(p)$, $j = 1, 2, \ldots$ can be reconstructed from the set of numbers $E_j(0)$, $j = 1, 2, \ldots$, that is, from the periodic eigenvalues, or, more generally, from the set $E_j(p_0)$, $j = 1, 2, \ldots$, provided that $\cos(p_0 \cdot \gamma) \neq 0$ for any $\gamma \in \Gamma$. Using the asymptotic expansions of the heat kernel (see §12) of such an operator in various one-dimensional directions, these authors proved that all the band functions of the 'reduced operators' with potentials

$$q_\gamma(x) = \int_0^1 q(x+s\gamma)\,ds = \sum_{\{\delta:\delta\cdot\gamma=0\}} a_\delta e^{i\delta\cdot x}$$

can be reconstructed from the band functions of A if

$$q(x) = \sum_{\delta\in\Gamma'} a_\delta e^{i\delta\cdot x}.$$

This makes it possible to reconstruct the band functions of the one-dimensional operators

$$-|\delta|^2 \frac{d^2}{ds^2} + Q_\delta(s),$$

where

$$Q_\delta(s) = \sum_{n\in\mathbb{Z}} a_{n\delta} e^{ins},$$

and use the theory of the one-dimensional inverse problem. The same authors demonstrated that for a generic analytic potential $q(x)$ there are only finitely many potentials with the same periodic eigenvalues as the corresponding Schrödinger operator.

We remark that it also proves useful to study the band functions $E_j(p)$ for complex values of the quasimomentum p, i.e., to consider their analytic continuation. By studying the analytic continuation of the function $f(\lambda) = E_j(\lambda p_0)$ of one variable, Avron and Simon (1978) proved that, in the case of the Schrödinger operator, the branch points are the only possible isolated singularities of this continuation, and if there are no singularities at all (i.e., $E_j(\lambda p_0)$ can be extended to an entire function) and $p_0 \in \Gamma$, then $f(\lambda) = (p_1+\lambda p_0)^2+C$ with $p_1 \in \Gamma'$ (this means that the band is the same as for $q \equiv c$). For a general elliptic operator A with periodic coefficients Kuchment (1982) considered the set Λ of pairs $(p,\lambda) \in \mathbb{C}^{n+1}$ such that the equation $A\psi = \lambda\psi$ has a Bloch solution with quasimomentum p (the intersection $\Lambda \cap \mathbb{R}^{n+1}$ is the union of graphs of all the functions $E_j(p)$ in the self-adjoint case). He proved that Λ is the set of all zeros of an entire function of order n (generally speaking, of an infinite type) in $\mathbb{C}^{n+1}$. Hence, in the case when A is self-adjoint, one can deduce that any irreducible component Λ_α of the analytic set Λ can be represented as the graph of an analytic function of p in the complement of an analytic subset $\Lambda'_\alpha \subset \Lambda_\alpha$ (the codimension of which in $\mathbb{C}^{n+1}$ is not less than two). Besides, in this case the projection $\Lambda \to \mathbb{C}^n$ is dense, and if K is the complement of this projection, then the intersection of K with any complex straight line $\{a\lambda+b\}$ in $\mathbb{C}^n$ (not lying in K) has capacity zero. In particular, K does not divide $\mathbb{C}^n$. In the case of the one-dimensional Schrödinger operator all the functions $E_j(p)$ can be obtained from one another by analytic continuation (i.e., Λ is irreducible). A much more detailed study of these functions is possible, but we shall not deal with this subject.

17.2. The Character of the Spectrum of an Operator with Periodic Coefficients

The expansion in terms of Bloch eigenfunctions described in the previous section yields the spectral expansion of any given self-adjoint elliptic differential operator A with periodic coefficients or any Γ-periodic difference operator A. Namely, if we introduce a space M with measure $d\mu$ being the union of countably many disjoint copies B_j, $j = 1, 2, \ldots$ of the Brillouin zone B with measure dp in the continuous case and finitely many copies B_j of B in the discrete case, and we define in $L_2(M, d\mu)$ the multiplication operator by the function $a = a(m)$ equal to $E_j = E_j(p)$ on $B_j = B$, then the given operator A defined in $L_2(\mathbb{R}^n)$ will be unitarily equivalent to the multiplication operator by a in $L_2(M, d\mu)$. It is therefore easy to give a complete description of the character of the spectrum of A in terms of the band functions $E_j(p)$. For example, we observe that the point spectrum $\sigma_{\mathrm{p}}(A)$ of the multiplication operator by an arbitrary real-valued function a in $L_2(M, d\mu)$ can be described as follows:

$$\lambda \in \sigma_{\mathrm{p}}(A) \Longleftrightarrow \mu\{m : a(m) = \lambda\} > 0.$$

In terms of the band functions of the given operator A, this means that

$$\lambda \in \sigma_{\mathrm{p}}(A) \Longleftrightarrow \mathrm{mes}\{p : \exists j,\ E_j(p) = \lambda\} > 0. \tag{17.9}$$

Taking into account that the functions $E_j(p)$ are piecewise analytic, we can see that the latter condition is satisfied if and only if $\lambda \in \sigma(A_p)$ for all p, or, equivalently, $E_j(p) \equiv \lambda$ for some j, given a suitable (not necessarily monotone) numbering of the eigenvalues $E_j(p)$. With the aid of perturbation theory, one can demonstrate that this is impossible, for example, in the case when A is the Schrödinger operator in $\mathbb{R}^3$ with a potential from $L_2(\mathbb{R}^3/\mathbb{Z}^3)$ (Thomas 1973; see also Kuchment 1982) or in the more general case when A is the Schrödinger operator with a periodic potential in $\mathbb{R}^n$, the Fourier coefficients of which belong to $l_2(\Gamma')$ for $n = 2, 3$ and to $l_\beta(\Gamma')$ with $\beta < (n-1)/(n-2)$ for $n > 3$ (see Reed and Simon 1978, Vol. 4, Theorem XIII.100). Moreover, if there is no point spectrum, then it follows easily that the whole spectrum is absolutely continuous, since the functions E_j are piecewise analytic (in particular, this is so in the above-mentioned cases for the Schrödinger operator). Kuchment proved that for a general elliptic operator A with smooth periodic coefficients, $\lambda \in \sigma_{\mathrm{p}}(A)$ if and only if the equation $(A - \lambda I)u = 0$ has a solution such that $|u(x)| \le c_a \exp(-a|x|)$ for any $a > 0$. He also extended the assertion on the non-existence of the point spectrum to more general operators of the form $L(D) + q(x)$.

The spectrum and the band functions of the Hill operator $A = -d^2/dx^2 + q(x)$ (with $q(x+a) = q(x)$) can be expressed in terms of the trace of the monodromy matrix, which is also called the *Hill discriminant*. Namely, we introduce the *monodromy operator* $M(\lambda)$, i.e., the translation operator by a in the two-dimensional space of all solutions of the equation $A\psi = \lambda\psi$. If we choose the two standard solutions $c(x)$ and $s(x)$ defined by the initial conditions $c(0) = s'(0) = 1$ and $c'(0) = s(0) = 0$, then the matrix of $M(\lambda)$ in the basis formed by these solutions reads

$$\begin{pmatrix} c(a) & s(a) \\ c'(a) & s'(a) \end{pmatrix}$$

and has determinant one because, being the Wronskian of $c(x)$ and $s(x)$, the determinant is independent of x by virtue of the Liouville theorem and equal to one for $x = 0$. The trace $D(\lambda) = c(a) + s'(a)$ of this matrix is called the *Hill discriminant*. The Bloch eigenfunctions with eigenvalue λ are eigenvectors of $M(\lambda)$ with eigenvalues $e^{\pm ipa}$, where p is the quasimomentum. The numbers $e^{\pm ipa}$ are therefore the solutions of the characteristic equation

$$\mu^2 - D(\lambda)\mu + 1 = 0$$

of the monodromy matrix, which implies that $D(\lambda) = 2\cos(pa)$. Furthermore, it follows easily that if $|D(\lambda)| \leq 2$, then the solutions $\mu_{1,2}$ of the equation have modulus one, which means that there exists a bounded Bloch eigenfunction. But if $|D(\lambda)| > 2$, then $|\mu_1| > 1$ and $|\mu_2| < 1$, given the appropriate numbering, and there exist two solutions of the equation $A\psi = \lambda\psi$, one of which decays exponentially at $-\infty$ and grows exponentially at $+\infty$, while the other one grows exponentially at $-\infty$ and decays exponentially at $+\infty$. The Green function can be easily constructed from these two solutions, so that $\lambda \notin \sigma(A)$ in the case at hand. At the same time, using the standard cut-off procedure, one can easily obtain a sequence of almost-eigenfunctions with compact support from a bounded Bloch solution of the equation $A\psi = \lambda\psi$, which implies that $\lambda \in \sigma(A)$. Thus we can see that

$$\lambda \in \sigma(A) \Longleftrightarrow |D(\lambda)| \leq 2. \tag{17.10}$$

It can be proved (see, for example, Coddington and Levinson 1955) that the graph of $D(\lambda)$ has the form presented in Fig. 6. In this figure λ_j are the eigenvalues of the periodic problem and μ_j are the eigenvalues of the antiperiodic problem, i.e., the values of λ for which there exists a periodic solution with period a (or, respectively, an antiperiodic solution, i.e., such that $\psi(x+a) = -\psi(x)$) of the equation $A\psi = \lambda\psi$. According to the theorems on the zeros of the solutions of a second-order equation, the eigenvalues can be ordered as follows:

$$\lambda_0 < \mu_0 \leq \mu_1 < \lambda_1 \leq \lambda_2 < \mu_2 \leq \mu_3 < \lambda_3 \leq \lambda_4 < \dots ,$$

the intervals $(-\infty, \lambda_0)$, (μ_0, μ_1), (λ_1, λ_2), (μ_2, μ_3), (λ_3, λ_4), ... being the forbidden zones (some of them may disappear; for example, there is no (μ_2, μ_3) in the figure). The lengths of the forbidden zones usually converge to zero. For example, if $q \in C^\infty$, then they converge to zero faster than any power of their number:

$$|\lambda_{2j} - \lambda_{2j-1}| \leq c_N j^{-N}, \quad |\mu_{2j+1} - \mu_{2j}| \leq c_N j^{-N} \tag{17.11}$$

(see, for example, Marchenko (1977); we remark that Gordon (1979) proved that the lengths of the gaps tend to zero in the case of an arbitrary, not necessarily periodic, bounded measurable potential q). The derivative $D'(\lambda)$ has zeros only in the intervals $[\mu_0, \mu_1]$, $[\lambda_1, \lambda_2]$, $[\mu_2, \mu_3]$, $[\lambda_3, \lambda_4]$, ... (i.e., the closures of the gaps and the points to which the vanishing gaps are reduced), in each of which there is precisely one zero (all the zeros of the derivative being non-degenerate). In particular, $D(\lambda)$ is strictly monotone in each portion $[\lambda_0, \mu_0]$, $[\mu_1, \lambda_1]$, $[\lambda_2, \mu_2]$, $[\mu_3, \lambda_3]$, ... of the spectrum.

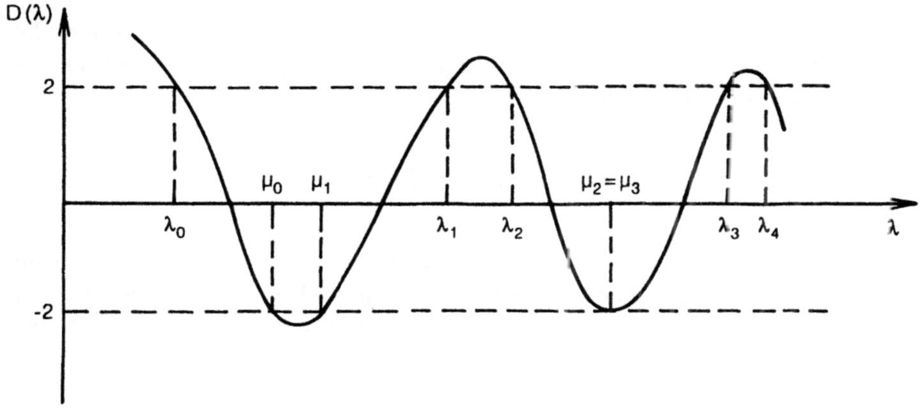

Fig. 6

A complete description of possible zones and gaps for potentials belonging to a fixed Sobolev smoothness class was given by Marchenko and Ostrovskij (1975) and Garnett and Trubowitz (1984, 1987). Presented in (Marchenko 1974) is a formulation and solution of the inverse spectral problem for the Hill operator, which can be used to study and solve nonlinear equations of Korteweg-de Vries type (on various aspects of the study of such equations, including connections with spectral theory, see the survey (Dubrovin et al. 1985)).

On the other hand, the article (Novikov 1983) deals with a number of concrete aspects of the study of the two-dimensional Schrödinger operator with a periodic magnetic field (the magnetic potential appearing in the Schrödinger operator is not necessarily periodic). In particular, a two-dimensional analogue of finite gap operators is presented in this article, along with a case in which the eigenfunctions of the ground state can be found explicitly.

17.3. Quantitative Characteristics of the Spectrum: Global Quasimomentum, Rotation Number, Density of States, and Spectral Function

The behaviour of $D(\lambda)$ described above makes it possible to introduce the *global quasimomentum* $p = p(\lambda)$ as a non-decreasing continuous function of $\lambda \in \mathbb{R}$ equal to zero on $(-\infty, \lambda_0]$, constant in each gap (μ_0, μ_1), (λ_1, λ_2), (μ_2, μ_3), ..., and satisfying the equation $2\cos(p(\lambda)a) = D(\lambda)$ for $\lambda \in \sigma(A)$ (such a function $p(\lambda)$ exists and is unique). In particular, it follows that $p(\lambda) = \pi/a$ for $\lambda \in [\mu_0, \mu_1]$, $p(\lambda) = 2\pi/a$ for $\lambda \in [\lambda_1, \lambda_2]$, and, in general, $p(\lambda) = l\pi/a$ in the closure of the l-th gap (counting the gaps which turn into points). For every $\lambda \in \sigma(A)$ the equation $A\psi = \lambda\psi$ has Bloch solutions ψ_λ and $\bar{\psi}_\lambda$ with quasimomentum $\pm p(\lambda)$ (if λ is neither a periodic nor a quasiperiodic eigenvalue, then the solutions are linearly independent), which we shall assume to be normalized by the relation

$$\frac{1}{a}\int_0^a |\psi_\lambda(x)|^2\, dx = 1.$$

(This normalization is convenient because it does not change if the problem is considered as a periodic one with multiple period la for an integer $l > 0$.) The spectral function $e(\lambda; x, y)$ of the Hill operator can be expressed in terms of these solutions and the global quasimomentum:

$$e(\lambda; x, y) = \frac{1}{\pi}\int_{-\infty}^{\lambda} \operatorname{Re}\left[\psi_\lambda(x)\overline{\psi_\lambda(y)}\right] dp(\lambda). \tag{17.12}$$

(This follows easily from the expansion in terms of Bloch eigenfunctions described in Sect. 17.1.)

Another notion of global quasimomentum, which is useful in the study of the inverse problem for the Hill operator perturbed by a decreasing potential, was employed by Firsova (for example, see Firsova 1986). Incidentally, we remark that the articles by Rofe-Beketov (1984 and references therein), Zheludev (1970), Malozemov (1988), and other authors are also devoted to the study of such a perturbed Hill operator.

The global quasimomentum of the Hill operator has the following two interpretations: to within a multiplier it coincides with the rotation number $w(\lambda)$ and with the integrated density of states $N(\lambda)$, which are defined as follows.

The *rotation number* $w(\lambda)$ can be defined with the aid of any non-trivial real solution ψ of the equation $A\psi = \lambda\psi$ by the formula

$$w(\lambda) = -\lim_{x\to+\infty} \frac{1}{2\pi x} \operatorname{Arg}\left(\psi(x) + i\psi'(x)\right), \tag{17.13}$$

in which any branch of the argument that is continuous with respect to x can be chosen. It can be proved (Johnson and Moser 1982) that this limit exists and is independent of the choice of ψ. Moreover, $w(\lambda) = 0$ if $\lambda < \inf \sigma(A)$. The function w is continuous and non-decreasing. It is constant in each gap of the spectrum of A, i.e., has properties similar to those of the global quasimomentum $p = p(\lambda)$. The simple relationship

$$w(\lambda) = (2\pi)^{-1} p(\lambda) \tag{17.14}$$

between $w(\lambda)$ and $p(\lambda)$ is therefore hardly surprising. Finally, the *integrated density of states* or the *limiting spectral distribution function* $N(\lambda)$ can be defined by

$$N(\lambda) = \frac{1}{a} \int_0^a e(\lambda; x, x)\, dx, \tag{17.15}$$

$e(\lambda; x, y)$ being the spectral function of A. We remark that since the spectral projection E_λ commutes with the translation by a, it follows that

$$e(\lambda; x + a, y + a) = e(\lambda; x, y),$$

which is an important property of the spectral function. Thus, in particular, the function $x \mapsto e(\lambda; x, x)$ is periodic with period a, and $N(\lambda)$ is the mean value of this function. From (17.15) and the positive definiteness of the kernel $e(\lambda; x, y)$ it is clear that $N(\lambda)$ is a non-decreasing function, which is equal to zero for $\lambda < \inf \sigma(A)$ and constant in each gap. Moreover, the spectrum $\sigma(A)$ is equal to the set of growth points of $N(\lambda)$, that is,

$$\sigma(A) = \left\{\lambda : N(\lambda + \varepsilon) - N(\lambda - \varepsilon) > 0 \text{ for any } \varepsilon > 0\right\}. \tag{17.16}$$

The derivative $\rho(\lambda) = dN(\lambda)/d\lambda$ is usually called the *density of states*. However, this term is sometimes used to refer to the measure on $\mathbb{R}$ determining the distribution function $N(\lambda)$ (in our case it can be demonstrated that $N(\lambda)$ is absolutely continuous, so that $\rho(\lambda)$ is the density of the measure) and, in some publications, to the function $N(\lambda)$ itself. The meaning of the term 'density of states' becomes clear from the formula

$$N(\lambda) = \lim_{L\to\infty} L^{-1} N_L(\lambda), \tag{17.17}$$

where $N_L(\lambda)$ is the ordinary distribution function of the spectrum of A on the interval $[0, L]$ or any other interval of length L with fixed self-adjoint boundary conditions at the end-points. An outline of the proof of the existence of the limit in (17.17) and the formula itself will be presented below in a much more general context. The limit also exists for the one-dimensional periodic difference Schrödinger operator, in which case it defines $N(\lambda)$.

We observe that $N(\lambda)$ is connected with the global quasimomentum by

$$N(\lambda) = \frac{p(\lambda)}{\pi}, \tag{17.18}$$

which follows from (17.12) and (17.15). Formulae (17.18) and (17.14) imply that $N(\lambda)$ is connected with the rotation number by

$$N(\lambda) = 2w(\lambda). \tag{17.19}$$

By differentiating the formula $D(\lambda) = 2\cos(p(\lambda)a)$ (with $\lambda \in \sigma(A)$), we obtain the following expression for $\rho(\lambda) = N'(\lambda) = \pi^{-1}p'(\lambda)$ in terms of $D(\lambda)$ valid everywhere, except for the end-points of the forbidden zones:

$$\rho(\lambda) = \frac{|D'(\lambda)|}{\pi a\sqrt{4 - D^2(\lambda)}}\,\theta\big(4 - D^2(\lambda)\big), \tag{17.20}$$

θ being the *Heaviside function* ($\theta(\mu) = 1$ for $\mu > 0$ and $\theta(\mu) = 0$ for $\mu \le 0$). In view of the above-mentioned properties of $D(\lambda)$, the graph of $\rho(\lambda)$ has the form presented in Fig. 7.

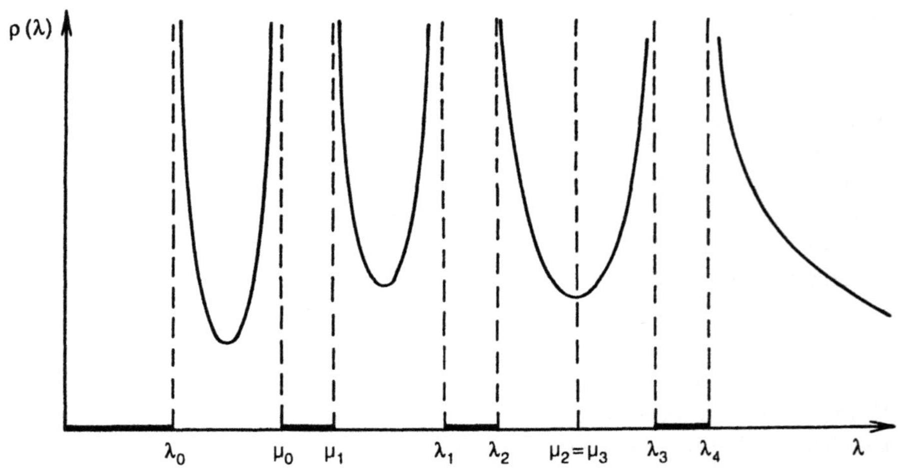

Fig. 7

At each end-point $\tilde{\lambda}$ of any forbidden zone there is a singularity of the type $|\lambda - \tilde{\lambda}|^{-1/2}$. If the forbidden zone vanishes, then there is no singularity at the corresponding point $\tilde{\lambda}$ and $\rho(\tilde{\lambda}) > 0$ (see, for example, the point $\tilde{\lambda} = \mu_2 = \mu_3$ in Fig. 7; it is also easy to prove that $N \in C^\infty$ everywhere, including neighbourhoods of all such points $\tilde{\lambda}$, but excluding the end-points of all non-vanishing forbidden zones).

We shall now discuss the asymptotic behaviour of the objects introduced above as $\lambda \to +\infty$. To this end, we first observe that if ψ is an arbitrary solution of the equation $A\psi = \lambda\psi$, then the logarithmic derivative $u = \psi'/\psi$ satisfies the Riccati equation

$$u' + u^2 = q - \lambda. \tag{17.21}$$

Setting $\lambda = \mu^2$, one can find a formal asymptotic solution of this equation of the form

$$u = i\mu\left(v_0 + \frac{v_1}{\mu} + \frac{v_2}{\mu^2} + \dots\right), \quad v_j = v_j(x), \tag{17.22}$$

where $v_0 \equiv 1$, $v_1 = 0$, $v_2 = q/2$, and

$$v_k = \frac{1}{2}\left[iv'_{k-1} - \sum_{r=1}^{k-1} v_r v_{k-r}\right], \quad k \geq 3.$$

Hence all v_k can be found by iteration as polynomials of q and its derivatives. It follows that the equation $A\psi = \lambda\psi$ has a formal asymptotic solution

$$\psi(x,\mu) = \exp\left[i\mu x + \sum_{k=1}^{\infty} \mu^{-k} \int_0^x v_k(t)\,dt\right]. \tag{17.23}$$

Replacing the infinite sum by

$$\sum_{k=1}^{N} \mu^{-k} \int_0^x v_k(t)\,dt + \mu^{-N} \int_0^x \tilde{v}_N(t,\mu)\,dt$$

and going over to an integral equation for $\tilde{v}_N$, one can prove by the successive approximation method that there exists a solution of this equation such that $\tilde{v}_N(t,\mu) = O(\mu^{-1})$ uniformly in $t \in [0,B]$ for any fixed $B > 0$ (Marchenko 1977). Consequently, for any integer $N > 0$ there exists an exact solution $\psi_N(x,\mu)$ that satisfies (17.23) to within terms of the form $O(\mu^{-N})$. Since all the functions v_k are periodic (with period a), we can see that ψ_N is an 'almost Bloch' function, i.e., more precisely,

$$\psi_N(x+a,\mu) = e^{ia\left[p_N(\mu) + O\left(\mu^{-N}\right)\right]}\psi_N(x,\mu),$$

where

$$p_N(\mu) = \mu - \sum_{k=1}^{N} c_k \mu^{-k}, \quad c_k = \frac{i}{a}\int_0^a v_k(t)\,dt, \tag{17.24}$$

and where $O\left(\mu^{-N}\right)$ may depend on x (being uniform in x for $x \in [0,B]$). It is easy to prove that all the numbers c_k are real (it can also be proved that $c_k = 0$ for odd numbers k). The solutions ψ_N and $\bar{\psi}_N$ are linearly independent and the actual monodromy matrix can be represented in the basis $\{\psi_N, \bar{\psi}_N\}$. Thereupon the actual quasimomentum $p(\lambda)$ and Bloch eigenfunctions $\psi_\lambda(x), \overline{\psi_\lambda(x)}$ are close to the functions $p_N(\lambda)$ and $\psi_N(x, \sqrt{\lambda}), \bar{\psi}_N(x, \sqrt{\lambda})$, which serve as their models. As a result, we find that $p(\lambda)$ has a complete asymptotic expansion as $\lambda \to +\infty$. Consequently, this is also the case for $w(\lambda)$ and $N(\lambda)$. For example,

$$N(\lambda) = \pi^{-1}\sqrt{\lambda} + \lambda^{-1/2}\sum_{k=0}^{\infty} d_k \lambda^{-k}. \tag{17.25}$$

(Here d_k differ from the coefficients c_k from (17.24) by multiplicative constants and by the numbering.) This asymptotic expression can be differentiated any number of times with respect to λ on any set of the form

$$M_{\varepsilon,l} = \{\lambda : |\lambda - \hat{\lambda}_k| \le \varepsilon\hat{\lambda}_k^{-1} \text{ for all } k\},$$

where $\varepsilon > 0$ and $l > 0$ are arbitrary, and where the points $\hat{\lambda}_k$ are chosen in such a way that each gap of the spectrum of A contains one of them.

Using the asymptotics of the Bloch eigenfunctions described above, one can obtain the following complete asymptotics of the spectral function (Shenk and Shubin 1985):

$$e(\lambda; x, y) = \frac{\sin\left[\sqrt{\lambda}(x-y)\right]}{\pi(x-y)} + \sum_{k=1}^{N}\left[f_k(x,y)\int_\lambda^\infty \frac{\cos\left[\sqrt{\lambda}(x-y)\right]}{\lambda^{k+1/2}}\,d\lambda\right.$$
$$\left. + g_k(x,y)\int_\lambda^\infty \frac{\sin\left[\sqrt{\lambda}(x-y)\right]}{\lambda^k}\,d\lambda\right] + O(\lambda^{-N}), \tag{17.26}$$

where $f_k, g_k \in C^\infty(\mathbb{R}\times\mathbb{R})$. For any fixed $B > 0$ this asymptotics is uniform in $x, y \in \mathbb{R}$ such that $|x-y| \le B$. In particular, for $x = y$ we find that

$$e(\lambda; x, x) \sim \pi^{-1}\sqrt{\lambda} + \sum_{k=0}^{\infty} h_k(x)\lambda^{-k-1/2} \tag{17.27}$$

uniformly with respect to all $x \in \mathbb{R}$. It is clear that formula (17.25) can be obtained from the latter by integrating with respect to x. Away from the diagonal, the asymptotic formula

$$e(\lambda; x, y) = \sum_{k=0}^{N}\{\lambda^{-k}p_k(x,y)\sin\left[\sqrt{\lambda}(x-y)\right]$$
$$+\lambda^{-k-1/2}q_k(x,y)\cos\left[\sqrt{\lambda}(x-y)\right]\} + O(\lambda^{-N-1}), \tag{17.28}$$

in which $p_k, q_k \in C_\infty$ for $x \ne y$, can be easily obtained from (17.26) by integration by parts for any $N = 1, 2, \ldots$ and $\varepsilon \le |x-y| \le A$, given arbitrary fixed $\varepsilon > 0$ and $A > 0$.

We shall also turn our attention to the description of the integrated density of states $N(\lambda)$ of a many-dimensional self-adjoint operator with periodic coefficients and positive principal symbol. As in the one-dimensional case, the spectral function $e(\lambda; x, y)$ of such an operator A is periodic, i.e.,

$$e(\lambda; x+\gamma, y+\gamma) = e(\lambda; x, y), \quad \gamma \in \Gamma,$$

Γ being the lattice of periods of the coefficients. In particular, the function $x \mapsto e(\lambda; x, x)$ is Γ-periodic and $N(\lambda)$ can be defined as the mean value

$$N(\lambda) = \frac{1}{\operatorname{mes} E_\Gamma} \int\limits_{E_\Gamma} e(\lambda; x, x)\, dx, \tag{17.29}$$

where E_Γ is an elementary cell of Γ. It is readily seen that $N(\lambda)$ is a non-decreasing function, which is equal to zero for $\lambda < \inf \sigma(A)$ and constant in each gap of the spectrum, $\sigma(A)$ being equal to the set of growth points of this function. It is possible to give a description of $N(\lambda)$ of the type (17.17) with the aid of a limit over domains which blow up. Such a description will be presented below in a more general context along with some information on the asymptotic behaviour of $N(\lambda)$ as $\lambda \to +\infty$ not restricted to the periodic case. Right now we state an expression for $N(\lambda)$ in terms of the band functions $E_j(p)$ (see Shubin 1979):

$$\begin{aligned} N(\lambda) &= (2\pi)^{-n} \sum_{j=1}^{\infty} \operatorname{mes} \{p : p \in B, E_j(p) \le \lambda\} \\ &= (2\pi)^{-n} \int\limits_B N_p(\lambda)\, dp, \end{aligned} \tag{17.30}$$

where $N_p(\lambda)$ is the ordinary distribution function for the discrete spectrum of the operator $A_p = a_p(x, p + D_x)$ on the torus $\mathbb{R}^n/\Gamma$. This formula can be easily derived from the following many-dimensional analogue of (17.12):

$$e(\lambda; x, y) = (2\pi)^{-n} \int\limits_B \sum_{E_j(p) \le \lambda} \psi_{j,p}(x) \overline{\psi_{j,p}(x)}\, dp, \tag{17.31}$$

$\psi_{j,p}$ being the Bloch eigenfunction of A with quasimomentum p normalized in the usual way:

$$\frac{1}{\operatorname{mes} E_\Gamma} \int\limits_{E_\Gamma} \left|\psi_{j,p}(x)\right|^2 dx = 1.$$

Moreover, $\psi_{j,p}$ must be chosen as a measurable function of p.

We also mention that for a semi-bounded self-adjoint elliptic operator $A = a(D)$ with constant coefficients, which can be regarded as an operator with periodic coefficients and an arbitrary lattice of periods, the explicit formula

$$e(\lambda; x, y) = (2\pi)^{-n} \int\limits_{\{\xi : a(\xi) < \lambda\}} e^{i(x-y)\cdot\xi}\, d\xi \tag{17.32}$$

for $e(\lambda; x, y)$ yields the explicit formula

$$N(\lambda) = (2\pi)^{-n} \operatorname{mes}\{\xi : a(\xi) < \lambda\} \tag{17.33}$$

for $N(\lambda)$. In particular, for $A = -\Delta$ we find that $N(\lambda) = (2\pi)^{-n} v_n \lambda^{n/2} \theta(\lambda)$, where v_n is the volume of the unit ball in $\mathbb{R}^n$ and θ is the Heaviside function.

§18 Operators with Almost Periodic Coefficients

18.1. General Definitions. Essential Self-Adjointness

(See Shubin 1974, 1978c.) We recall that a continuous function $f : \mathbb{R}^n \to \mathbb{C}$ is called *almost periodic in the sense of Bohr* or *uniformly almost periodic* if for any $\varepsilon > 0$ there is a compact set $K \subset \mathbb{R}^n$ such that every translation $x + K$ of K contains an *ε-almost period* of f, i.e., a vector $\tau \in \mathbb{R}^n$ such that $\sup_x |f(x + \tau) - f(x)| < \varepsilon$. An equivalent definition is that the family $\{f(\cdot + \tau), \tau \in \mathbb{R}^n\}$ of all translations of f must be precompact in the uniform convergence topology on $\mathbb{R}^n$. The definition of an almost periodic function on $\mathbb{Z}^n$ is exactly the same.

Operators with almost periodic coefficients can be used to model the quantum-mechanical motion of electrons in media with certain deviations from periodicity, for example in some liquids and alloys. Moreover, questions of the spectral theory of such operators emerge in a number of mechanical problems, for example, when considering the linearization of systems with conditionally periodic motions. In particular, there is a model in which the structure of the spectra of one-dimensional operators with almost periodic coefficients is responsible for the structure of the rings of Saturn (Avron and Simon 1981).

We denote by CAP $(\mathbb{R}^n)$ the set of all almost periodic functions on $\mathbb{R}^n$ in the sense of Bohr. Every almost periodic function f in the sense of Bohr is bounded. If the usual norm

$$\|f\|_\infty = \sup_x |f(x)|$$

is introduced in CAP $(\mathbb{R}^n)$, then the latter becomes a commutative Banach algebra (with the usual addition and multiplication). The set Trig $(\mathbb{R}^n)$ of all trigonometric polynomials, i.e., finite sums of the form

$$f(x) = \sum_\xi f_\xi e^{i\xi\cdot x},$$

is dense in CAP ($\mathbb{R}^n$). The latter can therefore be thought of as the set of all functions in $\mathbb{R}^n$ being uniform limits of trigonometric polynomials. The space of maximal ideals of the Banach algebra CAP ($\mathbb{R}^n$) is called the *Bohr compact* and denoted by $\mathbb{R}^n_B$. There is a natural continuous embedding $\mathbb{R}^n \subset \mathbb{R}^n_B$, under which $\mathbb{R}^n$ becomes a dense subset of $\mathbb{R}^n_B$. The addition in $\mathbb{R}^n$ can be extended by continuity to an operation on $\mathbb{R}^n_B$, which turns the latter into a topological group, $\mathbb{R}^n$ being of measure zero in $\mathbb{R}^n_B$ relative to the Haar measure on $\mathbb{R}^n_B$. The mean value of an almost periodic function f in the sense of Bohr is defined by

$$M\{f\} = \lim_{R\to+\infty} \frac{1}{R^n} \int\limits_{|x_j|\le R/2} f(x)\,dx. \tag{18.1}$$

(It can be proved that the limit exists for any function $f \in$ CAP ($\mathbb{R}^n$); to this end it suffices to consider trigonometric polynomials.) The mean value of $f \in$ CAP ($\mathbb{R}^n$) can also be written as

$$M\{f\} = \int\limits_{\mathbb{R}^n_B} \hat{f}\,d\mu_n, \tag{18.2}$$

where $d\mu_n$ is the Haar measure on $\mathbb{R}^n_B$ (normalized in such a way that the measure of the whole group $\mathbb{R}^n_B$ is equal to one), and where $\hat{f}$ is the extension of f from $\mathbb{R}^n$ to $\mathbb{R}^n_B$ by continuity. Using the mean value, one can introduce the scalar product

$$(f, g)_B = M\{f \cdot \bar{g}\}$$

on CAP ($\mathbb{R}^n$). The completion of CAP ($\mathbb{R}^n$) in the corresponding norm is the *Bezikowich space* $B^2(\mathbb{R}^n)$, which, by (18.2), is canonically isomorphic to $L_2(\mathbb{R}^n_B) = L_2(\mathbb{R}^n_B, d\mu_n)$. The space $B^2(\mathbb{R}^n)$ is a non-separable Hilbert space, in which the exponents $\{e^{2\pi i\xi\cdot x},\ \xi \in \mathbb{R}^n\}$ form an orthonormal basis. Expanding any function $f \in$ CAP ($\mathbb{R}^n$) in this basis, we obtain the series

$$f(x) = \sum_{k=1}^{\infty} f_k e^{2\pi i\xi_k\cdot x}, \quad f_k = M_x\{e^{-2\pi i\xi_k\cdot x} f(x)\}.$$

called the *Fourier series* of f (we remark that $M_x\{e^{-2\pi i\xi\cdot x} f(x)\} = 0$ for all but at most countably many vectors ξ). Here the vectors ξ_k are called *frequencies* and the smallest $\mathbb{Z}$-submodule in $\mathbb{R}^n$ generated by the set of all frequencies is called the *frequency module* of f. The frequency module can also be defined using the notion of the hull. By the *hull* $H(f)$ of f we mean the closure of the set $\{f(\cdot+\tau),\ \tau \in \mathbb{R}^n\}$ of translations in the uniform convergence topology. There is a canonical transformation $\mathbb{R}^n \to H(f)$ mapping τ into $f(\cdot+\tau)$. Under this transformation, the image of $\mathbb{R}^n$ is dense in $H(f)$. The addition in $\mathbb{R}^n$ can be extended by continuity to an operation inducing the structure of a compact Abelian group on $H(f)$, whose dual group (the group of characters) can be identified with the frequency module if one observes that the restriction of any character to $\mathbb{R}^n$ is a character of the group $\mathbb{R}^n$, and so has the form $e^{2\pi i\xi\cdot x}$, the character being uniquely defined by the vector-valued frequency ξ. The hull can also be defined as the set of translations $\{\hat{f}(\cdot+\tau),\ \tau\in\mathbb{R}^n_B\}$, where $\hat{f}$ is the extension of f to $\mathbb{R}^n_B$ by continuity. This being the case, $H(f)$ is the quotient group of $\mathbb{R}^n_B$ relative to the subgroup of all $\tau\in\mathbb{R}^n_B$ such that the translation by τ leaves $\hat{f}$ unchanged.

We also introduce the space

$$\mathrm{CAP}^\infty(\mathbb{R}^n)=\{f: f\in C^\infty(\mathbb{R}^n),\ \partial^\alpha f\in \mathrm{CAP}\,(\mathbb{R}^n)\ \text{for any}\ \alpha\}.$$

We consider a formally self-adjoint uniformly elliptic operator A of the form (17.1) with coefficients $a_\alpha\in\mathrm{CAP}^\infty(\mathbb{R}^n)$. For brevity, we shall refer to such an operator as an *almost periodic self-adjoint elliptic operator* (it is often sufficient to assume merely that $a_\alpha\in\mathrm{CAP}\,(\mathbb{R}^n)$, but we shall confine ourselves to the smooth case for simplicity).

The operator A can be regarded as an unbounded operator in the Hilbert space $L_2(\mathbb{R}^n)$ with domain $C_0^\infty(\mathbb{R}^n)$ or $S(\mathbb{R}^n)$, or as an unbounded operator in $B^2(\mathbb{R}^n)$ with domain $\mathrm{Trig}\,(\mathbb{R}^n)$ or $\mathrm{CAP}^\infty(\mathbb{R}^n)$. These operators are both essentially self-adjoint (in the former case it does not matter that the operator is almost periodic). To prove that this is so one can use the scale of Sobolev spaces $H^s(\mathbb{R}^n)$ in the former case and the analogous scale of spaces $H^s(\mathbb{R}^n_B)$ in the latter, $H^s(\mathbb{R}^n_B)$ being defined as the completion of $\mathrm{Trig}\,(\mathbb{R}^n)$ in the norm

$$\Big\|\sum_\xi f_\xi e^{i\xi\cdot x}\Big\|_{s,B}=\Big(\sum_\xi\big(1+|\xi|^2\big)^{s/2}\big|f_\xi\big|^2\Big)^{1/2}.$$

Theorems on elliptic regularity of the form

$$\{u\in H^{-\infty},\ Au=f\in H^s\}\Longrightarrow u\in H^{s+m} \tag{18.3}$$

(here $H^{-\infty}=\bigcup_{s\in\mathbb{R}}H^s$) are valid in either of these scales. The theorems can be obtained, for example, with the aid of the technique of pseudodifferential operators. It follows from (18.3) that H^m is the domain of the maximal operator associated with A. On the other hand, it is clear that H^m is contained in the domain of the minimal operator. Hence the minimal and maximal operators coincide, which is equivalent to essential self-adjointness.

18.2. General Properties of the Spectrum and Eigenfunctions

We shall now consider self-adjoint operators in $L_2(\mathbb{R}^n)$ and $B^2(\mathbb{R}^n)$ obtained by taking the closure of either of the operators described above defined by the same expression A of the form (17.1) on $C_0^\infty(\mathbb{R}^n)$ and $\mathrm{Trig}\,(\mathbb{R}^n)$, respectively. We denote by $\sigma(A)$ the spectrum of the former operator and by $\sigma_B(A)$ the spectrum of the latter.

Theorem 18.1 (on the equality of the spectra; Shubin 1976a):

$$\sigma(A) = \sigma_B(A). \tag{18.4}$$

We remark that the theorem is concerned only with the equality of the spectra as subsets of $\mathbb{R}$. The spectrum of A in $L_2(\mathbb{R}^n)$ can have a completely different character than that in $B^2(\mathbb{R}^n)$. For example, if $A = a(D)$ is an operator with constant coefficients, then its spectrum in $L_2(\mathbb{R}^n)$ is absolutely continuous, while it has pure point spectrum in $B^2(\mathbb{R}^n)$. By analogy, for operators with periodic coefficients an orthonormal basis of eigenfunctions in $B^2(\mathbb{R}^n)$ can be constructed from the Bloch eigenfunctions of A. However, in general, the spectrum of the same operator in $L_2(\mathbb{R}^n)$ is no longer a point spectrum (moreover, the point spectrum may not exist at all, as in the case of the Schrödinger operator with a periodic potential; see Sect. 17.2).

Theorem 18.1 can be proved by approximating functions from $\mathrm{CAP}^\infty(\mathbb{R}^n)$ by functions from $C_0^\infty(\mathbb{R}^n)$ (which can be achieved by standard cut-offs) and, conversely, approximating each function $\varphi \in C_0^\infty(\mathbb{R}^n)$ by functions $\psi_k \in \mathrm{CAP}^\infty(\mathbb{R}^n)$ such that

$$\lim_{k\to\infty} \|\psi_k\|_B^2 = \|\varphi\|^2, \quad \lim_{k\to\infty} \|(A-\lambda I)\psi_k\|_B^2 = \|(A-\lambda I)\varphi\|^2.$$

The latter can be achieved by taking ψ_k to be the convolution

$$\psi_k(x) = \int \chi_k(x-y)\varphi(y)\,dy, \tag{18.5}$$

where $\chi_k \in \mathrm{CAP}^\infty(\mathbb{R}^n)$ is a sequence of functions selected in a special way depending on the coefficients of A. Another approach, which makes it possible to prove Theorem 18.1 with the aid of the theory of C^*-algebras, was proposed by Biktashev and Mishchenko (1980) as well as Baaj Saad (1988).

Theorem 18.1 is also valid for pseudodifferential (not necessarily self-adjoint) operators in $\mathbb{R}^n$ whose symbols are almost periodic in x. The ellipticity condition can also be relaxed. (Ellipticity is used only in the proof of essential self-adjointness. Instead of ellipticity, it suffices to adopt a certain hypoellipticity condition. In the case of pseudodifferential operators of order zero the assertion is valid without any additional conditions.)

The whole spectrum of an almost periodic elliptic operator A is essential, i.e., does not contain any isolated eigenvalues of finite multiplicity. Indeed, by shifting any function $\varphi \in C_0^\infty(\mathbb{R}^n)$ such that $\|\varphi\| = 1$ and $\|(A-\lambda I)\varphi\| < \varepsilon/2$ by sufficiently large vectors being δ-almost periods of the functions a_α and their derivatives up to the m-th order for a sufficiently small $\delta > 0$, we can obtain an orthonormal system of functions $\{\varphi_1, \varphi_2, \ldots\}$ such that $\|(A-\lambda I)\varphi_j\| < \varepsilon$ for all $j = 1, 2\ldots$. It follows immediately that the spectrum of A in $L_2(\mathbb{R}^n)$ is essential. Besides, if $\lambda \in \sigma(A)$, then the convolution (18.5) with φ_j from the orthonormal system just constructed with k large enough yields a system of functions $\psi_j \in \mathrm{CAP}^\infty(\mathbb{R}^n)$ that is 'almost orthogonal' in $B^2(\mathbb{R}^n)$ and such that $\|\psi_j\|_B^{-1}\|(A-\lambda I)\psi_j\|_B < \varepsilon$, which proves that the spectrum of A in $B^2(\mathbb{R}^n)$ is essential.

For $n = 1$ no eigenvalue can have infinite multiplicity (the multiplicity cannot exceed $\operatorname{ord} A$), so that $\sigma(A)$ is a perfect set in this case (a closed set without isolated points). Subject to certain more specialized assumptions, it is possible to obtain much sharper results, which will be described below.

The notion of the hull of an almost periodic operator, which is analogous to that of a function, is useful in the study of the spectrum of an almost periodic elliptic operator A. Namely, the *hull* $H(A)$ of an operator $A = a(x, D_x)$ is defined to be the set of all operators $\tilde{A} = \tilde{a}(x, D_x)$ whose symbols $\tilde{a}(x,\xi)$ are uniform limits (with respect to x) of translations $a(x+\tau,\xi)$ of the symbol of A. In other words, $\tilde{A} \in H(A)$ if

$$\tilde{A} = \sum_{|\alpha|\le m} \tilde{a}_\alpha(x) D^\alpha$$

and there exists a sequence $\{\tau_k,\ k = 1,2,\ldots\}$ such that $\tilde{a}_\alpha(x) = \lim_{k\to+\infty} a(x+\tau_k)$ uniformly in x for all α with $|\alpha| \le m$ (here a_α are the coefficients of A). It is easily seen that $\sigma(\tilde{A}) = \sigma(A)$ for any $\tilde{A} \in H(A)$. As in the case of a single function, the hull $H(A)$ has the structure of a compact topological group, on which it is convenient to consider the Haar measure. Given this measure, we can talk of a property satisfied almost everywhere on $H(A)$ or for almost all $\tilde{A} \in H(A)$. Here is an example of using the hull: if A is an almost periodic elliptic operator, then $\lambda \in \sigma(A)$ if and only if there exists an operator $\tilde{A} \in H(A)$ such that the equation $\tilde{A}u = \lambda u$ has a bounded solution (Shubin 1978c). Let us also mention the following important property of mobility of the point spectrum of the one-dimensional Schrödinger operator with an almost periodic potential pointed out by Pastur (1980):

$$\mu\{\tilde{A} : \tilde{A} \in H(A),\ \lambda \in \sigma_p(\tilde{A})\} = 0$$

for any fixed $\lambda \in \mathbb{R}$, where μ is the Haar measure on $H(A)$. We shall explain this property in the more general context of random operators (this is the case proved by Pastur). In this situation Kirsch and Martinelli (1982) proved that there are fixed closed subsets S_1, S_2, and S_3 of $\mathbb{R}$ such that for almost all $\tilde{A} \in H(A)$ the closure of the point spectrum is equal to S_1, the absolutely continuous spectrum is equal to S_2, and the singular continuous spectrum is equal to S_3. It is now a consequence of the above-mentioned result belonging to Pastur that S_1 is locally uncountable, that is, if $\lambda_0 \in S_1$, then the set $S_1 \cap (\lambda_0 - \varepsilon, \lambda_0 + \varepsilon)$ is uncountable for any $\varepsilon > 0$. A. Marchenko (1979) proved that the spectrum $\sigma(A)$ of an almost periodic self-adjoint elliptic operator can be approximated by the spectra $\sigma(A_k)$ of operators in bounded domains with suitable boundary conditions in the sense that

$$\sigma(A) = \bigcap_m \overline{\bigcup_{k>m} \sigma(A_k)},$$

(i.e., $\sigma(A)$ is the limit of the sequence $\sigma(A_k)$). One should not think that the presence of a pure point spectrum in $B^2(\mathbb{R}^n)$, which is the case for operators with constant or periodic coefficients, is also typical of operators with almost periodic coefficients. In what follows we shall see that even for the one-dimensional Schrödinger operator with an almost periodic potential there are various possibilities (for example, a pure point spectrum in $L_2(\mathbb{R})$ with exponentially decaying eigenfunctions, which rules out the possibility of a pure point spectrum in $B^2(\mathbb{R})$). A few general results on the structure of the spectrum and the behaviour of eigenfunctions are known for many-dimensional almost periodic operators. We shall present one of these results concerned with the Schrödinger operator with quasiperiodic coefficients.

A function $f : \mathbb{R}^n \to \mathbb{C}$ is said to be *quasiperiodic* if $f \in \mathrm{CAP}\,(\mathbb{R}^n)$ and the Fourier series of f has the form

$$f(x) = \sum_{m_1,\dots,m_N \in \mathbb{Z}} c_{m_1\dots m_N} e^{2\pi i(m_1\alpha_1+\dots+m_N\alpha_N)\cdot x}, \tag{18.6}$$

where $\alpha_1, \dots, \alpha_N \in \mathbb{R}^n$. This means that f has a finitely generated frequency module $\{m_1\alpha_1 + \dots + m_N\alpha_N : m_1, \dots, m_N \in \mathbb{Z}\}$. A quasiperiodic function f of this form can be obtained as a composition $f = \hat{f} \circ \pi$, where $\pi : \mathbb{R}^n \to \mathbb{T}^N$ is a continuous homomorphism from $\mathbb{R}^n$ into the torus $\mathbb{T}^N = \mathbb{R}^N/\mathbb{Z}^N$ having the form $p \circ j$ with $j : \mathbb{R}^n \to \mathbb{R}^N$ a linear mapping, $p : \mathbb{R}^N \to \mathbb{T}^N$ the canonical projection, and $\hat{f} \in C(\mathbb{T}^N)$. The expansion (18.6) can now be obtained with the aid of the standard Fourier series of $\hat{f}$. If $\hat{f}$ is sufficiently smooth, then the series (18.6) will be absolutely and uniformly convergent. We shall write $f \in \hat{C}^r$ if $\hat{f} \in C^r(\mathbb{T}^N)$. The *arithmetic* (or *Diophantine*) condition

$$\exists\, k > 0,\ \exists\, c > 0 : \left|\sum_{j=1}^{N} m_j\alpha_j\right| \geq c|m|^{-k},$$
$$m = (m_1, \dots, m_N) \in \mathbb{Z}^N \setminus \{0\} \tag{18.7}$$

plays an important role. It means that the vectors $\alpha_1, \dots, \alpha_N$ are linearly independent over the field $\mathbb{Q}$ of rational numbers in a slightly sharpened sense. It is easily demonstrated that the condition is satisfied for almost all systems of vectors $\alpha_1, \dots, \alpha_N$ (in the sense of the Lebesgue measure on $\mathbb{R}^{nN}$) for any given N.

The following theorem is practically the only significant result on eigenfunctions in the many-dimensional case.

Theorem 18.2 (Kozlov 1983). *Let $A = -\Delta + q(x)$, where q is a quasiperiodic function of class $\hat{C}^r$ on $\mathbb{R}^n$. Suppose that the arithmetic condition (18.7) is satisfied, r is large enough (depending on k in (18.7)), and the norm of q in $\hat{C}^r$ (which is equal to the norm of $\hat{q}$ in $C^r(\mathbb{T}^N)$) is small enough. This being the case, if $\lambda_0 = \inf \sigma(A)$, then the equation $A\psi = \lambda_0\psi$ has a non-zero quasiperiodic solution ψ that is positive everywhere, has the same frequency module as q, and belongs to $\hat{C}^s$, where $s = s(r) \to +\infty$ as $r \to \infty$.*

The proof of this theorem employs the Kolmogorov-Arnol'd-Moser accelerated convergence method, which makes it possible to use the successive approximation method to find a number λ_0 and a quasiperiodic function ψ having the properties specified in the theorem such that

$$\psi(A - \lambda_0 I)\psi = -\sum_{j=1}^{n} \frac{\partial}{\partial x_j} \psi^2 \frac{\partial}{\partial x_j}.$$

Thus, it follows easily that the assertion of the theorem is true. Moreover, it turns out that $\lambda_0 = \inf \sigma(A)$. We remark that a quasiperiodic (and even an almost periodic) solution ψ of the equation $(A - \lambda_0 I)\psi = 0$ is unique.

If $q(x)$ is not required to be small, it may turn out that the equation $(A - \lambda_0 I)\psi = 0$ has no quasiperiodic or even almost periodic solutions. Even in the one-dimensional case all the solutions of this equation may belong to $L_2(\mathbb{R})$ (see below).

18.3. The Spectrum of the One-Dimensional Schrödinger Operator with an Almost Periodic Potential

We consider the one-dimensional Schrödinger operator $A = -d^2/dx^2 + q(x)$ with an almost periodic potential $q \in \mathrm{CAP}(\mathbb{R})$. As opposed to operators with periodic potentials, the spectrum of this operator may no longer have a zone structure, and, as we shall now see, it is natural to expect it to be a *perfect Cantor set* (not necessarily of measure zero), that is, a closed subset of $\mathbb{R}$ without isolated points, the complement of which is everywhere dense. Besides, in this case the spectrum may have both the singular continuous and point components, but may also remain absolutely continuous.

Following Simon (1982), we shall explain why the spectrum tends to be a perfect Cantor set. We consider the doubly periodic function

$$f(x, y) = \sum_{n_1, n_2 \in \mathbb{Z}} a_{n_1, n_2} \exp\left[2\pi i(n_1 x + n_2 y)\right],$$

where the coefficients a_{n_1,n_2} decrease fast enough as $|n| \to \infty$ (here $n = (n_1, n_2)$) and $a_{-n} = \bar{a}_n$. Now let us consider the Schrödinger operator A with potential $q(x) = f(x, \alpha x)$. To begin with, let $\alpha = r/s$, where r and s are relatively prime positive integers. Then the potential $q(x)$ is periodic with period $2\pi s$. In general, the gaps of this potential lie near the periodic and antiperiodic eigenvalues, which, for small q, are close to the corresponding eigenvalues of $A_0 = -d^2/dx^2$ equal to $(\pi l/s)^2$, where $l = 0, 1, 2, \ldots$, or, equivalently, to $[\pi(n_1 + n_2\alpha)]^2$, where $n_1, n_2 \in \mathbb{Z}$. It is natural to expect that the same will also be true for any irrational number α, in which case the points $[\pi(n_1 + n_2\alpha)]^2$, where $n_1, n_2 \in \mathbb{Z}$, are everywhere dense on the half-axis $[0, +\infty)$, so that if each of these points lies inside a gap, then the complement of the spectrum will be an open everywhere dense set on $[0, +\infty)$. The following more precise argument is based on perturbation theory: if $a_{n_1,n_2} > 0$ for all n_1 and n_2, then, for the operator $A_\lambda = -d^2/dx^2 + \lambda q(x)$ with the same $q(x)$ as above (with $\alpha = r/s$), the length of the gap containing the point $(2\pi l/s)^2$ is

$$2\lambda\Bigg(\sum_{n_1+n_2\alpha=l/s} a_{n_1,n_2}\Bigg) + O(\lambda^2)$$

for small λ. Thus, judging by the first approximation of perturbation theory, all the gaps are non-empty in this case. Moreover, in the same approximation the total length of all the gaps is equal to $O\big(\lambda\Sigma|a_n|\big)$, so that the Lebesgue measure of the spectrum would have to be infinite, even though the spectrum is nowhere dense.

In order to state precise assertions let us fist consider the case of *limit periodic potentials*, i.e., almost periodic potentials that are uniform limits of periodic functions (it can be demonstrated that this is so if and only if the frequency module has one generator over the field $\mathbb{Q}$). In this case it was proved independently by Chulaevskij (1981), Moser (1981), and Avron and Simon (1982) that the spectrum is a Cantor set. Namely, the following result is true.

Theorem 18.3 (Avron and Simon 1982). *In the space of all limit periodic potentials (with the standard uniform metric) there exists a dense subset of type G_δ consisting of potentials q such that the spectrum of the corresponding Schrödinger operator $A = -d^2/dx^2 + q(x)$ is a perfect Cantor set. The same is true in the space of potentials q of the special form*

$$q(x) = \sum_{n=0}^{\infty} a_n \cos(x/2^n), \quad \sum_n |a_n| < +\infty.$$

The proof can be obtained by using perturbation theory to trace any new gaps emerging in the spectrum as one goes over from a periodic potential to its perturbation with multiple period. At the same time one can establish that there exists a dense set of limit periodic potentials (or potentials of the special form specified above) with absolutely continuous spectrum (of multiplicity two) being simultaneously a Cantor set. In particular, the spectrum becomes absolutely continuous if $q(x)$ admits a rapidly convergent approximation by periodic potentials. For example, the following result is true.

Theorem 18.4 (Chulaevskij 1981). *Suppose that the potential q has the form*

$$q(x) = \sum_{n=0}^{\infty} \varepsilon_n q_n(x/T_n), \tag{18.8}$$

where each function q_n is periodic with period one, $q_n \in C^2$, $\max|q_n| \leq 1$, $T_{n+1}/T_n \in \mathbb{N} \setminus \{1\}$, and

$$\sum_{n=N}^{\infty} |\varepsilon_n| < \beta_N / T_N$$

with $\beta_N \leq \exp(-c_N T_N)$, *where* $c_N \to +\infty$ *as* $N \to \infty$. *Then the spectrum is absolutely continuous and there exist eigenfunctions* $\psi(x, \lambda)$ *(solutions of the equation* $A\psi = \lambda\psi$*) of the form*

$$\psi(x, \lambda) = e^{ip(\lambda)x}\chi(x, \lambda), \tag{18.9}$$

where χ *is a limit periodic function in* x *(with the same numbers* T_n*), and where* $p(\lambda)$ *is the boundary value of an analytic function having holomorphic branches on the complex plane* $\mathbb{C} \setminus \{\lambda : \lambda \in \mathbb{R},\ \lambda \geq \min q(x)\}$ *with a cut. The branches of* $p(\lambda)$ *are continuous up to the boundary of the specified domain,* $p(\lambda) \in \mathbb{R}$ *for every point of the spectrum, and* $p(\lambda) \in i\mathbb{R} \setminus \{0\}$ *in the complement of the spectrum on the real axis (that is, inside the gaps). Moreover, there exists an everywhere dense subset* $\mathcal{L}$ *of* $C^2(\mathbb{R}/\mathbb{Z})$ *such that if* $q_n \in \mathcal{L}$ *for all but finitely many* n, *then the spectrum is a Cantor set.*

However, the spectrum does not always have to be absolutely continuous (or even simply continuous), even in the case of a limit periodic potential. For example, Chulaevskij and Molchanov (1984) demonstrated that there are limit periodic potentials with pure point Cantor spectrum of Lebesgue measure zero (with eigenfunctions decaying faster than any power of $|x|$ as $|x| \to \infty$, although not exponentially). Examples of such potentials can be obtained with the aid of the probabilistic technique, which will be discussed later on in §19.

A situation similar to Theorem 18.4 can also take place for quasiperiodic potentials, as can be seen from the earliest significant results on one-dimensional almost periodic operators due to Dinaburg and Sinai (1975) and Belokolos (1975, 1976). In the case of one independent variable the fact that $q(x)$ is quasiperiodic means that it has the form

$$q(x) = f(\alpha_1 x, \alpha_2 x, \ldots, \alpha_N x), \tag{18.10}$$

where $f = f(y_1, \ldots, y_N)$ is a continuous periodic function with period one with respect to each variable, and where $\alpha_1, \alpha_2, \ldots, \alpha_N$ are arbitrary real numbers. It is clear that in this case q can be uniformly approximated by trigonometric polynomials of the form

$$\sum_m c_m e^{2\pi i(m,\alpha)x},$$

where $m \in \mathbb{Z}^N$ and $(m, \alpha) = m_1\alpha_1 + \ldots + m_N\alpha_N$, and the frequency module is equal to $\{(m, \alpha), m \in \mathbb{Z}^N\}$.

Theorem 18.5 (Dinaburg and Sinai 1975). *Suppose that* q *has the form (18.10) with a real analytic function* f *and with a system* $(\alpha_1, \ldots, \alpha_N)$ *satisfying the Diophantine condition (18.7). Then for any* $\varepsilon > 0$ *there exist* $C'(\varepsilon), C''(\varepsilon) > 0$ *such that every neighbourhood*

$$O_m = \left\{ \lambda : \left|\sqrt{\lambda} - \pi(m, \alpha)\right| < \frac{C'(\varepsilon)}{|(m, \alpha)| + 1} \right\}$$

contains a neighbourhood

$$\tilde{O}_m = \left\{ \lambda : \left|\sqrt{\lambda} - \lambda_m\right| < C''(\varepsilon) \exp\left[-\frac{|m|}{\ln^{1+\varepsilon}|m|} \right] \right\}$$

such that if $\lambda \neq \bigcup_m \tilde{O}_m = \mathfrak{M}$, *then the equation* $A\psi = \lambda\psi$ *has two linearly independent solutions* ψ *and* $\bar{\psi}$, *where* ψ *has the form (18.9) with a quasiperiodic function* χ *(of the form (18.10) with the same vector* $(\alpha_1, \dots, \alpha_N)$*) and with* $p(\lambda)$ *such that* $|p(\lambda) - \sqrt{\lambda}| \leq c'''/\sqrt{\lambda}$, *where* c''' *is a constant.*

This means that A has Bloch type eigenfunctions (with periodicity replaced by quasiperiodicity) outside exponentially small intervals (each of which can either be a gap or can contain a part of the spectrum).

Hence one can easily deduce that the spectrum of A has an absolutely continuous component, the density of the absolutely continuous measure $d(E_\lambda f, f)$ on $[0, +\infty) \setminus \mathfrak{M}$ being almost everywhere equal to

$$\frac{|\rho(\lambda)|^2}{2\pi\sqrt{\lambda}(1 + o(1))},$$

where

$$\rho(\lambda) = \int_{-\infty}^{+\infty} f(x)\overline{\psi(x, \lambda)}\, dx, \quad o(1) \to 0 \text{ as } \lambda \to +\infty.$$

The proof of Theorem 18.5 is based on the reduction of a system of two first-order differential equations to which one can reduce the equation $A\psi = \lambda\psi$ to a system with constant coefficients by successive modifications of the unknown functions, which can be found with the aid of the KAM-theory (the Kolmogorov-Arnol'd-Moser accelerated convergence method). Certain refinements of Theorem 18.5 are contained in (Rüssmann 1980; Moser and Pöschel 1984). Following (Dinaburg and Sinai 1975), many authors used the KAM-theory in various problems of spectral theory. For example, it was used in (Craig 1983; Pöschel 1983) to construct examples of almost periodic difference Schrödinger operators with pure point spectrum (§19 will be concerned with other examples of this kind). Sinai (1985) proved that for the one-dimensional difference Schrödinger operator arising in the linearization of the difference analogue of the sine-Gordon equation there exist Bloch eigenfunctions near the left end of the spectrum.

Theorem 18.5 provides information on the spectrum only in a Cantor set of sufficiently large measure, but it does not say anything about what happens inside the gaps of this set. In particular, the theorem does not indicate whether or not the spectrum is a Cantor set. In this connection, we remark that, using the known description of finite gap potentials and perturbing the given finite gap potential, Levitan and Savin (1984) demonstrated that for any fixed in advance arbitrary frequency modulus that is not isomorphic to $\mathbb{Z}$ there exists an almost periodic potential with this modulus such that the corresponding Schrödinger operator has a Cantor spectrum (a frequency modulus isomorphic to $\mathbb{Z}$ corresponds to the periodic case).

Using the methods of KAM-theory, Sinai (1987) considered the one-dimensional difference Schrödinger operator $H_g = -\Delta + gV(z)$ with $V(z) = \cos 2\pi(\omega z + \alpha)$ for $z \in \mathbb{Z}$. This operator is often called the *Harper operator* or *almost Mathieu operator*. It emerges in the study of the Schrödinger operator with a constant magnetic field and periodic electric field by the methods of perturbation theory. In (Sinai 1987) it is assumed that the coupling constant g is sufficiently small and ω is typical enough, i.e., it is poorly approximated by rational numbers (these conditions for ω, which are satisfied for almost all ω, can be stated explicitly in terms of an expansion of ω as a continued fraction). Any 1-periodic Morse C^2-function on $\mathbb{R}$ with exactly two non-singular critical points in $\mathbb{R}/\mathbb{Z}$ can also appear in place of $\cos 2\pi x$. It has been proved that H_g has pure point spectrum and the eigenfunctions decay exponentially (this is an example of the so-called *Anderson localization*, a more detailed discussion of which will be presented in §19).

The similar difference operator

$$P_0 = \cos hD_x + \cos x$$

considered in $L_2(\mathbb{R})$ can be represented as the direct integral of the operators H_g of the form described above (the integral is with respect to α). The spectrum of P_0 and the spectra of the analogous operators with Weyl symbols $p(x, h\xi)$, where $p = p(x, \eta)$ is a 2π-periodic function in x and ξ, were studied in detail by Helffer and Sjöstrand (for example, see Helffer and Sjöstrand 1987, 1988 and references therein), who used microlocal analysis to investigate the tunnelling effect between various potential wells. In particular, they proved that if P is a suitable analytic perturbation of P_0 and $h/2\pi$ can be represented as the continued fraction

$$\frac{h}{2\pi} = \cfrac{1}{q_1 + \cfrac{1}{q_2 + \cfrac{1}{q_3 + \dots}}}, \quad q_j \in \mathbb{Z}, \quad 1 \le j < \infty,$$

where $|q_j| \ge c_0$, c_0 is large enough, and $|h| < 2/c_0$, then the spectrum of P is a Cantor set of Lebesgue measure zero.[8]

Among other qualitative results on the spectrum, let us mention the following theorem, which provides a sufficient condition for the absence of the point spectrum.

[8] (Added in the English edition.) There is a lot of recent literature on the almost Mathieu operator; see e.g. Shubin (1993) and references therein.

Theorem 18.6 (Gordon 1976). *Let q be a potential such that there exists a sequence of periodic potentials q_m with periods $T_m \to \infty$ for which*

$$\sup_{x\in[-2T_m,2T_m]} \left|q(x) - q_m(x)\right| \le cm^{-T_m}.$$

This being the case, if ψ is a solution of the equation $A\psi = \lambda\psi$ (here $A = -d^2/dx^2 + q(x)$), then

$$\overline{\lim_{|x|\to\infty}} \frac{|\psi'(x)|^2 + |\psi(x)|^2}{|\psi'(0)|^2 + |\psi(0)|^2} \ge \frac{1}{4}.$$

In particular, under the assumptions of this theorem, A has no point spectrum, for if $\psi \in L_2(\mathbb{R})$ is a solution of the equation $A\psi = \lambda\psi$, then it can be proved that $\psi(x) \to 0$ and $\psi'(x) \to 0$ as $|x| \to +\infty$.

On the other hand, Gordon (1976) also demonstrated that for any system $(\alpha_1, \dots, \alpha_N)$ of real numbers linearly independent over the field $\mathbb{Q}$ of rational numbers there exists a quasiperiodic function q of the form (18.10) such that the point spectrum of $A = -d^2/dx^2 + q$ is non-empty (moreover, for some λ there exists an eigenfunction ψ such that $|\psi(x)| \le C\exp(-|x|^{1/2})$).

18.4. The Density of States of an Operator with Almost Periodic Coefficients

Let $A = a(x, D_x)$ be a uniformly elliptic self-adjoint operator of order m in $\mathbb{R}^n$ (which implies that m is an even number) with coefficients in $\mathrm{CAP}^\infty(\mathbb{R}^n)$. We shall always assume that the principal symbol $a_m = a_m(x,\xi)$ of A is positive (for $\xi \ne 0$). Consequently, A is a lower semi-bounded operator. For such operators one can define the integrated density of states, i.e., the distribution function $N(\lambda)$, by the limit

$$N(\lambda) = \lim_{k\to+\infty} (\operatorname{mes} V_k)^{-1} N_{V_k}(\lambda), \tag{18.11}$$

where V_k is a bounded domain in $\mathbb{R}^n$, $\operatorname{mes} V_k$ is the Lebesgue measure of V_k, and $N_{V_k}(\lambda)$ is the standard distribution function of the discrete spectrum of A in V_k with self-adjoint boundary conditions

$$B_j u\big|_{\partial V_k} = 0, \quad j = 1, 2, \dots, m/2 \tag{18.12}$$

satisfying the ellipticity condition with a parameter for $A - \lambda I$ (see, for example, Agranovich and Vishik 1964) uniformly in k (for example, the Dirichlet conditions are suitable for any A). For simplicity, we shall assume that either V_k has smooth boundary, or A has periodic coefficients, V_k is a parallelepiped formed by the elementary cells of the lattice of periods, and the periodicity conditions (with the lattice of periods for which V_k is an elementary cell) serve as the boundary conditions. Furthermore, we assume that the domains blow up in a sufficiently regular way, for example, homothetically, as $k \to +\infty$. Under these conditions, the limit (18.11) exists for all but perhaps countably many λ (to be more precise, it can fail to exist only at the points of discontinuity of $N(\lambda)$, which is a non-decreasing function and can be defined to be right-continuous at every point of discontinuity). Moreover, the limit does not depend on the choice of the sequence of domains V_k or boundary conditions (18.12) (Shubin 1978b). It follows that $N(\lambda)$ depends only on A.

To prove that the limit in (18.11) exists one can, for example, use Helly's selection theorems (see, for example, Shilov 1961), which imply that it suffices, firstly, to demonstrate that there exists λ_0 independent of k such that $\lambda_0 < \inf \sigma(A_k)$, where A_k is the self-adjoint operator in $L_2(V_k)$ defined by A and the boundary conditions (18.12) (its domain consists of all functions $u \in H^m(V_k)$ satisfying the boundary conditions), secondly, to estimate each of the functions $N_k(\lambda) = (\operatorname{mes} V_k)^{-1} N_{V_k}(\lambda)$ by $C(1+|\lambda|)^N$ uniformly with respect to k, and, thirdly, to prove the existence of the limit of the Laplace transform $\tilde{N}_k(t)$ as $k \to \infty$. Here the *Laplace transform* of any distribution function $N(\lambda)$ is defined by

$$\tilde{N}(t) = \int_{-\infty}^{\infty} e^{-\lambda t}\, dN(\lambda), \quad t > 0. \tag{18.13}$$

Then

$$\tilde{N}_k(t) = (\operatorname{mes} V_k)^{-1} \operatorname{Tr}\left(e^{-tA_k}\right) = (\operatorname{mes} V_k)^{-1} \int_{V_k} G_k(t,x,x)\, dx,$$

where G_k is the kernel of the operator e^{-tA_k}, or, equivalently, the Green function of the parabolic operator $\partial/\partial t - A$ in the domain $[0,+\infty) \times V_k$ with boundary conditions (18.12). Comparing the Green function G_k with the fundamental solution $G = G(t,x,y)$ of the Cauchy problem, one can easily find with the aid of the results of Ehjdel'man and Ivasishen (1970) that the limit $\lim_{k\to\infty} \tilde{N}_k(t)$ exists and is equal to

$$\tilde{N}(t) = M_x\{G(t,x,x)\}. \tag{18.14}$$

(The right-hand side makes sense, since it can be demonstrated by analysing the construction of the fundamental solution by Levy's method that $G(t, x+z, y+z) \in \operatorname{CAP}(\mathbb{R}^n_z)$ for any $t > 0$ and $x, y \in \mathbb{R}^n$; in particular, $G(t,x,x) \in \operatorname{CAP}(\mathbb{R}^n_x)$.) It can be seen from (18.14) that $N(\lambda)$ does not depend on the choice of the sequence of V_k or the boundary conditions.

In this discussion one could use the resolvent $(A - \lambda I)^{-1}$ and the Stieltjes transform, rather than the exponent e^{-tA} and the Laplace transform.

By analogy, one can prove the existence of other limits, for example,

$$F_L(\lambda) = \lim_{k\to\infty} (\operatorname{mes} V_k)^{-1} \sum_{\lambda_l < \lambda} (L\psi_l, \psi_l),$$

where $\{\psi_l,\ l = 1, 2, \ldots\}$ is the orthonormalized system of all eigenfunctions of the operator A_k, λ_l are the corresponding eigenvalues, L is an arbitrary differential operator with almost periodic coefficients (for $L = I$ we obtain $F_L(\lambda) = N(\lambda)$), and the scalar product is taken in $L_2(V_k)$. Here is another example:

$$F_{f,g}(\lambda) = \lim_{k\to\infty} (\operatorname{mes} V_k)^{-1} \sum_{\lambda_l<\lambda} (f, \psi_l)(\psi_l, g),$$

where $f, g \in \mathrm{CAP}(\mathbb{R}^n)$ (this example corresponds to a rank one operator L for the function F_L considered above). If A' and A'' are two given operators of the type described above, ψ' and ψ'' are their eigenfunctions, λ_l' and λ_l'' are the corresponding eigenvalues, and L' and L'' are differential operators with almost periodic coefficients, then the limit

$$F_{L',L''}(\lambda, \mu) = \lim_{k\to+\infty} (\operatorname{mes} V_k)^{-1} \sum_{\substack{\lambda_l'<\lambda \\ \lambda_r''<\mu}} (L'\psi_l', \psi_r'')(L''\psi_r'', \psi_l')$$

exists in the sense of weak convergence of the corresponding measures on the (λ, μ)-plane. Hence, for $A' = A'' = A$, $L' = \partial/\partial x_i$, and $L'' = \partial/\partial x_j$, one obtains a tensor $S^{ij}(\lambda, \mu)$, which has physical interpretation as the conductivity tensor. By considering the inverse functions, one can prove the existence of the *Fermi energy*

$$E^F(\rho) = \lim_{\substack{p\to\infty \\ p/\operatorname{mes} V_k \to \rho}} \frac{\lambda_1 + \lambda_2 + \ldots + \lambda_p}{p}, \quad \rho > 0,$$

where $\lambda_1 \le \lambda_2 \le \ldots$ are the eigenvalues of A_k arranged into an increasing sequence with their multiplicity taken into account (Shubin 1979).

18.5. Interpretation of the Density of States with the Aid of von Neumann Algebras and Its Properties

The theory of von Neumann algebras (see, for example, Dixmier 1969) makes it possible to interpret the integrated density of states $N(\lambda)$ of an almost periodic self-adjoint operator in the same way as the standard distribution function of the discrete spectrum, but using the generalized trace and dimension. We consider the Hilbert space $\mathcal{H} = L_2(\mathbb{R}^n_B \times \mathbb{R}^n) = B^2(\mathbb{R}^n) \otimes L_2(\mathbb{R}^n)$ and denote by e_λ the multiplication operator by $e_\lambda(x) = e^{i\lambda x}$ ($\lambda \in \mathbb{R}^n$) in $L_2(\mathbb{R}^n)$ and $B^2(\mathbb{R}^n)$. We denote by T_λ the translation operator by λ, that is, $T_\lambda u(x) = u(x - \lambda)$. Let $\mathfrak{A}$ be the algebra of bounded linear operators on $\mathcal{H}$ obtained as the weak closure of the algebra spanned by the set of operators

$$\{e_\lambda \otimes e_\lambda,\ \lambda \in \mathbb{R}^n\} \cup \{I \otimes T_\lambda,\ \lambda \in \mathbb{R}^n\}. \tag{18.15}$$

For a bounded linear operator A on $\mathcal{H}$ one can prove that $A \in \mathfrak{A}$ if and only if the operator commutes with every operator from the set

$$\{T_{-\lambda} \otimes T_\lambda,\ \lambda \in \mathbb{R}^n\} \cup \{e_\lambda \otimes I,\ \lambda \in \mathbb{R}^n\}. \tag{18.16}$$

An unbounded operator A in $\mathcal{H}$ is said to be *associated* with $\mathfrak{A}$ (in which case we write $A\eta\mathfrak{A}$) if it commutes with every operator from the set (18.16). In the case of a self-adjoint operator this is so if and only if either $(A - \lambda I)^{-1} \in \mathfrak{A}$ for every $\lambda \notin \sigma(A)$ or for one such λ, or $E_\lambda \in \mathfrak{A}$ for every spectral projection E_λ of A.

Coburn, Moyer, and Singer (1973) were the first to use $\mathfrak{A}$ in the theory of almost periodic operators in order to construct the index theory of such operators. The algebra was obtained as a special case of the general construction due to Murray and von Neumann (1936), which is applicable to any dynamical system. The basic fact we need is that $\mathfrak{A}$ is a II_∞*-factor*. This means that the *exact normal semi-finite trace* $\mathrm{Tr}_B : \mathfrak{A}^+ \to [0, +\infty]$, unique up to a scalar factor, exists in $\mathfrak{A}^+ = \{A : A \in \mathfrak{A}, A \geq 0\}$ and takes all the values from $[0, +\infty]$ on the set $\mathrm{Proj}\,(\mathfrak{A}) = \{P : P \in \mathfrak{A},\ P^2 = P = P^*\}$ consisting of all the orthogonal projections that belong to $\mathfrak{A}$. We recall (see Dixmier 1969) that, by definition, this trace has the following properties:

a) $\mathrm{Tr}_B(\lambda_1 A_1 + \lambda_2 A_2) = \lambda_1 \mathrm{Tr}_B A_1 + \lambda_2 \mathrm{Tr}_B A_2,\ \ A_j \in \mathfrak{A}^+,\ \ \lambda_j \geq 0$;
b) $\mathrm{Tr}_B(A^*A) = \mathrm{Tr}_B(AA^*)$ for any $A \in \mathfrak{A}$;
c) $A \in \mathfrak{A}^+,\ \mathrm{Tr}_B A = 0 \Longrightarrow A = 0$ (*exactness*);
d) if $A_\gamma \in \mathfrak{A}^+$ and $A_\gamma \nearrow A$ (i.e., A_γ is a monotone directed set of operators from $\mathfrak{A}^+$ strongly convergent to A), then $\mathrm{Tr}_B A_\gamma \to \mathrm{Tr}_B A$ (*the property of being normal*);
e) for any $A \in \mathfrak{A}^+$ we have $\mathrm{Tr}_B A = \sup\limits_{\substack{C \in \mathfrak{A}^+ \\ C \leq A \\ \mathrm{Tr}_B C < +\infty}} \mathrm{Tr}_B C$ (*semi-finiteness*).

To fix the normalization we define the trace of any operator of the form $I \otimes a(D)$, where $0 \leq a \in L_\infty(\mathbb{R}^n)$, by the formula

$$\mathrm{Tr}_B(I \otimes a(D)) = (2\pi)^{-n} \int\limits_{\mathbb{R}^n} a(\xi)\, d\xi \tag{18.17}$$

(one can prove that this is possible). The trace Tr_B is now uniquely defined.

The trace Tr_B can be extended by linearity to the ideal $\mathfrak{S}_1(\mathfrak{A})$ of $\mathfrak{A}$ consisting of $A \in \mathfrak{A}$ such that $\mathrm{Tr}_B|A| < +\infty$. For a projection $P \in \mathrm{Proj}(\mathfrak{A})$ the trace $\mathrm{Tr}_B P$ can be regarded as the generalized dimension $\dim_B(P\mathcal{H})$ of the space $P\mathcal{H}$. The above-mentioned properties of the spectrum imply that the generalized dimension has the same properties as the standard one, except that it is defined only for those subspaces $L \in \mathcal{H}$ for which the corresponding orthogonal projection P_L belongs to $\mathfrak{A}$. It follows that $\dim_B L = \mathrm{Tr}_B P_L$.

Next, let A be an almost periodic self-adjoint elliptic operator in $\mathbb{R}^n$. We introduce an operator $A^\#$ in $\mathcal{H}$ by the formula

$$A^{\#}u(x,y) = a(x+y, D_y)u(x,y).$$

(The elements of $\mathcal{H}$ are classes of functions $u = u(x,y)$ of two variables $x \in \mathbb{R}^n_B$ and $y \in \mathbb{R}^n$, the translation $a(x+y, D_y)$ being understood in the sense of extending the coefficients of A to $\mathbb{R}^n_B$ by periodicity.) It follows that, for every fixed $x \in \mathbb{R}^n_B$, $A^{\#}$ is the direct integral of all those operators $A_x = a(x+y, D_y)$ that are almost periodic elliptic operators in $L_2(\mathbb{R}^n)$. It is easily seen that $\sigma(A_x) = \sigma(A)$ for any $x \in \mathbb{R}^n_B$, which implies that $\sigma(A^{\#}) = \sigma(A)$.

Using the invariance of the trace (that is, the property that $\mathrm{Tr}_B(V^{-1}AV) = \mathrm{Tr}_B A$ for any $A \in \mathfrak{S}_1(\mathfrak{A})$ and any invertible $V \in \mathfrak{A}$), one can obtain explicit trace formulae for many important classes of operators. Namely, let A be a given operator in $L_2(\mathbb{R}^n)$ with kernel $K_A = K_A(x,y)$ such that $K_A(x+z, y+z)$ is a continuous function of x, y with values in CAP $(\mathbb{R}^n_z)$ and

$$|K_A(x,y)| \le C(1+|x-y|)^{-N}$$

for some $N > n$ (in particular, almost periodic pseudodifferential operators of order $m < -n$ have this property). We introduce the operator

$$A^{\#}u(x,y) = \int K_A(x+y, x+t)u(x,t)\,dt$$

(as in the case of a differential operator A, this is the direct integral of $A_x = T_{-x}AT_x$). Since $(I \otimes T_{-x})A^{\#}(I \otimes T_x) = (A_x)^{\#}$, it follows that $\mathrm{Tr}_B(A_x)^{\#} = \mathrm{Tr}_B A^{\#}$. Taking the mean value with respect to x, we find that $\mathrm{Tr}_B A^{\#} = \mathrm{Tr}_B A^{\#}_{av}$, where $A_{av} = M_x\{A_x\}$ is the operator with kernel $K_{A_{av}}(y,t) = M_x\{K_A(x+y, x+t)\}$ depending only on $y - t$, so that it is a convolution operator, and, consequently, $A^{\#}_{av}$ has the form $I \otimes a(D)$. Using formula (18.17), one can now easily prove that

$$\mathrm{Tr}_B A^{\#} = M_x\{K_A(x,x)\}. \tag{18.18}$$

For an almost periodic pseudodifferential operator $A = a(x, D_x)$ of order $m < -n$ this yields

$$\begin{aligned}\mathrm{Tr}_B A^{\#} &= (2\pi)^{-n} \int M_x\{a(x,\xi)\}\,d\xi = \\ &= (2\pi)^{-n} \int\limits_{\mathbb{R}^n_B \times \mathbb{R}^n} a(x,\xi)\,d\mu_n(x)\,d\xi,\end{aligned} \tag{18.19}$$

where $d\mu_n(x)$ is the Haar measure on $\mathbb{R}^n_B$.

Let $\{\tilde{E}_\lambda\}$ be the family of spectral projections of $A^{\#}$. We remark that, in general, $\tilde{E}_\lambda \ne (E_\lambda)^{\#}$ for any spectral projection E_λ of A. Moreover, $(E_\lambda)^{\#}$ may be undefined, while it is clear that $\tilde{E}_\lambda \in \mathfrak{A}$, since $\mathfrak{A}$ is weakly closed. It turns out (Shubin 1978b) that

$$N(\lambda) = \mathrm{Tr}_B \tilde{E}_\lambda, \quad \lambda \in \mathbb{R}. \tag{18.20}$$

To prove this equality it suffices to go over to the Laplace transform and use formulae (18.14) and (18.18) as well as the fact that $(e^{-tA})^{\#} = e^{-tA^{\#}}$. In particular, it follows from (18.20) that $\sigma(A)$ is equal to the set of growth points of $N(\lambda)$.

In the case of a positive almost periodic elliptic operator A (i.e., in the case when $\sigma(A) \subset (0, +\infty)$) the trace Tr_B can be used to define the *ζ-function*

$$\zeta_A(z) = \mathrm{Tr}_B(A^z)^{\#}. \tag{18.21}$$

Using the theory of complex powers of elliptic operators (see e.g. Egorov and Shubin 1988b, Sect. 1.7), one can prove that the right-hand side is defined and finite if $\operatorname{Re} z < -n/m$ (here m is the order of A) and the following analogue of the Seeley theorem on the meromorphic continuation of the ζ-function is valid: ζ_A admits a meromorphic continuation to the whole complex plane $\mathbb{C}$ with possible simple poles at the points $z_j = (j-n)/m, \;\; j = 0, 1, 2, \ldots$ only (excluding the points $z = 0, 1, 2, \ldots,$ at which there are no poles). The residues of the poles and the values at $z = 0, 1, 2, \ldots$ can be expressed in terms of the symbol of A by the standard formulae, except that the integrals with respect to x must be replaced by the mean values (Shubin 1976b). In particular, by Ikehara's Tauberian theorem, this yields the following asymptotics of $N(\lambda)$ as $\lambda \to +\infty$:

$$\begin{gathered} N(\lambda) = N_0(\lambda)(1 + o(1)), \\ N_0(\lambda) = (2\pi)^{-n} M_x \{ \operatorname{mes} \{ \xi : a_m(x, \xi) < \lambda \} \}, \end{gathered} \tag{18.22}$$

where mes denotes the Lebesgue measure in $\mathbb{R}^n_\xi$. By the hyperbolic equation method, one can refine the estimate of the remainder to obtain Hörmander's estimate (Kiselev 1978) with $o(1)$ replaced by $O(\lambda^{-1/m})$. In the case of the Schrödinger operator $A = -\Delta + q(x)$ with $q \in \mathrm{CAP}\,(\mathbb{R}^n)$ a still better asymptotics is possible (Shubin 1979):

$$N(\lambda) = (2\pi)^{-n} v_n \lambda^{n/2} (1 + O(\lambda^{-1})) \tag{18.23}$$

(here v_n is the volume of the unit sphere in $\mathbb{R}^n$). This asymptotics can be obtained by letting $k \to +\infty$ in the inequalities

$$N^0_{V_k}(\lambda - M) \le N_{V_k}(\lambda) \le N^0_{V_k}(\lambda + M)$$

divided by $\operatorname{mes} V_k$ (here $M = \sup_x |q(x)|$ and N_V and N^0_V are the distribution functions of the discrete spectra of A and $A_0 = -\Delta$ in a domain $V \subset \mathbb{R}^n$), followed by using formula (17.33), which gives $N(\lambda)$ for A_0.

For the one-dimensional Schrödinger operator $A = -d^2/dx^2 + q(x)$ with $q \in \mathrm{CAP}^\infty(\mathbb{R})$ Savin (1988) obtained the complete asymptotic expansion (17.25) for $N(\lambda)$, the coefficients of which can be obtained from q by means of the same formulae with

$$\frac{1}{a} \int\limits_0^a f(x)\, dx$$

replaced by $M_x\{f(x)\}$. Incidentally, we remark that the rotation number $w(\lambda)$ of the one-dimensional Schrödinger operator with an almost periodic potential can also be defined by (17.13) and is connected with $N(\lambda)$ by the same formula (17.19) as in the periodic case (Johnson and Moser 1982).

There is a variational principle for $N(\lambda)$ similar to the ordinary Glazman lemma (cf. Sect. 1.11):

$$N(\lambda) = \sup_{\substack{P\in \mathrm{Proj}\,(\mathfrak{A}) \\ P(A^{\#}-\lambda I)P\leq 0}} \mathrm{Tr}_B P. \tag{18.24}$$

This can be used in the usual way to study $N(\lambda)$ and, in particular, to obtain the asymptotics of $N(\lambda)$ as $\lambda \to +\infty$. For example, with the aid of the variational principle, using a suitable modification of the approximate spectral projection method (see §15), Bezyaev (1978) found the asymptotics of $N(\lambda)$ as $\lambda \to +\infty$ with an estimate of the remainder for almost periodic hypoelliptic operators (in this case $N(\lambda)$ is defined by (18.20)).

In some cases one can find the asymptotics of $N(\lambda)$ as $\lambda \to \lambda_0 = \inf \sigma(A)$ as well as that of $N_V(\lambda)$ as V is inflated in a suitable way and, simultaneously, $\lambda \to +\infty$ at various rates (the asymptotics of $N(\lambda)$ as $\lambda \to \infty$ is then a special case, in which the diameter of V increases much faster than λ). Such asymptotic formulae are connected with averaging problems and were studied by Kozlov (1982), who considered a uniformly elliptic operator (of order $2m$) of the divergent form

$$A = \sum_{|\alpha|,|\beta|\leq m} D^\alpha a_{\alpha\beta}(x) D^\beta \tag{18.25}$$

with coefficients $a_{\alpha\beta} = \bar{a}_{\beta\alpha} \in \mathrm{CAP}\,(\mathbb{R}^n)$ and with the Dirichlet boundary conditions. Suppose that V is a bounded domain in $\mathbb{R}^n$ with smooth boundary, $V_l = lV$, $\lambda_j^{(l)}$ are the eigenvalues of A in V_l with the Dirichlet boundary conditions, and

$$N_l(\lambda) = \frac{\#\{j : \lambda_j^{(l)} \leq \lambda\}}{\mathrm{mes}\, V_l}.$$

We already know that $N_l(\lambda) \to N(\lambda)$ as $l \to +\infty$. Now let us set

$$N_l(\rho, \delta; \lambda) = l^\delta N_l(l^\rho \lambda)$$

and consider the limits of this function as $l \to +\infty$ for various ρ (δ is chosen depending on ρ so as to ensure that the limit is finite and different from zero). Firstly, it turns out that

$$\lim_{l\to+\infty} l^{-n\rho/2m} N_l(l^\rho\lambda) = N_0(\lambda) \quad \text{for } \rho > 0, \tag{18.26}$$

where $N_0(\lambda)$ is defined in terms of the principal symbol (of order $2m$) as in (18.22). For $\rho < 0$ one must assume, in addition, that A has no lower-order terms (i.e., $a_{\alpha\beta} \equiv 0$ for $|\alpha| + |\beta| < 2m$) and is strictly positive in the sense that

$$\sum_{|\alpha|=|\beta|=m} a_{\alpha\beta}\lambda_\alpha\bar{\lambda}_\beta \geq \varepsilon \sum_{|\alpha|=m} |\lambda_\alpha|^2 \tag{18.27}$$

for some $\varepsilon > 0$ and any set of numbers $\{\lambda_\alpha : |\alpha| \leq m\}$. Then it turns out that

$$\lim_{l\to\infty} N_l(\rho, \delta; \lambda) = 0 \quad \text{for } \rho < -2m \text{ and any } \delta, \tag{18.28}$$

$$\lim_{l\to\infty} N_l(-2m, n; \lambda) = \hat{N}_V(\lambda), \tag{18.29}$$

where $\hat{N}_V$ is the distribution function of the eigenvalues of the Dirichlet problem in V for a special operator $\hat{A} = \hat{a}(D)$ with constant coefficients called the *averaged operator*. Moreover, we set

$$\hat{N}(\lambda) = (2\pi)^{-n}\text{mes}\left\{\xi : \hat{a}(\xi) \leq \lambda\right\}.$$

The following asymptotic formula is then valid for the limiting function $N(\lambda)$ (also subject to the condition (18.27) and in the absence of lower-order terms):

$$N(\lambda) = \hat{N}(\lambda) + o\big(\lambda^{n/2m}\big), \quad \lambda \to +0. \tag{18.30}$$

Finally, if $m = 1$ and there are no lower-order terms (in this case (18.27) is a consequence of uniform ellipticity), then

$$\lim_{l\to\infty} N_l(\rho, -\rho n/2; \lambda) = \hat{N}(\lambda), \quad -2 < \rho < 0. \tag{18.31}$$

We also remark that, under the assumptions of Theorem 18.2, each of the asymptotic formulae (18.26) and (18.28)–(18.31) just described is valid for the Schrödinger operator $A = -\Delta + q(x)$ (with a sufficiently smooth quasiperiodic potential under the Diophantine condition for the frequency) if A is replaced by $A - \lambda_0 I$, where $\lambda_0 = \inf \sigma(A)$, since, by Theorem 18.2, A can be reduced to the required divergent form.

In addition, we shall turn our attention to the properties of $N(\lambda)$ for the one-dimensional Schrödinger operator. First of all, it is easily seen that $N(\lambda)$ is continuous in this case. The existence of a discontinuity at λ_0 would mean that $\lambda_0 \in \sigma_p(\tilde{A})$ for almost every $\tilde{A} \in H(A)$, which, as we have already mentioned, is impossible (we remark that $N(\lambda)$ is even Hölder continuous with index $1/2$ in the periodic case (Avron and Simon 1982)).

Theorem 18.8 (Johnson and Moser 1982). *If $A = -d^2/dx^2 + q$, where $q \in \mathrm{CAP}(\mathbb{R})$ and $\lambda \in \mathbb{R} \setminus \sigma(A)$, then $N(\lambda) \in \Omega_q$, where Ω_q is the frequency module of q.*

The theorem is important because it makes it possible to label the gaps of the spectrum (which can be a Cantor set!) by the elements of the frequency module. Elliott (1982, 1986) proposed another proof based on the theory of C^*-algebras.[9]

Finally, of interest is the relationship between $N(\lambda)$ and the Lyapunov exponent $\gamma(\lambda)$ defined for every operator $\tilde{A} \in H(A)$ by

$$\gamma(\lambda) = \lim_{L\to\infty} |L|^{-1} \ln \|T_L\|, \tag{18.32}$$

where T_L is a linear operator in $\mathbb{R}^2$ mapping $(\psi(0), \psi'(0))$ into $(\psi(L), \psi'(L))$, ψ being a solution of the equation $\tilde{A}\psi = \lambda\psi$. For any fixed $\lambda \in \mathbb{R}$ the limit in (18.32) exists for almost all $\tilde{A} \in H(A)$ and is independent of $\tilde{A}$ (this follows from Oseledets' multiplicative ergodic theorem (Oseledets 1968)). It is easily seen that $\gamma(\lambda) \geq 0$, since $\det T_L = 1$. Besides, if $\lambda \notin \sigma(A)$, then $\gamma(\lambda) > 0$, because in this case $\lambda \notin \sigma(\tilde{A})$ for any $\tilde{A} \in H(A)$, and the equation $\tilde{A}\psi = \lambda\psi$ has an exponentially increasing solution. However, it is also possible that $\gamma(\lambda) > 0$ for $\lambda \in \sigma(A)$. This means that for almost all $\tilde{A} \in H(A)$ every solution ψ of the equation $\tilde{A}\psi = \lambda\psi$ either increases or decreases exponentially as $x \to +\infty$ (the same is true for $x \to -\infty$). The set $\sigma(A) \cap \{\lambda : \gamma(\lambda) > 0\}$ can therefore be expected to consist of the point spectrum and the singular continuous spectrum. Indeed, this turns out to be the case to within a set of measure zero: if $\psi \in \mathcal{H}_{\mathrm{ac}}$, then $\{\lambda : \gamma(\lambda) > 0\}$ is a set of measure zero relative to the absolutely continuous spectral measure $d(E_\gamma\psi, \psi)$ (Ishii 1973; Pastur 1974). This result was substantially refined by Kotani (1982), who demonstrated that $\{\lambda : \gamma(\lambda) = 0\}$ is the essential support of the absolutely continuous spectrum $\sigma_{\mathrm{ac}}(A)$, i.e., it coincides with it to within a set of Lebesgue measure zero.

The relationship between $N(\lambda)$ and $\gamma(\lambda)$ is given by the Thouless formula (Thouless 1972; for the proof see Avron and Simon 1983)

$$\gamma(\lambda) = \gamma_0(\lambda) + \int_{-\infty}^{\infty} \ln|\lambda - \lambda'|\, d\big[N(\lambda') - N_0(\lambda')\big], \tag{18.33}$$

where $\gamma_0(\lambda) = \big[\max(0, -\lambda)\big]^{1/2}$ and $N_0(\lambda) = \pi^{-1}\big[\max(0, \lambda)\big]^{1/2}$ correspond to the case $q \equiv 0$. Formula (18.33), in which the integral is to be understood as the principal value, is valid for almost all λ (with respect to the Lebesgue measure in $\mathbb{R}$). Introducing the function

$$f(z) = \sqrt{-z} + \int_{-\infty}^{\infty} \ln(z - \lambda')\, d\big[N(\lambda') - N_0(\lambda')\big],$$

which is analytic on the complex half-plane $\{z : \operatorname{Im} z > 0\}$, we observe that $\operatorname{Im} f(z)$ tends to the limit $\operatorname{Im} f(\lambda + i0) = \pi N(\lambda)$ on the real axis, and, by (18.33), $\operatorname{Re} f(\lambda + i0) = \gamma(\lambda)$. It follows that $\gamma(\lambda) + i\pi N(\lambda)$ is the limit of

[9] (Added in the English edition.) An extensive theory of gap labelling evolved from these results; see e.g. Bellissard et al. (1985), Bellissard (1989, 1992), and references therein.

an analytic function defined on the upper half-plane, so that, in principle, $\pi N(\lambda)$ and $\gamma(\lambda)$ can be obtained from one another by the Hilbert transform. Following the review (Simon 1982), the origin of this fact can be elucidated by observing that if $\psi(x) \sim e^{i\alpha x}$ at infinity, then the Lyapunov exponent $\gamma(\lambda)$ is connected with $\operatorname{Im}\alpha$, while the rotation number, which is proportional to $N(\lambda)$, is connected with $\operatorname{Re}\alpha$.

Let us also mention the following inequality due to Deift and Simon (1983), which is valid for the one-dimensional Schrödinger operator:

$$\frac{d[N^2(\lambda)]}{d\lambda} \geq \frac{1}{\pi^2} \quad \text{for almost all } \lambda \text{ such that } \gamma(\lambda) = 0, \tag{18.34}$$

where, for any monotone function $f = f(\lambda)$, the derivative $df/d\lambda$ is understood as the limit

$$\lim_{\varepsilon \to +0} \frac{f(\lambda + \varepsilon) - f(\lambda - \varepsilon)}{2\varepsilon},$$

which exists for almost all λ. We remark that (18.34) becomes an exact equality for the zero potential. In particular, it follows from (18.34) that if $M = \{\lambda : \gamma(\lambda) = 0\}$, then

$$N^2(b) - N^2(a) \geq \pi^{-2}\mathrm{mes}\big[M \cap (a,b)\big]$$

for any interval $(a, b) \in \mathbb{R}$. Kotani's inequality

$$4\pi w(\lambda)\gamma(\lambda) \geq \operatorname{Im}\lambda, \tag{18.35}$$

where $w(\lambda)$ is the rotation number, is satisfied for $\operatorname{Im}\lambda \geq 0$ (Kotani 1982).

On other aspects of the spectral theory of almost periodic elliptic operators see also the surveys by Shubin (1979), Simon (1982), Johnson (1983), and Pastur (1987b), as well as the books by Cycon et al. (1987), Carmona and Lacroix (1990), and Figotin and Pastur (1992).

§19 Operators with Random Coefficients

Many publications have been devoted to operators with random coefficients. Among them are the general surveys (Pastur 1973, 1987a, 1987b; Carmona 1985, 1986; Kirsch 1985; Martinelli and Scoppola 1986; Souillard 1986; Spencer 1986).[10] In the present section we do not attempt to give a complete account of all results in this area, and we confine ourselves only to the most

[10](Added in the English edition.) See also the books by Carmona and Lacroix (1990) and Figotin and Pastur (1992).

characteristic and mainly many-dimensional results, referring the reader for more details to the above-mentioned surveys and references therein.

19.1. Translation Homogeneous Random Fields

We shall briefly explain the physical context which gives rise to random fields and operators (see Ziman 1979; Gredeskul et al. 1982). Random functions, fields, or operators arise from the desire to describe the behaviour of the corresponding objects, which are known inaccurately and can be expected to exhibit a statistically stable behaviour. A rough surface in $\mathbb{R}^{n+1}$, which can be described as a graph $y = f(x)$ with $x \in \mathbb{R}^n$, serves as a typical example. Failing to know f accurately, we can assume that it depends on a random parameter $\omega \in \Omega$, where Ω is a *probability space* (a set with a σ-algebra $\mathcal{F}$ of subsets and a probability measure μ on $\mathcal{F}$). Thus $f = f(\omega, x)$. So far f cannot have any additional properties as compared to the deterministic situation, since, in particular, one can take Ω consisting of a single point. However, we shall now impose the following *translation homogeneity* condition: we assume that all the probability characteristics of the surface remain unchanged under any translation on the x-plane. If we consider a large piece of the surface and the irregularities are small, then this assumption is natural. Mathematically, it means that for any $\omega \in \Omega$ and $z \in \mathbb{R}^n$ the shifted function $x \mapsto f(\omega, x+z)$ belongs the same family f, but now with a different random parameter ω, which we denote by $T_z\omega$, the given probability measure μ being preserved under the transformation $T_z : \Omega \to \Omega$ (this means precisely that the probability characteristics are preserved: the measure of any set of random functions f must be the same as that of the set obtained by shifting all the functions by z).

We observe that the translations by z acting on functions form a group isomorphic to $\mathbb{R}^n$. A natural requirement is that the transformations T_z should have the same group property. This will be satisfied automatically if there are as many functions $f(\omega, \cdot\,)$ as 'labels' ω, that is, if for any two different elements ω the corresponding functions on $\mathbb{R}^n$ are different. It is often assumed that the probability measure is defined directly in the space of functions, i.e., Ω is a subset of the space of functions on $\mathbb{R}^n$. This is, however, unsuitable for our purposes, since it makes it necessary to change Ω all the time (for example, when going over from f to $f+1$ or f^2). Finally, f can be assumed to be a vector-valued function (with values in $\mathbb{C}^N$).

This motivation gives rise to the following definition.

Definition 19.1. A system consisting of the following objects is called a *homogeneous random field* (with values in $\mathbb{C}^N$):

a) a probability space $(\Omega, \mathcal{F}, \mu)$ and a dynamical system defined on it with n-dimensional time $\{T_z, z \in \mathbb{R}^n\}$; here $T_z : \Omega \to \Omega$, the family $\{T_z\}$ is *measurable* in the sense that that the transformation $\Omega \times \mathbb{R}^n \to \Omega$ assigning

$T_z\omega$ to (ω, z) is measurable (the σ-algebra of Borel measurable sets should be considered on $\mathbb{R}^n$), and the group properties $T_0 = \mathrm{id}$ and $T_{z+w} = T_z T_w$ are satisfied for all $z, w \in \mathbb{R}^n$;

b) a measurable function $f : \Omega \times \mathbb{R}^n \to \mathbb{C}^N$ such that

$$f(\omega, x+z) = f(T_z\omega, x) \tag{19.1}$$

for almost all $\omega \in \Omega$ and all $x, z \in \mathbb{R}^n$, that is, for all $\omega \in \Omega'$ and $x, z \in \mathbb{R}^n$, where Ω' is a subset of full measure in Ω.

Fixing $\omega \in \Omega$, we obtain a function $f(\omega, \cdot\,)$ on $\mathbb{R}^n$ called a *realization* of the given random field. The realizations of a homogeneous random field can no longer be arbitrary. They almost surely, i.e., for almost all ω, satisfy a number of properties which follow from homogeneity. For example, if Ω consists of a single point ω_0, then homogeneity implies that $f(\omega_0, x) = \mathrm{const}$. Furthermore, if $f(\,\cdot\,, 0) \in L_1(\Omega)$ (for example, this is the case when f is bounded), then, by Birkhoff's ergodic theorem (see, for example, Cornfel'd et al. 1980), the mean value

$$\bar{f}(x) = M_x\{f(\omega, x)\} = \lim_{R\to\infty} \frac{1}{R^n} \int\limits_{|x_j|\le R/2} f(\omega, x)\, dx, \tag{19.2}$$

which is analogous to the mean value of an almost periodic function (see (18.1)), exists for almost all ω. Being a random variable (a function on Ω), this mean value is invariant under the translations $\{T_z,\ z \in \mathbb{R}^n\}$, that is,

$$\bar{f}(T_z\omega) = \bar{f}(\omega) \tag{19.3}$$

for almost all ω and all $z \in \mathbb{R}^n$. An important and frequently encountered case is that of an ergodic dynamical system $\{T_z,\ z \in \mathbb{R}^n\}$, i.e., such that every invariant measurable function on Ω is almost surely constant. In this case the field itself is often called *ergodic* or *metrically transitive*. This being so,

$$\bar{f}(\omega) = \mathrm{const} = \int\limits_{\Omega} f(\omega, 0)\, d\mu(\omega) = \int\limits_{\Omega} f(\omega, x)\, d\mu(\omega) \tag{19.4}$$

for almost all ω (the right-hand side is independent of $x \in \mathbb{R}^n$), i.e., $\bar{f}(\omega)$ is the mathematical expectation of the random variable $f(\,\cdot\,, 0)$. In the general (non-ergodic) case $\bar{f}(\omega)$ is the conditional expectation of $f(\,\cdot\,, 0)$ given the σ-algebra of T-invariant sets. In the case when $f(\,\cdot\,, 0) \in L_2(\Omega)$ this means that $\bar{f}(\,\cdot\,) = Pf(\,\cdot\,, 0)$, where P is the orthogonal projection onto the space of T-invariant functions in $L_2(\Omega)$. This formula is also meaningful in the general case, since P can be extended to a continuous linear operator $P : L_1(\Omega) \to L_1(\Omega)$.

Let us present examples of homogeneous random fields.

Example 19.1. Let Γ be a lattice in $\mathbb{R}^n$ (see §17), let the torus $\Omega = \mathbb{R}^n/\Gamma$ be equipped with the normalized Lebesgue measure $d\mu$ (so that $\mu(\Omega) = 1$), and let $\hat{g}$ be a measurable function on Ω. We set

$$f(\omega, x) = \hat{g}(\omega + x), \quad \omega \in \Omega,\ x \in \mathbb{R}^n, \tag{19.5}$$

where $\omega + x$ is understood in the natural way as the sum of ω and the class of x modulo Γ. It is easily seen that f is a homogeneous random field, all of whose realizations are Γ-periodic functions, which can be obtained from one another by translations.

Example 19.2. Let $g = g(x)$ be a quasiperiodic function on $\mathbb{R}^n$ (see §18) that can be expressed as the composition $g = \hat{g} \circ \pi$, where $\pi : \mathbb{R}^n \to \mathbb{T}^N$ is a continuous homomorphism from $\mathbb{R}^n$ to $\mathbb{T}^N = \mathbb{R}^N/\mathbb{Z}^N$ and $\hat{g}$ is a measurable function on $\mathbb{T}^N$. Let $\Omega = \mathbb{T}^N$ and let $d\mu$ be the Lebesgue measure on Ω. Then formula (19.5) defines a homogeneous random field if $\omega + x$ is understood as the translation of ω by x given the natural action of $\mathbb{R}^n$ on Ω defined by π. The quasiperiodic function g is then a realization of the random field f. All the remaining realizations form the hull $H(g)$ of g.

Example 19.3. Let $\Omega = \mathbb{R}^n_B$ be the Bohr compact of $\mathbb{R}^n$ with the Haar measure $d\mu$ (see §18). The embedding $\mathbb{R}^n \subset \mathbb{R}^n_B$ defines the action of $\mathbb{R}^n$ on $\mathbb{R}^n_B$ by translations. For any almost periodic function $g \in \mathrm{CAP}\,(\mathbb{R}^n) = C(\mathbb{R}^n)$ we can introduce a homogeneous random field f by (19.5). Then g will be a realization of f, while all the remaining realizations will belong to the hull of g.

It follows that periodic and almost periodic functions are realizations of homogeneous random fields (although of a rather special form). Below we present other examples of such fields.

Example 19.4. Let $\{\xi_z(\omega),\ z \in \mathbb{Z}^n\}$ be identically distributed independent random variables, let $\omega \in \Omega_0$, where Ω_0 is a probability space with measure $d\mu_0$, and let $\chi = \chi(x)$ be a measurable function on $\mathbb{R}^n$ with support in the cube $[0,1]^n$. We set

$$f(\omega, x) = \sum_{z \in \mathbb{Z}^n} \xi_x(\omega)\chi(x - z). \tag{19.6}$$

The space Ω_0 can be assumed to be the direct product

$$\Omega_0 = \prod_{x \in \mathbb{Z}^n} \Omega_{1z}$$

of identical probability spaces Ω_1 with measures $d\mu_1(\omega)$ and with (the same) random variable $\xi_z(\omega_1)$, $\omega_1 \in \Omega_1$ defined on each of them, so that $\xi_z(\omega)$ can be obtained by lifting $\xi_z(\omega_1)$ to Ω_0 with the aid of the natural projection $\pi_z : \Omega_0 \to \Omega_{1z}$. It proves convenient to think of Ω_0 as the set of functions on $\mathbb{Z}^n$ with values in Ω_1. Given this interpretation, there are natural translations $T_z : \Omega_0 \to \Omega_0$ for $z \in \mathbb{Z}^n$ transforming any function $\omega : \mathbb{Z}^n \to \Omega_1$ of this kind into the function $(T_z\omega)(z') = \omega(z' + z)$. The invariance property (19.1) under translations by vectors with integral coordinates follows from the fact that the random variables $\xi_z(\omega)$ have the same distribution, which is sufficient for the majority of applications (for example for the existence of the mean value (19.2) in the case when $\xi_0(\,\cdot\,) \in L_1(\Omega_1)$ and χ is a bounded function). Besides, if required, homogeneity with respect to all translations can be achieved by means of the Smale suspension (see Anosov et al. 1985) while leaving the given random field essentially unchanged. Namely, we replace Ω_0 by $\Omega = \Omega_0 \times K_1$, where $K_1 = \{x = (x_1, \dots, x_n),\ 0 \le x_j < 1,\ j = 1, \dots, n\}$ is the unit cube with the Lebesgue measure defined on it, and set

$$\hat{f}(\omega, x) = \hat{f}((\omega_0, t), x) = f(\omega_0, x + t)$$

for $\omega = (\omega_0, t)$, $\omega_0 \in \Omega_0$, and $t \in K_1$. It remains to set $T_y\omega = (T_{z+z_0}\omega_0, t')$ for $y \in \mathbb{R}^n$, $y = z + \tau$, $z \in \mathbb{Z}^n$, $\tau \in K_1$, and $\omega = (\omega_0, t)$, where $t + \tau = z_0 + t'$ for $z_0 \in \mathbb{Z}^n$ and $t' \in K_1$. This defines a dynamical system on Ω (with n-dimensional time), with respect to which $\hat{f}$ is a homogeneous random field.

Remark 19.1. Example 19.4 has a more natural discrete version. The notion of a homogeneous random field makes sense in the case of a discrete argument $z \in \mathbb{Z}^n$, provided $\mathbb{R}^n$ is replaced by $\mathbb{Z}^n$ everywhere in Definition 19.1. In this case any system $\{\xi_z(\omega),\ z \in \mathbb{Z}^n\}$ of identically distributed independent random variables can serve as an example of a homogeneous random field (one should set $f(\omega, z) = \xi_z(\omega)$).

Example 19.5. Let $\varphi = \varphi(x)$ be a fixed bounded measurable function with compact support in $\mathbb{R}^n$. We set

$$f(\omega, x) = \sum_i \varphi(x - x_i), \tag{19.7}$$

where the points $\{x_i\}$ form a random set with *Poisson distribution* relative to the Lebesgue measure in $\mathbb{R}^n$, which means that the probability that N of these points belong to a set $X \subset \mathbb{R}^n$ of finite Lebesgue measure $\operatorname{mes} X$ can be expressed as

$$P\{\operatorname{card}(\{x_i\} \cap X) = N\} = \frac{e^{-c \operatorname{mes} X}(c \operatorname{mes} X)^N}{N!}, \tag{19.8}$$

where the constant c can be interpreted as the mean concentration of the points x_i (the average number of points x_i in X is equal to $c \operatorname{mes} X$). It must also be assumed that if X and Y are disjoint subsets of $\mathbb{R}^n$, then the distributions of points in X and Y are independent. It is easily verified that the points $\{x_i\}$ have the same distribution (in the probabilistic sense) in X and $X + a$ for any subset $X \subset \mathbb{R}^n$ and translation vector $a \in \mathbb{R}^n$. In this situation a field f of the form (19.7) is called a *Poisson-generated field* or simply a *Poisson field* (with potential φ). We remark that, in general, $f(\omega, x)$ may turn out to be undefined for some x (or even for all x), because the sum in (19.7) has infinitely many terms. The probability of such a distribution of $\{x_i\}$ is, however, equal to zero in view of (19.8).

Example 19.6. In the case when $n = 1$ homogeneous random fields are usually called *stationary stochastic processes*. The theory of such processes (see, for example, Cramer and Leadbetter 1967) provides numerous examples of one-dimensional homogeneous random fields. Among them there are *Markov processes* along with their discrete versions, i.e., *Markov chains* (see, for example, Venttsel' 1975), as well as *diffusion processes* (see Itô and McKean 1965; Venttsel' 1975). An important example of such a situation is connected with the operator $A = -\Delta + h$ on a compact Riemannian manifold M (without boundary), Δ being the Laplace-Beltrami operator on M and h a vector field on M. Let $p(t, x, y)$ be the kernel of the operator $\exp(-tA)$ (the fundamental solution of the Cauchy problem for the parabolic operator $\partial/\partial t - A$ on M). Then $p(t, x, y)$ can be regarded as the probability density of the transition from x to y in time t for a random walk $Q(\omega, t)$ on M (a Markov diffusion process on M). By the ergodic theorem, $p(t, x, y) \to \pi(y)$ as $t \to \infty$. We take $\pi(y)$ to be the probability density of the initial distribution. This gives rise to a probability distribution on arbitrary trajectories $Q : [0, +\infty) \to M$. Since π defines a stationary probability distribution, we obtain a probability measure on the set of paths $Q : \mathbb{R} \to M$ (i.e., paths $Q(t)$ defined for all t). Now let us set $f(t) = F(Q(t))$, where F is a (vector-valued) function on M (the dependence on ω is implicit). Then $f(t) = f(\omega, t)$ is a stationary stochastic process (or a homogeneous random field on $\mathbb{R}^1$), which we shall call a *diffusion-generated* process.

The finite-dimensional distributions

$$P\{\omega : f(\omega, x_1) \in B_1, \ldots, f(\omega, x_k) \in B_k\}, \tag{19.9}$$

where $x_1, \ldots, x_k \in \mathbb{R}^n$ and $B_1, \ldots, B_k$ are Borel subsets of $\mathbb{C}^N$ (we assume that f is a field with values in $\mathbb{C}^N$) are important objects characterizing homogeneous random fields. Formula (19.9) defines a probability measure on $(\mathbb{R}^n)^k$ depending on $x_1, \ldots, x_k$, which play the role of parameters. For various values of k these measures are consistent with one another in the obvious sense. Conversely, if this consistency condition is satisfied, then the well-known Kolmogorov theorem (see, for example, Venttsel' 1975) ensures that there is a random field with the given finite-dimensional distributions (the field being homogeneous if the distributions (19.9) are preserved under the translation of each of the points $x_1, \ldots, x_k$ by the same arbitrary vector $\tau \in \mathbb{R}^n$). Various classes of random fields can be defined in terms of finite-dimensional distributions. For example, a field is said to be *Gaussian* if all the measures (19.9) are Gaussian measures on the corresponding Euclidean spaces, i.e., have densities $\exp(-Q(y))$, where Q is a positive definite quadratic form.

19.2. Random Differential Operators

We shall now consider a *random differential operator*

$$A = A_\omega = a(\omega, x, D_x) = \sum_{|\alpha| \le m} a_\alpha(\omega, x) D_x^\alpha \tag{19.10}$$

whose coefficients $\{a_\alpha(\omega, x) : |\alpha| \le m\}$ form a homogeneous random field. For simplicity, we shall consider only those cases when the estimates

$$\left|\partial_x^\beta a_\alpha(\omega, x)\right| \le c_\beta \tag{19.11}$$

are satisfied for any multi-indices α and β with $|\alpha| \le m$ (the constant c_β being independent of ω), or A is the Schrödinger operator

$$A = A_\omega = -\Delta + q(\omega, x) \tag{19.12}$$

(in the latter case the potential q must be a homogeneous random field).

Along with the differential operators (19.10), it proves useful to consider *random pseudodifferential operators* $A = A_\omega = a(\omega, x, D_x)$ whose symbols satisfy the estimates

$$\left|\partial_\xi^\alpha \partial_x^\beta a(\omega, x, \xi)\right| \le c_{\alpha,\beta}(1 + |\xi|)^{m-|\alpha|}, \tag{19.13}$$

$a(\omega, x, \xi)$ being a homogeneous random field for every fixed $\xi \in \mathbb{R}^n$. We say that $A = a(\omega, x, D_x)$ is a *random elliptic operator* if it is uniformly elliptic with respect to $\omega \in \Omega$ and $x \in \mathbb{R}^n$ (this is automatically true for the Schrödinger operator (19.12)). In the ergodic case the operators themselves (as well as the fields formed by their coefficients or symbols) are often said to be *ergodic* or *metrically transitive*.

We shall consider the operators (19.10) or (19.12) as random operators in $L_2(\mathbb{R}^n)$ with domain $C_0^\infty(\mathbb{R}^n)$. In order that the Schrödinger operator (19.12) be defined on $C_0^\infty(\mathbb{R}^n)$ for almost all ω, we shall assume from now on that

$$q(\,\cdot\,, 0) \in L_2(\Omega). \tag{19.14}$$

The homogeneity condition can be written in the operator form

$$A_{T_z\omega} = U_z A_\omega U_{-z}, \tag{19.15}$$

where U_z is the translation operator by $z \in \mathbb{R}^n$, i.e., $(U_z f)(x) = f(x + z)$ for any $f \in L_2(\mathbb{R}^n)$. The equality (19.15) must be satisfied on $C_0^\infty(\mathbb{R}^n)$ for almost all $\omega \in \Omega$. The homogeneity condition (19.15) can be carried over to arbitrary functions of A such as, for example, polynomials of A (with constant coefficients), the resolvent of the closure of A, and the spectral projections (provided A is essentially self-adjoint). We remark that for an integral operator B with a continuous (in x, y) kernel $K_B = K_B(\omega, x, y)$ the property (19.15) means that

$$K_B(T_z\omega, x, y) = K_B(\omega, x+z, y+z) \tag{19.16}$$

for almost all ω and all $x, y, z \in \mathbb{R}^n$. In particular, the restriction $K_B(\omega, x, x)$ of the kernel to the diagonal is a homogeneous random field and the *random trace*

$$\mathrm{Tr}_R B = M_x\{K_B(\omega, x, x)\} \tag{19.17}$$

can be defined if $K_B(\,\cdot\,, 0, 0) \in L_1(\Omega)$. In the ergodic case the random trace is independent of ω. In (Fedosov and Shubin 1978a, 1978b) it was proved that it has properties similar to those of the ordinary trace. Namely, $\mathrm{Tr}_R B$ is linear in B and $\mathrm{Tr}_R B \geq 0$ for every non-negative operator B (in this case the equality $\mathrm{Tr}_R B = 0$ is satisfied only for $B = 0$). Furthermore, let A and B be two *homogeneous random Carleman operators*, i.e., let A and B be homogeneous random integral operators with kernels $K = K_A$ or K_B such that

$$\sup_{y,\omega} \int |K(\omega, x, y)|^2\, dx < \infty, \qquad \sup_{x,\omega} \int |K(\omega, x, y)|^2\, dy < \infty. \tag{19.18}$$

This being the case, if the operators AB and BA have kernels which are continuous in x, y (for almost all ω), then

$$\mathrm{Tr}_R(AB) = \mathrm{Tr}_R(BA). \tag{19.19}$$

The kernels of AB and BA are continuous, for example, if the estimates (19.18) are valid for both kernels K_A and K_B with K replaced by $\partial_x^\alpha \partial_y^\beta K$ for any multi-indices α and β. Formula (19.19) is also valid if B satisfies the specified estimates of the derivatives and A is merely an arbitrary homogeneous random pseudodifferential operator (with the estimates (19.13) of the symbol), or if A and B are two homogeneous random pseudodifferential operators of order m_1 and m_2 such that $m_1 + m_2 < -n$.

By the construction presented in Example 19.3, for almost periodic operators, which can be represented as random ones, Tr_R is equal to the trace Tr_B described in §18.

In what follows we shall also require that the dynamical system $\{T_z,\ z \in \mathbb{R}^n\}$ should define a *strongly continuous group of unitary operators* in $L_2(\Omega)$. This means that if V_z are the unitary operators introduced in $L_2(\Omega)$ by $V_z f(\omega) = f(T_z\omega)$, then the vector-valued function $z \mapsto V_z f$ on $\mathbb{R}^n$ with values in $L_2(\Omega)$ is continuous for any $f \in L_2(\Omega)$. We remark that this condition needs only to be verified for a dense set of functions f, and, by the well-known von Neumann theorem, it is automatically satisfied if $L_2(\Omega)$ is a separable space (Reed and Simon 1972, Theorem VIII.9), or if Ω is a compact topological space (with a Borel measure μ) and T_z is a continuous group of homeomorphisms of Ω. Subject to the above continuity condition, the action of a homogeneous random differential (or pseudodifferential) operator $A = a(\omega, x, D_x)$ can also be defined in $L_2(\Omega)$, which can be interpreted as the space of homogeneous random fields. Namely, we introduce the space $C_b^\infty(\Omega)$

consisting of $u \in L_2(\Omega)$ such that the function $x \mapsto v(\omega, x) = u(T_x\omega)$ has bounded derivatives of any order satisfying the inequalities $\left|\partial_x^\alpha u(T_x\omega)\right| \le c_\alpha$ with non-random constants c_α for any multi-index α and almost all ω. It can be demonstrated (Dedik and Shubin 1980) that $C_b^\infty(\Omega)$ is dense in $L_2(\Omega)$. Let us now set

$$(A_\Omega u)(\omega) = a(\omega, x, D_x)u(T_x\omega)\big|_{x=0} \tag{19.20}$$

for $u \in C_b^\infty(\Omega)$.

On identifying u with a homogeneous random field v, we can write (19.20) in the form

$$A_\Omega v(\omega, x) = a(\omega, x, D_x)v(\omega, x) \tag{19.20$'$}$$

(the right-hand side is a homogeneous random field, which is equal to the right-hand side of (19.20) for $x = 0$). If $a = a(\omega, x, \xi)$ admits the estimates (19.13), then A_Ω transforms the space $C_b^\infty(\Omega)$ into itself. If A is a random Schrödinger operator with a potential that satisfies (19.14), then A_Ω transforms $C_b^\infty(\Omega)$ into $L_2(\Omega)$. In both cases A_Ω can be regarded as an (unbounded) operator in $L_2(\Omega)$.

Remark 19.2. Along with random differential and pseudodifferential operators, one often considers *random difference operators*, i.e., operators A in $l^2(\mathbb{Z}^n)$ depending on a random parameter $\omega \in \Omega$, along with a dynamical system $\{T_z,\ z \in \mathbb{Z}^n\}$ acting on the probability space Ω such that the matrix elements $K_A(\omega, x, y)$, where $x, y \in \mathbb{Z}^n$, satisfy the homogeneity condition (19.16) for $z \in \mathbb{Z}^n$. As an example we mention the discrete Schrödinger operator (19.12), where Δ is to be understood as the Laplace operator on the lattice $\mathbb{Z}^n$ and q is a homogeneous random field in the sense of Remark 19.1. In particular, if $\{q(\,\cdot\,, x),\ x \in \mathbb{Z}^n\}$ is a system of identically distributed independent random variables, then we are dealing with the so-called *Anderson model*.

19.3. Essential Self-Adjointness and Spectra

To study the spectrum of a random elliptic operator in $L_2(\mathbb{R}^n)$ one must go over to the closure of that operator for every fixed ω. Let the given random operator A_ω be symmetric on $C_0^\infty(\mathbb{R}^n)$ for almost all ω. Then there arises the question of whether or not it is self-adjoint in $L_2(\mathbb{R}^n)$ for almost all ω. This question can be answered most easily if sufficient conditions for essential self-adjointness are satisfied for the non-random operator obtained for almost every fixed ω. For example, this is the case for random elliptic operators whose symbols satisfy the estimates (19.13), since any non-random uniformly elliptic formally self-adjoint operator with such estimates of the symbol is essentially self-adjoint (see, for example, Shubin 1975). The properties of almost all realizations of a random field can often be spelt out explicitly. It is, nevertheless, more convenient to state the conditions for essential self-adjointness in terms of the probabilistic characteristics of the coefficients. We present two results of this kind, which apply to the Schrödinger operator (19.12) in the ergodic case.

Theorem 19.1 (Figotin 1983). *Let p be the least integral number strictly greater than $n/2$ and let $q(\cdot, 0) \in L_{p+2}(\Omega)$. Then the Schrödinger operator (19.12) is essentially self-adjoint in $L_2(\mathbb{R}^n)$ for almost all ω.*

The condition imposed on q means that the $(p+2)$-nd moment of the random variable $q(\cdot, 0)$ exists and is finite. In particular, the condition is satisfied for any Gaussian potential q, any Poisson potential q defined in the same way as the field f in Example 19.5 (by (19.7)) with an arbitrary function $\varphi \in L_{p+2}(\mathbb{R}^n)$, and for any potential (19.6), where χ is a bounded function and ξ_0 is a random variable, the $(p+2)$-nd moment of which is finite.

The following theorem provides a condition of a different kind (imposing a restriction on the correlations, rather than on q at a fixed point):

Theorem 19.2 (Grenkova 1981). *Suppose that q has* finite correlation radius, *i.e., there exists $r_0 > 0$ such that if $V_1, V_2 \subset \mathbb{R}^n$ are two domains such that* $\operatorname{dist}(x, y) \geq r_0$ *for any $x \in V_1$ and $y \in V_2$, then the σ-algebras $\mathcal{F}_j$, $j = 1, 2$, generated by the random variables $\{q(\cdot, x), x \in V_j\}$ are statistically independent. Then the Schrödinger operator (19.12) is essentially self-adjoint in $L_2(\mathbb{R}^n)$ for almost all ω.*

Let us pose the problem of essential self-adjointness for the operator A_Ω defined on C_b^∞ by (19.20).

Theorem 19.3 (Dedik and Shubin 1980). *If A is a random formally self-adjoint elliptic operator for which the estimates (19.13) of the symbol are satisfied, then A_Ω is essentially self-adjoint in $L_2(\Omega)$.*

We shall now turn our attention to the spectra $\sigma(A)$ and $\sigma(A_\Omega)$ of the operators $A = A_\omega$ and A_Ω. In what follows each of them will always be understood as the spectrum of the closure of the corresponding operator in $L_2(\mathbb{R}^n)$ or $L_2(\Omega)$. Everywhere below we shall assume that A is a random elliptic operator of the form (19.10) or (19.12) and, for simplicity, we shall confine ourselves to the ergodic case. We observe that, in view of (19.15), the operators A_ω and $A_{T_z\omega}$ are similar, and so have the same spectrum. It follows that $\sigma(A_\omega)$ is an invariant function on Ω (the values of this function being closed subsets of $\mathbb{C}$). It can be easily deduced that the spectrum is the same for almost all ω. It will be denoted by $\sigma(A)$. The same is true (in the self-adjoint case) for the absolutely continuous part $\sigma_{ac}(A_\omega)$ of the spectrum, the singular continuous part $\sigma_{sc}(A_\omega)$ of the spectrum, and the closure $\overline{\sigma_p(A_\omega)}$ of the point spectrum (the set of eigenvalues), each of these parts of the spectrum being independent of ω for almost all ω (Pastur 1974, 1987b; Kunz and Souillard 1980; Kirsch 1985). We remark that, in general, this not so for $\sigma_p(A_\omega)$, i.e., the set of eigenvalues itself. For if we consider the orthogonal projection $E_\omega(\{\lambda\})$ of $\operatorname{Ker}(A_\omega - \lambda I)$ for a given number λ, then $E_\omega(\{\lambda\})$ will be a homogeneous random operator in the sense of (19.15) and its ordinary trace $\operatorname{Tr} E_\omega(\{\lambda\}) = \dim \operatorname{Ker}(A_\omega - \lambda I)$ will be non-random (for almost all ω). It is easily demonstrated that this trace is equal to 0 or $+\infty$ (this is obvious, for example, if the random trace (19.17) is defined and finite). In particular, for the one-dimensional Schrödinger operator A_ω we have $\dim \operatorname{Ker}(A_\omega - \lambda I) \leq 2$, which implies that $\operatorname{Tr} E_\omega(\{\lambda\}) = 0$, so that the probability that the given fixed point λ is an eigenvalue is equal to zero (Pastur 1974). However, this does not mean at all that there is no point spectrum. On the contrary, as we shall see below, in many one-dimensional situations the whole spectrum is a *pure point* one, i.e., there exists an orthonormal basis of eigenfunctions. It follows that the eigenvalues, being very sensitive to perturbations, depend essentially on ω in this case.

We shall now discuss the relationship between the spectra $\sigma(A)$ and $\sigma(A_\Omega)$ of the same operator A in $L_2(\mathbb{R}^n)$ and $L_2(\Omega)$. One can easily verify that

$$\sigma(A_\Omega) \subset \sigma(A). \tag{19.21}$$

However, the inclusion in the opposite direction is not always true. For example, let $\Omega = \mathbb{R}^n/\Gamma$, where Γ is a lattice in $\mathbb{R}^n$ (so that Ω is an n-dimensional torus). Suppose that $\mathbb{R}^n$ acts on Ω by natural translations. Then, by homogeneity, the random elliptic operator A_ω has Γ-periodic coefficients for almost all ω. In this case $\sigma(A_\Omega)$ is just the spectrum of A_ω on Ω, which means that it is discrete. At the same time $\sigma(A_\omega)$ has a zone structure (see §17), which implies that it cannot be equal to $\sigma(A_\Omega)$. It turns out that the existence of non-trivial periods for the dynamical system $\{T_x\}$ is the only obstacle to $\sigma(A)$ and $\sigma(A_\Omega)$ being equal. Namely, in the general case we introduce the group of periods of the dynamical system by the formula

$$\Gamma = \{z \in \mathbb{R}^n,\ T_z\omega = \omega \text{ for a.e. } \omega\}, \tag{19.22}$$

or, equivalently, we define Γ as the group of those $z \in \mathbb{R}^n$ for which $V_z = I$.

Theorem 19.4 (Kozlov and Shubin 1984). *Let A be a random elliptic operator of order $m > 0$ whose symbol satisfies the estimates (19.13) and let $\{T_z\}$ be an* aperiodic *dynamical system, i.e., let $\Gamma = \{0\}$. Then $\sigma(A_\Omega) = \sigma(A)$. In the general case, for an arbitrary group of periods Γ, the spectrum $\sigma(A_\Omega)$ is for almost all ω equal to that defined by the operators A_ω in $L_2(\mathbb{R}^n/\Gamma)$.*

Obviously, this theorem generalizes Theorem 18.1 on the equality of the spectra of an almost periodic operator in $L_2(\mathbb{R}^n)$ and $B^2(\mathbb{R}^n)$. In order to prove the theorem, one can establish that if A is self-adjoint and $\varphi \in S(\mathbb{R})$, then $\varphi(A)$ is a random integral operator in $L_2(\mathbb{R}^n)$ whose kernel $K(\omega, x, y)$ decreases rapidly away from the diagonal, i.e., as $|x - y| \to +\infty$, making it possible to define $\varphi(A_\Omega)$ with the aid of the same kernel in $L_2(\Omega)$, which can be identified with the space of homogeneous random fields. The aperiodicity condition enables one to establish that $\varphi(A) = 0$ if and only if $\varphi(A_\Omega) = 0$, which is equivalent to the equality of the spectra.

Subject to certain conditions other than those described above, a result similar to Theorem 19.4 can be obtained with the aid of the technique of von Neumann algebras (Baaj Saad 1988), which also covers operators of order $m \leq 0$.[11]

19.4. Density of States

The *limiting spectral distribution function* or the *integrated density of states* $N(\lambda)$ can be defined for a random self-adjoint elliptic differential operator A in the same way as in the almost periodic case (see Sect. 18.4), i.e., by formula (18.11). In the case of second-order operators the existence of the limit in (18.11) for almost all ω, which is sometimes referred to as *self-averagibility*, was proved in (Slivnyak 1966; Pastur 1971, 1973) and, in the case of higher-order operators A, in (Gusev 1977), the scheme of the proof described in Sect. 18.4 being also applicable in this case. Moreover, the invariance under translations in x implies immediately that $N(\lambda)$ as a function of the random parameter ω is invariant under the transformations from $\{T_z,\ z \in \mathbb{R}^n\}$. Therefore $N(\lambda)$ is a non-random function in the ergodic case, to which we confine ourselves in what follows for the sake of simplicity. Moreover, in the case of operators of the form (19.10) (whose coefficients satisfy the estimates (19.11)) $N(\lambda)$ can be expressed in terms of the spectral projection E_λ of A by the formula

$$N(\lambda) = \mathrm{Tr}_R E_\lambda = M_x\{e(\lambda, x, x)\} = \bar{e}(\lambda, 0, 0), \tag{19.23}$$

where $\bar{e}(\lambda, 0, 0)$ denotes the mathematical expectation of $e(\lambda, 0, 0)$ (see (19.2) and (19.17)). The scheme of the proof of self-averagibility remains the same as for almost periodic operators (see Sect. 18.4). In the case of the Schrödinger operator one can use the Feynman-Kac representation of the fundamental solution of the Cauchy problem and the Green function in terms of the Wiener integral (see Kac 1959; Dynkin 1963). In this way we obtain, in particular, the following expression for the Laplace transform of $N(\lambda)$ (see (18.13):

$$\tilde{N}(t) = (2\pi t)^{-n/2} W \left\{ M\left\{ \exp\left(- \int_0^t q(\omega, x(s))\, ds \right) \right\} \middle| x(t) = 0 \right\}, \tag{19.24}$$

where W denotes the Wiener integral over the trajectories $x(s)$ of a Brownian motion in $\mathbb{R}^n$ such that $x(0) = x(t) = 0$, and M denotes the mean value with respect to the random parameter ω appearing in the potential.

Other results on self-averagibility were presented in Sect. 18.4. All of them can be carried over to operators with random coefficients (Pastur 1973; Gusev

[11](Added in the English edition.) Recently Jingbo Xia (1993) has offered an analytical approach, which also covers operators of order $m \leq 0$.

1977). This is not so for Kozlov's results on the operators (18.25) described in Sect. 18.5.

As in the almost periodic case, the spectrum $\sigma(A_\omega)$ of a random operator A_ω in $L_2(\mathbb{R}^n)$ is equal to the set of growth points of $N(\lambda)$ for almost all ω (see, for example, Gusev and Shubin 1977; Pastur 1980).

We also remark that formula (19.23) makes it possible to define $N(\lambda)$ for a larger class of operators, for example, for random self-adjoint elliptic or hypoelliptic pseudodifferential operators. Moreover, the following variational principle, which is analogous to (18.24), can be established for $N(\lambda)$ (Bogorodskaya and Shubin 1986):

$$N(\lambda) = \sup_{P \in \mathcal{P}_\lambda(A)} \operatorname{Tr}_R P, \tag{19.25}$$

where $\mathcal{P}_\lambda(A)$ is the set of infinitely smoothing (in the standard Sobolev scale of spaces $H_s(\mathbb{R}^n)$) homogeneous random orthogonal projections P such that $P(A - \lambda I)P \leq 0$. The index theory of random elliptic operators constructed by Fedosov and Shubin (1978a, 1978b) is used in the proof of (19.25) in an essential way.

For random elliptic operators the asymptotic properties of $N(\lambda)$ as $\lambda \to +\infty$ are basically the same as in the almost periodic case. With the aid of the trace (19.17), one can introduce the *ζ-function of a positive random self-adjoint elliptic operator* A of the form (19.10) (whose coefficients satisfy (19.11)) and prove that it admits a meromorphic continuation to $\mathbb{C}$ with the same analytic properties as in the almost periodic case (Sect. 18.5). In particular, it follows that the Weyl asymptotics (18.22) is valid for $N(\lambda)$. Moreover, Gusev (1977) carried over Hörmander's estimate of the remainder in this asymptotic formula to random elliptic operators (for example, the estimate can be obtained by establishing that the asymptotics of the spectral function is uniform in x, followed by taking the mean value with respect to x). The asymptotic formula (18.23) is valid for the Schrödinger operator with a bounded random potential (the proof of this formula can be carried over without any modifications).

We also consider the physically interesting question concerning the behaviour of $N(\lambda)$ at the lower end of the spectrum (also called the *fluctuation boundary*). Sometimes, in the almost periodic case, it is also possible to find such an asymptotics of $N(\lambda)$ (see formula (18.30) and the description of how to extend (18.30) to the case of Schrödinger operators that follows formula (18.31)), however, the multitude of feasible models of random operators enables one to obtain a number of new results. Let us present examples of such results (Pastur 1973, 1977) for random Schrödinger operators (19.12).

a) Let q be a Gaussian potential (see Sect. 19.1) whose correlation function $B(x) = M\{q(\omega, x)q(\omega, 0)\}$ satisfies the following conditions: $B \in C^2(\mathbb{R})$, $|B(x)| \leq c|x|^{-\alpha}$, where $\alpha > 0$, and $|\tilde{B}(0) - \tilde{B}(x)| \leq c|\ln|x||^{-a}$ for $|x| < 1$, where $a > 1$ and $\tilde{B}(x) = -\Delta B(x)$. Then

$$\ln N(\lambda) = -\frac{\lambda^2}{2b}(1 + o(1)), \tag{19.26}$$

as $\lambda \to -\infty$, where $b = B(0)$.

b) Let q be a Poisson potential (Example 19.5) with a continuous function $\varphi \leq 0$ (*attracting centres*) having a strict minimum at zero. Then $\inf \sigma(A) = -\infty$ and

$$\ln N(\lambda) = -\frac{\lambda}{\varphi(0)} \ln \frac{\lambda}{\varphi(0)}(1 + o(1)) \tag{19.27}$$

as $\lambda \to -\infty$.

c) Let q be a Poisson potential with a function $\varphi \geq 0$ such that $\varphi(x) = \varphi_0 |x|^{-\alpha}(1 + o(1))$ as $|x| \to \infty$, where $n < \alpha < n + 2$ (*repelling long-range centres*). Then $\inf \sigma(A) = 0$ and

$$\ln N(\lambda) = -c(\varphi_0, \alpha, n)\lambda^{\frac{n}{\alpha}-1}(1 + o(1)) \tag{19.28}$$

as $\lambda \to +0$.

d) Let q be a Poisson potential with a function $\varphi \geq 0$ such that $\varphi(x) = o(|x|^{-n-2})$ as $|x| \to \infty$ (*repelling short-range centres*). Then $\inf \sigma(A) = 0$ and

$$\ln N(\lambda) = -c(\gamma_n \lambda^{-1})^{n/2}(1 + o(1)) \tag{19.29}$$

as $\lambda \to +0$, where c is the same constant as in (19.8) and γ_n is the first eigenvalue of the operator $(-\Delta)$ with Dirichlet conditions on an n-dimensional ball of volume one.

Assertions a)–c) can be obtained by estimating the Wiener integral (19.24) for $\tilde{N}(t)$ at large t, followed by applying the appropriate Tauberian theorems. To obtain assertion d) it is also necessary to analyse large deviations of the Wiener process (Donsker and Varadhan 1975; Freidlin and Venttsel' 1979). Other results of this kind along with relevant comments can be found in (Kirsch and Martinelli 1983; Kirsch 1985; Pastur 1987a, 1987b; Simon 1987; Grenkova and Molchanov 1988).

Bovier et al. (1988) used the supersymmetric cluster expansion of the resolvent in the Anderson model to obtain the essential information on the local smoothness of $N(\lambda)$ with respect to λ. These results imply, in particular, that if $q(\cdot, x)$, where $x \in \mathbb{Z}^n$, are random variables uniformly distributed on a finite interval in the Anderson model, then $N \in C^\infty(\mathbb{R})$. For other results on the local smoothness of $N(\lambda)$ see (Craig and Simon 1983a; Simon and Taylor 1985).

We also remark that, by virtue of Oseledets' multiplicative ergodic theorem (Oseledets 1968), the Lyapunov exponent $\gamma(\lambda)$ can also be defined by (18.32) for one-dimensional random Schrödinger operators ($\tilde{A}$ must be replaced by A_ω). In the ergodic case it is independent of ω for almost all ω. Moreover, all the properties of the exponent mentioned at the end of §18 (the relationship with the absolutely continuous spectrum, Thouless' formula (18.33), Deift and Simon's inequality (18.34)) can be carried over to this case.

19.5. The Character of the Spectrum. Anderson Localization

As we have seen in §18, the question concerning the character of the spectrum is very difficult and can hardly be solved completely for operators with almost periodic coefficients. This is even more so for general operators with random coefficients. However, in models that are much more random than the nearly deterministic almost periodic ones new additional possibilities emerge and several deep results can be obtained.

Anderson (1958) and Mott and Twose (1961) where the first to demonstrate heuristically that the spectrum and eigenfunctions of the one-dimensional Schrödinger operator can exhibit the following behaviour, provided the behaviour of the potential is sufficiently random:

a) the whole spectrum is a pure point one;

b) the eigenfunctions decay exponentially.

The two properties a) and b) are referred to as *Anderson localization*. In particular, if Anderson localization takes place, then the eigenvalues must form a dense subset of the spectrum $\sigma(A)$ of the operator A under consideration and there can be no absolutely continuous or singular continuous spectrum.

The first mathematical article in which it was proved that the spectrum is a pure point one (assertion a)) was that by Gol'dshejd et al. (1977), in which a diffusion-generated potential (see Sect. 19.6) was considered. Assertion b) on the exponential localization was proved by Molchanov (1978) in the same model. The proofs are based on a limiting passage from an interval $[-L, L]$ and use deep facts from the theory of degenerate parabolic equations satisfied by the transition densities of the Markov processes involved.

The exponential decay rate of eigenfunctions can be defined by means of the Lyapunov exponent (see Carmona 1982; Craig and Simon 1983b; Kotani 1986).

Subsequently a number of authors proposed new methods of establishing Anderson localization, making it possible to consider one-dimensional random Schrödinger operators (and more general second-order operators) of other kinds. For example, among them is the method put forward by Kunz and Souillard (1980) resting on the concepts of scattering theory, which was applied by them in the case of the discrete Schrödinger operator and later carried over by Royer (1980) and Carmona (1982) to the continuous case.

We shall now discuss the many-dimensional case. In this case a heuristic argument (see, for example, Martinelli and Scoppola 1986) indicates that, for sufficiently random potentials, Anderson localization should take place for $n = 2$, as before, and the spectrum should be continuous on a half-axis $\lambda > \lambda_0$ and pure point on the half-axis $\lambda < \lambda_0$ for $n \geq 3$ (λ_0 is then called the *mobility edge*).

In this case the rigorous results go back to the article (Fröhlich and Spencer 1983), in which Anderson's model (see Remark 19.2) was considered in the case of large disorder (in the sense that the potential is taken to be $gV(\omega, x)$, where g is a large parameter and V a fixed potential of the Anderson model) or low energy. Moreover, the distribution of the potential is assumed to have bounded density relative to the Lebesgue measure. Instead of localization, in was proved in (Fröhlich and Spencer 1983) that the kernel of the resolvent $(A-\lambda I+i\varepsilon)^{-1}$ decays exponentially as $|x-y| \to +\infty$ uniformly with respect to $\varepsilon > 0$. Later (see Fröhlich and Spencer 1984; Fröhlich et al. 1985; Martinelli and Scoppola 1986) it turned out that this implies Anderson localization. Moreover, the mechanism of localization was explained (see Martinelli and Scoppola 1986). It is based on the instability of quantum tunnelling discovered by Jona-Lasinio, Martinelli, and Scoppola. Namely, the lack of localization of eigenfunctions in the case of a periodic potential can be explained by the tunnelling effect and is connected with strong translation symmetry. A similar effect takes place, for example, for an even potential on $\mathbb{R}^1$ with two symmetric wells, in which case the eigenfunctions are either odd or even, i.e., *delocalized*. However, under a small perturbation that breaks evenness, the eigenfunctions become localized in one of the wells in the semiclassical approximation. For infinitely many wells a similar effect takes place in the Anderson model with large disorder or low energy.

In (Martinelli and Scoppola 1986) the following continuous model was also considered: the Schrödinger operator in $\mathbb{R}^n$ with potential

$$q(\omega, x) = \sum_i \xi_i \varphi(x - x_i),$$

where the points x_i have the Poisson distribution (see Example 19.5), ξ_i are identically distributed independent random variables with values in $[0, 1]$, the distribution of which has bounded density relative to the Lebesgue measure, and where φ is a bounded function with compact support such that $\varphi(x) \le 0$ for all x. In this case, if $\varphi \ne 0$, then the spectrum is unbounded from below and if the constant c in (19.8) (the mean concentration of admixture) is small enough, then the negative spectrum is a pure point one and the eigenfunctions decay exponentially, i.e., Anderson localization takes place for the negative spectrum.

For almost periodic potentials Anderson localization can also take place in the many-dimensional case. This is so, for example, in the so-called *'Maryland model,'* i.e., for the discrete Schrödinger operator $H_{\alpha,\theta,g}$ with potential

$$V(z) = V_{\alpha,\theta,g}(z) = g \tan\bigl[\pi(\alpha \cdot z) + \theta\bigr], \quad z \in \mathbb{Z}^n,$$

where $\alpha = (\alpha_1, \dots, \alpha_n) \in \mathbb{R}^n$ and $\theta \in [0, 2\pi]$ with $\theta \ne \pi(\alpha \cdot z) + \pi/2 \bmod 2\pi$, so that $V(z)$ is defined for all $z \in \mathbb{Z}^n$. It is assumed that $(1, \alpha_1, \dots, \alpha_n)$ is a system of frequencies satisfying the Diophantine condition (18.7). Then Anderson localization takes place for $H_{\alpha,\theta,g}$ (see Bellissard et al. 1983; Figotin and Pastur 1984; Simon 1985; Cycon et al. 1987, and references therein). Moreover, the spectrum is equal to $\mathbb{R}$ and has multiplicity one, and the integrated density of states can be evaluated explicitly.

A discussion of other aspects of localization theory can be found in (Delyon et al. 1985a, 1985b, 1985c; Simon et al. 1985; Simon and Wolff 1985b) as well as in the surveys mentioned at the beginning of the present section.[12]

§20
Non–Self-Adjoint Differential Operators that Are Close to Self-Adjoint Ones

20.1. Preliminary Remarks

The spectral theory of non–self-adjoint operators is much more complex than the theory of self-adjoint operators. To a large extent this is connected with the lack of a universal 'model' (similar to the multiplication operators in the self-adjoint theory – see Example 1.2 and Theorems 1.3 and 1.4). For various classes of non–self-adjoint operators it becomes necessary to use individual methods of some kind. The most important position among such methods is occupied by perturbation theory, which makes it possible to study operators that are close to self-adjoint ones. In the present section we shall, basically, deal with the latter class of operators.

We also remark that a technique based on studying the properties of the fundamental system of solutions of the equation $\mathcal{L}u - \lambda u = 0$ as an analytic function of λ is widely used in the spectral analysis of ordinary differential operators. This technique is applicable both in the self-adjoint and non–self-adjoint cases and makes it possible, in particular, to investigate the spectra of operators that fail to be close to self-adjoint ones. On this subject see (Naimark 1969; Kostyuchenko and Sargsyan 1979). In the dissipative case the so-called 'Nagy-Foias functional model' proves to be very effective. It can also be applied to certain many-dimensional problems. On using this model in the spectral theory of differential operators see the article by Pavlov in the present series.

As opposed to the self-adjoint case, non–self-adjoint operators fail to be determined by their spectral characteristics, even if only because there are non–self-adjoint operators whose spectrum is empty.

Example 20.1. The spectrum (in $L_2(0,1)$) of the operator of the Cauchy problem

$$A = -i\frac{d}{dt} \upharpoonright \{u \in H^1(0,1) : u(0) = 0\}$$

is empty.

[12](Added in the English edition.) See also Carmona and Lacroix (1990) and Figotin and Pastur (1992). Recently a simple new approach to localization problems has been suggested by Aizenman and Molchanov (1993); see also (Aizenman, 1993).

If the spectrum is non-empty, there arise many new questions as compared with the self-adjoint case. First of all, if λ is an eigenvalue of A, then, along with the eigenvectors $x \in \operatorname{Ker}(A - \lambda I)$, the operator may have associated elements ($x \in \operatorname{Ker}(A - \lambda I)^p$ for some $p > 1$). The linear set $\mathfrak{H}(\lambda) = \bigcup_p \operatorname{Ker}(A - \lambda I)^p$ is called the *root lineal* (it is an analogue of a Jordan cell in linear algebra). It may turn out not to be closed.

Example 20.2. For the operator

$$A = -i\frac{d}{dt} \restriction H^1(0,1)$$

every $\lambda \in \mathbb{C}$ is an eigenvalue. The set $\mathcal{P} \exp(i\lambda t)$, where $\mathcal{P}$ is the set of all polynomials, is the corresponding root lineal. It follows that every root lineal is non-empty and dense in $L_2(0,1)$.

In what follows we shall confine ourselves to closed operators in a Hilbert space $\mathfrak{H}$ such that the natural embedding of $D(A)$ (with the A-metric (1.1)) in $\mathfrak{H}$ is compact and, moreover, the resolvent set $\rho(A)$ is non-empty. The abstract theory of such operators with discrete spectrum is presented in the book by Gokhberg and Krein (1965). We remark that the operator from Example 20.1 belongs to the class under consideration, but the operator from Example 20.2 does not.

For definiteness, we assume that $0 \in \rho(A)$. Then the inverse operator A^{-1} is compact. The spectrum $\sigma(A)$ consists of a finite or infinite sequence of eigenvalues $\{\lambda_k\}$. If the sequence is infinite, then $|\lambda_k| \to \infty$ (hence the term 'operator with discrete spectrum'). The dimension of each root lineal $\mathfrak{H}(\lambda_k)$ is finite and is called the *algebraic multiplicity* of the corresponding eigenvalue λ_k, as opposed to the *geometric multiplicity* $\dim \operatorname{Ker}(A - \lambda_k I)$. In particular, one can see that the root lineals of an operator with discrete spectrum are closed. Because of this, they are usually called the *root subspaces*.

Let Γ be a positively oriented contour in $\mathbb{C}$, the intersection of which with the spectrum of A is empty (such a contour will be called *admissible* for A). If A is an operator with discrete spectrum and only one point λ of the spectrum of A lies inside Γ, then the Riesz formula

$$P(\lambda) = -(2\pi i)^{-1} \int_\Gamma (A - \zeta I)^{-1}\, d\zeta \tag{20.1}$$

defines the spectral projection (in general, not orthogonal) onto the root subspace $\mathfrak{H}(\lambda)$. Choosing a basis (for example, an orthonormal one) in each subspace $\mathfrak{H}(\lambda_k)$, where $\lambda_k \in \sigma(A)$, we obtain a system of root vectors of A. Important problems in non–self-adjoint spectral theory are connected with the study of such systems. The first one is the problem of completeness of a system of root vectors. Example 20.1 indicates that a system of root vectors of an operator with discrete spectrum is not always complete. Furthermore, even if the system of root vectors is compete, it fails, in general, to be a basis, that is, the decomposition

$$f \sim \sum_{\lambda \in \sigma(A)} P(\lambda) f \tag{20.2}$$

may fail to converge in $\mathfrak{H}$ for some $f \in \mathfrak{H}$. Decomposition theorems are concerned with the study of the convergence or summability of the series (20.2) in some sense. We recall that the completeness and decomposition properties are both valid for self-adjoint operators with discrete spectrum, because the eigenspaces are orthogonal to one another. The two properties are expressed by (1.5').

For many classes of non–self-adjoint operators it turns out that the expansion (20.2) converges, provided the terms are collected in a special way.

Definition 20.1. An operator A has a *root basis with parentheses* if the spectrum of A can be represented as a countable union $\sigma(A) = \Lambda_1 \cup \Lambda_2 \cup \dots$ of finite sets such that the series

$$\sum_j P_j f, \quad P_j = \sum_{\lambda \in \Lambda_j} P(\lambda), \tag{20.3}$$

where $P(\lambda)$ are the projections (20.1), converges to f for any $f \in \mathfrak{H}$.

This basis is a *Riesz basis with parentheses* if the series (20.3) converges unconditionally (i.e., if the sum is independent of the order of terms).

The expansion (20.3) is a particular averaged form of the ordinary root vector expansion (20.2). Information about the dimension of P_j (i.e., the frequency of the arrangement of parentheses) makes it possible to study the rate of convergence of the series (see Agranovich 1977, 1980). When there is no basis with parentheses, various summation methods can be applied to the series (20.2). Here we consider the *Abel method*.

For $\lambda \in \sigma(A)$ we introduce the operator

$$P(\alpha; \lambda, t) = -(2\pi i)^{-1} \int_\Gamma e^{-t\zeta^\alpha} (A - \zeta I)^{-1} \, d\zeta, \quad \alpha > 0,$$

where Γ is an admissible contour enclosing λ.

The Abel summation method of order α can be applied to (20.2) if the series

$$E_\alpha(f; t) = \sum_{\lambda \in \sigma(A)} P(\alpha; \lambda, t) f, \quad t > 0$$

converges for any $f \in \mathfrak{H}$ and $E_\alpha(f; t) \to f$ as $t \to +0$.

The quantitative spectral theory of non–self-adjoint operators gives rise to a new problem of estimating the distance between the eigenvalues and the real axis. The study of the spectral asymptotics involves the investigation of the distribution function of the absolute values or real parts of eigenvalues. For operators that are close to self-adjoint ones these two distribution functions are equivalent.

20.2. Basic Examples

We shall present typical examples of differential operators with discrete spectrum.

Example 20.3. A regular elliptic operator. Let X be a compact manifold without boundary, let $d\mu$ be a positive density on X, and let $\mathcal{L}$ be an elliptic differential expression of order m on X. The operator $\mathcal{L} \restriction H_m(X)$ is closed. If it has at least one regular point, then its spectrum is discrete. This follows directly from the compactness of the embedding of $H^m(X)$ in $L_2(X, d\mu)$.

It is essential to assume that there is at least one regular point. For instance, the operator $\mathcal{L}u = \exp(i\varphi)du/d\varphi$ in $L_2(S^1)$ is elliptic, but has no regular points: for every $\lambda \in \mathbb{C}$ the equation $\mathcal{L}u = \lambda u$ has a periodic solution $u(\varphi) = \exp\big(-\lambda e^{i\varphi}\big)$.

What has been said in Example 20.3 can be carried over to elliptic operators on vector bundles as well as the operators of regular elliptic boundary value problems on a manifold with boundary.

Example 20.4. The Schrödinger operator with a complex potential. We consider the differential expression

$$\mathcal{L}u = -\Delta u + Vu \tag{20.4}$$

in $L_2(\mathbb{R}^n)$ and assume that $V \in L_{1,\mathrm{loc}}(\mathbb{R}^n)$ is a complex-valued function. Suppose that the functions $V_0 = \operatorname{Re} V$ and $V_1 = \operatorname{Im} V$ satisfy the conditions

$$V_0(x) \geq c > 0, \quad |V_1(x)| \leq q + bV_0(x), \quad q, b \geq 0. \tag{20.5}$$

We shall use the method of quadratic forms to give a precise definition of the corresponding differential operator. Let $a_0[u] = \int \big(|\nabla u|^2 + V_0|u|^2\big)\, dx$ with the natural domain $d_0 = d[a_0]$ (see Example 3.6). Furthermore, let $A_0 = \operatorname{Op}(a_0)$ and let $a_1 = \int V_1 |u|^2\, dx$. Under the conditions (20.5), the variational triple $\{d_0; a_0, a_1\}$ defines a bounded self-adjoint operator T in d_0. For this operator

$$a_1[u, v] = a_0[Tu, v] = \Big(A_0^{1/2} Tu, A_0^{1/2} v\Big) \quad \forall\, u, v \in d_0. \tag{20.6}$$

The operator $S = A_0^{1/2} T A_0^{-1/2}$ is self-adjoint and bounded in $L_2(\mathbb{R}^n)$. It is natural to take the (closed) operator

$$A = A_0^{1/2}(I + iS)A_0^{1/2}, \quad D(A) = \Big\{u \in d_0 : A_0^{1/2} u - iSA_0^{1/2} u \in d_0\Big\} \tag{20.7}$$

as a realization of the differential expression (20.4). Indeed, by (20.6), we have $V_1 u = A_0^{1/2} S A_0^{1/2} u$ for $u \in C_0^\infty(\mathbb{R}^n)$.

The quadratic form

$$a[u] = a_0[u] + ia_1[u] = \int \left(|\nabla u|^2 + V|u|^2\right) dx, \quad u \in d_0 \tag{20.8}$$

is connected with A in a natural way. The values of this form (in the case when $q = 0$) lie in a sector $|\arg a[u]| < \theta$ in $\mathbb{C}$ with $\theta < \pi/2$. The operator A can be defined by means of (20.8) with the aid of the theory of sectorial quadratic forms (Kato 1966), which generalizes Friedrichs' construction. The resulting operator is the same as (20.7).

In addition, we assume that $V_0(x) \to \infty$ as $|x| \to \infty$ in the conditions of Example 20.4. Then $A_0^{-1/2}$ is a compact operator. It follows that so is A^{-1}, and, consequently, the spectrum of the Schrödinger operator (20.7) is discrete in this case.

20.3. Completeness Theorems

The application of abstract theorems on perturbations of self-adjoint operators is often an effective method of studying the spectral properties of non–self-adjoint operators with discrete spectrum. Theorems of this kind rest on formula (20.1). A detailed investigation of the resolvent of the operator is required to prove such theorems. Moreover, it is often necessary to use very deep results of function theory.

The following notions are needed to characterize the 'magnitude' of perturbations. Let A_0 be a self-adjoint operator in $\mathfrak{H}$. An operator B is said to be A_0-*compact* if $D(B) \supset D(A_0)$ and $B(A_0 - iI)^{-1}$ is a compact operator. This condition is satisfied if and only if B is compact as an operator from $D(A_0)$ (with the A_0-norm) to $\mathfrak{H}$. Under these conditions, the operator $(A_0 + B) \restriction D(A_0)$ is closed and, as we shall see later on, its spectral properties are close to the corresponding properties of A_0.

Furthermore, let $\{d; a_0, b\}$ be a triple that satisfies all the conditions of Sect. 1.10, except that the form $b[u, v]$ does not have to be Hermitian. As before, (1.10) defines a bounded operator $T = T(d; a_0, b)$ in d. If this operator is compact, then b is said to be an a_0-*compact* form. This being the case, if a_0 and b can be realized as the quadratic forms of operators A_0 and B in a Hilbert space $\mathfrak{H}$ (see Sect. 1.10), then B is said to be A_0-*compact in the sense of forms*. In this case $T(d; a_0, b)$ is compact if and only if $S = A_0^{1/2} T A_0^{-1/2}$ is compact in $\mathfrak{H}$. By definition, we set (cf. Example 20.4)

$$D(A_0 + B) = \left\{x \in d : (I + S)A_0^{1/2}x \in d\right\},$$
$$(A_0 + B)x = A_0^{1/2}(I + S)A_0^{1/2}x \tag{20.9}$$

('the sum in the sense of forms'). This definition is natural, since it follows easily form (1.10) that $A_0^{1/2} S A_0^{1/2} x = Bx$ for every x from the original domain of B. In what follows we shall write $S = A_0^{-1/2} B A_0^{-1/2}$ (even though this formula is not completely rigorous). As in the case of A_0-compactness in the operator sense, the spectral properties of (20.9) are close to the corresponding properties of A_0.

In applications to differential operators, A_0-compactness (in either sense) can be verified on the basis of embedding theorems. If A_0 is a regular elliptic operator, B is a differential operator with bounded coefficients, and $\operatorname{ord} B < \operatorname{ord} A_0$, then B is A_0-compact. Furthermore, if $a_0[u]$ and $b[u]$ are differential quadratic forms such that $a_0[u] \geq c\|u\|_{H^r}$ with $c > 0$ on $d[a_0]$ and $\operatorname{ord} b < r$, then b is an a_0-compact form.

Numerous completeness criteria for the system of root vectors of an operator that is close to a self-adjoint one are presented in the book by Gokhberg and Krein (1965). In the case of differential operators the following theorem (which is essentially due to Keldysh) turns out to be the most useful one.

Theorem 20.1. *Let A_0 be a self-adjoint operator with discrete spectrum and let B be an A_0-compact operator such that the series*

$$\sum_j s_n \left(B(A_0 - iI)^{-1}\right)^p$$

converges for some $p > 0$. Then $A = A_0 + B$ is an operator with discrete spectrum. Moreover, there are at most finitely many eigenvalues $\lambda_j(A)$ in the complement of any angle $|\operatorname{Im}\lambda| < \varepsilon|\operatorname{Re}\lambda|$, where $\varepsilon > 0$, and the system of root vectors of A is complete in $\mathfrak{H}$.

The results remain valid if A_0 is positive definite, B is an A_0-compact operator in the sense of forms, and the series

$$\sum_n s_n^p \left(A_0^{-1/2} B A_0^{-1/2}\right)$$

converges.

Results of other kind, allowing stronger deviation from self-adjointness, impose a condition on the numerical domain of A, that is, the set of values (Ax, x) of the quadratic form of A for $x \in D(A)$ such that $\|x\| = 1$.

Theorem 20.2 (Gokhberg and Krein 1965). *Let A be an invertible operator whose numerical domain is contained in a sector $|\arg z - \alpha| < \pi/2p$, where $p > 1$, and let*

$$s_n\left(A^{-1} - (A^{-1})^*\right) = o\left(n^{-1/p}\right).$$

Then the system of root vectors of A is complete.

20.4. Expansion and Summability Theorems. Asymptotic Behaviour of the Spectrum

Proceeding to questions concerned with the convergence of spectral expansions, we remark right at the very beginning that it is extremely rare for a system of root vectors to be a basis in the case of a non self-adjoint differential operator (examples of such operators in the one-dimensional case are presented in (Agranovich 1980)). On the other hand, there are many criteria for a system of root vectors to be a basis with parentheses. As suggested by the 'experimental material,' the bases with parentheses which arise naturally are usually unconditional.

Let $A_0 > 0$ be a self-adjoint operator. As a rule, conditions under which the system of root vectors of $A_0 + B$ is a basis with parentheses require that the 'perturbation' B be small relative to A_0 (exactly as in the case of the conditions ensuring the validity of spectral asymptotic formulae similar to those in §9).

Definition 20.2. Let $0 \le p < 1$. An operator B is said to be A_0^p*-bounded* if $D(B) \supset D(A_0^p)$ and BA_0^{-p} is bounded in $\mathfrak{H}$. We say that B is *p-subordinate* to A_0 if $D(B) \supset D(A_0)$ and $\|Bu\| \le c\|A_0u\|^p\|u\|^{1-p}$, where c is a constant. Finally, a quadratic form $b[u]$ is *p-subordinate* to $a_0 = \mathrm{QF}(A_0)$ (or to A_0) if

$$|b[u]| \le c(a_0[u])^{p/2}\|u\|^{2-p}, \quad u \in d[a_0]. \tag{20.10}$$

We observe that A_0^p-boundedness implies p-subordination, but p-subordination implies merely A_0^q-boundedness for any $q > p$. If A_0 is an operator with discrete spectrum and either B or its form is p-subordinate to A_0, then B is A_0*-compact* (in the appropriate sense). This makes it possible for $A_0 + B$ to be defined correctly.

Theorem 20.3 (see Agranovich 1977; Markus and Matsaev 1981, 1982). *Let A_0 be a positive definite self-adjoint operator and let $N(\lambda; A_0) = o(\lambda^\gamma)$ as $\lambda \to \infty$. Let us assume that B is p-subordinate to A_0 for some $p \in (0,1)$. Then, for $p + \gamma \le 1$, the system of root vectors of $A = A_0 + B$ is a Riesz basis with parentheses. If $p + \gamma > 1$, then the expansion (20.2) is summable by the Abel method of order α, where $\alpha > p + \gamma - 1$.*

Theorems on bases can usually be proved by comparing (20.2) with the analogous orthonormal eigenvector expansion for the self-adjoint operator A_0. Let Γ_j be a sequence of admissible contours for A and A_0 enclosing parts Λ_j and M_j of the spectra of A and A_0, and let P_j and Q_j be the corresponding spectral projections. Then the series (20.3) converges if and only if so does the series

$$\sum_j (Q_j - P_j) f$$
$$= -(2\pi i)^{-1} \sum_j \int_{\Gamma_j} ((A - \zeta I)^{-1} - (A_0 - \zeta I)^{-1}) f \, d\zeta. \quad (20.11)$$

It follows that the problem can be reduced to finding the desired system of admissible contours Γ_j and estimating integrals of the form (20.11).

Similarly, the proofs of the theorems on the asymptotic behaviour of the spectrum of a non-self-adjoint operator are based on estimates of the resolvent. In addition to Sect. 1.11, we denote by $\tilde{N}(r; A)$ the number of eigenvalues of A in the disc $|\lambda| < r$ with their algebraic multiplicity taken into account. Let Γ_r be an admissible contour for A and A_0 enclosing these eigenvalues. Then

$$\tilde{N}(r; A) - \tilde{N}(r; A_0)$$
$$= -(2\pi i)^{-1} \mathrm{Tr} \int_{\Gamma_r} ((A - \zeta I)^{-1} - (A - \zeta_0 I)^{-1}) \, d\zeta. \quad (20.12)$$

If $\sigma(A)$ and $\sigma(A_0)$ lie sufficiently far away from the contour Γ_r, i.e., there is a sufficiently large 'gap' both in $\sigma(A)$ and $\sigma(A_0)$, then the fact that B is 'small' implies that the resolvents $(A - \zeta I)^{-1}$ and $(A_0 - \zeta I)^{-1}$ are close to one another, and, as a result, the integral (20.12) is small. If there is no such gap in the spectrum, then it can be formed artificially by adding suitable finite-dimensional operators to A and A_0. For a detailed presentation of the technique in question and the results obtained see (Markus and Matsaev 1982).

Theorem 20.4 (Markus and Matsaev 1981, 1982). *Let $A_0 > 0$ be a self-adjoint operator with discrete spectrum and let B be an operator p-subordinate to A_0 in the sense of operators or forms with $0 \leq p < 1$. Then all but finitely many eigenvalues of $A = A_0 + B$ are contained inside the domain $|\mathrm{Im}\, \lambda| \leq c|\mathrm{Re}\, \lambda|^p$ and*

$$\tilde{N}(r; A) - N(r; A_0) = O\big(N(r + qr^p; A_0) - N(r - qr^p; A_0)\big) \quad (20.13)$$

for some $q > 0$.

The theorem can be extended to exponents $p < 0$, the condition for B being that the operator BA_0^{-p} must admit an extension to a bounded operator in $\mathfrak{H}$. Theorem 20.4 can also be carried over to non–semi-bounded operators A_0, in which case a formula of the type (20.13) turns out to be valid separately for the spectrum of A near the positive and negative rays.

These results are important because they provide practically complete solutions (at the abstract level!) to problems concerning the asymptotic behaviour of the spectra of operators being close to self-adjoint ones. Formula (20.10) can be used to find the leading term of the asymptotics of the spectrum of A. Also, if either an estimate of the remainder or the second term of the spectral asymptotics is known for A_0 and B is sufficiently small, then (20.10) yields an estimate of the remainder or the second term of $\tilde{N}(r; A)$, respectively.

If B is merely A_0-compact, then (20.13) is replaced by the inequality

$$\left|\tilde{N}(r;A) - N(r;A_0)\right| \le c_1\big(N(r(1+\varepsilon);A_0) \\ - N(r(1-\varepsilon);A_0)\big) + c_2, \qquad (20.14)$$

which is valid for any $\varepsilon > 0$ and all $r > r(\varepsilon)$. Besides, the spectrum of A is finite in the complement of any angle $|\arg\lambda| < \theta$. In particular, it follows from (20.14) that the relation

$$\frac{\tilde{N}(r;A)}{N(r;A_0)} \to 1$$

is valid, subject to the rather weak regularity condition

$$\lim_{\varepsilon\to 0}\left(\limsup_{r\to\infty}\frac{N(r(1+\varepsilon);A_0)}{N(r;A_0)}\right) = 1 \qquad (20.15)$$

for $N(r;A_0)$. In other words, subject to the condition (20.15), the leading term of the spectral asymptotics of a self-adjoint operator A_0 is invariant under compact (non–self-adjoint) perturbations.

Condition (20.15) is satisfied, for example, if $N(r;A_0) \sim cr^{\alpha}$. It is much weaker than, for example, the Tauberian condition in Theorem 12.2. As was demonstrated by Markus and Matsaev (1982), the asymptotics may no longer be preserved if the condition is violated.

In the conclusion let us mention that the results of this section can be extended without any substantial modifications to operators of the form $A_0 + B$ obtained by perturbing an invertible normal operator A_0, the spectrum of which is contained in a small neighbourhood of finitely many rays in $\mathbb{C}^1$ (see Markus and Matsaev 1982).

20.5. Application to Differential Operators

To apply the abstract results (in particular, Theorems 20.1–20.4) in a concrete situation one needs explicit criteria for the non–self-adjoint perturbation to be small, as required by the theorems. For the operators considered in Examples 20.3 and 20.4 such criteria can be obtained quite easily using, in particular, the results of §5 on the estimates of the spectrum. The results below are, basically, contained in (Agranovich 1977, 1980; Markus and Matsaev 1981, 1982).

Under the conditions of Example 20.3, the operator A^{-1} transforms $L_2(X)$ into $H^m(X)$. It follows that the s-numbers of A^{-1} are majorized by the s-numbers of the embedding operator from $H^m(X)$ into $L_2(X)$. Thus $s_j(A^{-1}) = O(j^{-m/n})$, where $n = \dim X$. The same estimate is valid for $s_j((A^{-1})^*) = s_j(A^{-1})$. Hence, by Theorem 20.2, if the numerical domain of A is contained in a sector of $\mathbb{C}^1$ whose angle is less than $\pi n/m$, then the system of root vectors of A is complete.

Let us now assume that A is close to a self-adjoint operator. To begin with, we consider an elliptic operator on a manifold without boundary. Let $A = A_0 + B$, where $A_0 = A_0^* > 0$, $\operatorname{ord} A_0 = m$, and B is a differential operator of order $m' < m$. Then A_0 is an isomorphism from $H^m(X)$ onto $L_2(X)$ and B is $A_0^{m'/m}$-bounded in $L_2(X)$. It follows that

$$s_j(BA_0^{-1}) = s_j(BA_0^{-m'/m}A_0^{-1+m'/m})$$
$$= O(s_j(A_0^{-1})^{1-m'/m}) = O(j^{-(m-m')/n}),$$

which ensures that the s-numbers of BA_0^{-1} decrease, as required by Theorem 20.1. It follows that A has a complete system of root vectors.

Furthermore, by Theorem 20.3, the root vectors of A form a Riesz basis with parentheses for $m - m' \geq \dim X$, and the expansion (20.2) is summable by the Abel method of order $\alpha > 1 - (m - m')/n$ for $m - m' < n$.

Next, applying Theorem 20.4 to the operator under consideration, we can see that the spectral asymptotics

$$\tilde{N}(\lambda; A) - a_0\lambda^{n/m} = O(\lambda^{(n-1)/m}), \quad \lambda \to +\infty$$

with an estimate of the remainder is valid for A, the coefficient a_0 being the same as for A_0. Moreover, all but finitely many points of the spectrum of A belong to the domain

$$|\operatorname{Im}\lambda| \leq c(\operatorname{Re}\lambda)^{m'/m}.$$

But if the two-term asymptotic formula

$$N(\lambda; A_0) = a_0\lambda^{n/m} + a_1\lambda^{(n-1)/m} + o(\lambda^{(n-1)/m})$$

is known for A_0 (for example, if the conditions of Theorems 13.4 and 13.9 are met) and $m' < m - 1$, then the same two-term asymptotic formula is also valid for A.

Subject to the natural modifications, these results can be carried over to the case of non–semi-bounded elliptic operators on X.

For an elliptic operator A acting on a manifold with boundary both the differential expression and the boundary condition can give rise to non–self-adjointness. Let $\mathcal{L}_0$ be a uniformly elliptic differential expression of order $m = 2k$ acting on functions in a domain $X \subset \mathbb{R}^n$. Let B_j, where $j \in \{1, \ldots, k\}$, be a system of boundary operators of order m_j with $0 \leq m_1 < \ldots < m_k < m$ involving the derivative of order m_j in the normal direction to the boundary. Suppose that the system $(\mathcal{L}, B_j)$ defines the operator of a regular elliptic problem and the operator

$$A_0 u = \mathcal{L}_0 u, \quad D(A_0) = \{u \in H^m(X) : B_j u|_{\partial X} = 0 \ \forall\, j\}$$

is self-adjoint and invertible in $L_2(X)$ (for simplicity, we assume that the boundary and all the coefficients are smooth).

Let A be an operator defined in the domain

$$D(A) = \{u \in H^m(X) : (B_j + B_j')u\big|_{\partial X} = 0 \ \forall j\}$$

by the expression $Au = \mathcal{L}u = (\mathcal{L}_0 + \mathcal{L}')u$, where $\mathcal{L}'$ is a differential expression of order $m' < m$ and B_j' are boundary operators of order $m_j' < m_j$.

In the case at hand the application of Theorems 20.1, 20.3, and 20.4 presents difficulties, because A_0 and A have different domains. However, as was demonstrated in (Markus and Matsaev 1981), one can construct a continuous operator Σ from $H^s(X)$ into $H^{s+1}(X)$, where $-1 \le s \le m-1$, such that $I - \Sigma$ is invertible in $L_2(X)$ and maps $D(A_0)$ onto $D(A)$. It follows that $I - \Sigma$ furnishes the similarity between A and $\tilde{A} = A_0 + \tilde{B}$, where $\tilde{B}$ is an operator from H^s into H^{s-m+1}, even though not a differential one. If the order m_j' of the boundary operator B_j' satisfies the inequality $m_j' \le m_j - \mu$ with $\mu > 1$ for each $j \in \{1, \dots k\}$, then the above construction yields a continuous operator Σ from $H^s(X)$ to $H^{s+\mu}(X)$, where $-\mu \le s \le m - \mu$. Then $I - \Sigma$ furnishes the similarity between A and $\tilde{A} = A_0 + \tilde{B}$, where $\tilde{B} : H^s \to H^{s-\tilde{m}}$ with $\tilde{m} = \max(m', m - \mu)$.

The spectra of similar operators are the same, and so are the convergence properties of their spectral expansions. It follows that the assertions stated above for an operator on a manifold without boundary can be carried over to the system of root vectors of the operator of an elliptic boundary value problem (with m' replaced by $\tilde{m}$).

In the conclusion we consider the Schrödinger operator with a complex-valued potential, the modulus of which grows at infinity. We assume that $V_1 = o(V_0)$ as $|x| \to \infty$ in the conditions of Example 20.4. Then the operator T defined by (20.6) is compact in $d[a_0]$. In other words, $B : u \mapsto V_1 u$ is A_0-compact in the sense of forms. Since $s_j(A_0^{-1/2} T A_0^{-1/2}) = O(s_j(A_0^{-1}))$, the conditions of Theorem 20.1 are met, for example, if $V_0(x) \ge c(1 + |x|^\varepsilon)$, where $\varepsilon > 0$. Under those conditions, the system of root vectors of A is complete. Furthermore, we assume that $V_1 = O(V_0^p)$, where $0 < p < 1$. Then

$$\begin{aligned}\int |V_1||u|^2\,dx &\le \Big(\int |V_1|^{1/p}|u|^2\,dx\Big)^p \Big(\int |u|^2\,dx\Big)^{1-p} \\ &\le c(a_0[u])^p \|u\|^{2-2p}.\end{aligned}$$

It follows that the quadratic form of the perturbation is p-subordinate to A_0. Now, applying Theorems 20.3 and 20.4, we can obtain concrete results on the convergence of the spectral expansions and the position of the spectrum of the Schrödinger operator.

We use the Schrödinger operator $A = -\Delta + V_0 + iV_1$ with $V_1 > 0$ and $V_1 \le bV_0$ as an example to demonstrate how to apply Theorem 20.2 in the situation at hand. The numerical domain of A is contained in a sector of angle π/p, where $p = \pi/\arctan b$. Thus, in order that the system of root vectors be complete, it is sufficient that the series

$$\sum_j s_j^p(A^{-1}) \leq c \sum_j s_j^p(A_0^{-1})$$

be convergent. We assume that $n \geq 3$. Then, as follows from (5.11),

$$\sum_j s_j^p(A_0^{-1}) = \int_0^\infty \lambda^{-p}\, dN(\lambda; A_0) \leq c \int (V_0(x))^{-p+n/2}\, dx, \quad 2p > n.$$

It follows that if

$$V_0^{-p+n/2} \in L_1(\mathbb{R}^n), \quad n \geq 3, \tag{20.16}$$

then the system of root vectors of the Schrödinger operator is complete. If the condition $V_1 > 0$ in (20.16) is relaxed, then p must be replaced by $p/2$.

Comments on the Literature

The literature concerned with various aspects of spectral theory is immense. Listed in the bibliography below are only the basic monographs, surveys, as well as articles containing the most fundamental results related to the problems touched upon in the present volume.

The results obtained up to the early 60-ties are reflected in the books (Titchmarsh 1946; Courant and Hilbert 1953; Dunford and Schwartz 1963; Glazman 1963; Berezanskij 1966). The books (Coddington and Levinson 1955; Naimark 1969; Levitan and Sargsyan 1970; Joergens and Rellich 1976; Marchenko 1977; Kostyuchenko and Sargsyan 1979) are concerned with the theory of ordinary differential operators. The methods of functional analysis used in spectral theory, in the first place the theory of operators in Hilbert space, are presented in (Maurin 1959; Dunford and Schwartz 1963; Gokhberg and Krein 1965; Akhiezer and Glazman 1966; Kato 1966; Reed and Simon 1972, 1975, 1978, 1979; Birman and Solomyak 1980). One can learn about the fundamental functional spaces and embedding theorems from (Nikol'skij 1977). Information on non-smooth and unbounded weights, domains, and the like can be found in (Maz'ya 1985; Mynbaev and Otelbaev 1988).

The classical theory of boundary value problems for partial differential equations is presented in (Lions and Magenes 1968). Recent advances in these problems are connected with the emergence of the technique of pseudodifferential operators and Fourier integral operators (FIO). The expositions of this material in (Shubin 1978a; Taylor 1981; Tréves 1982; Grubb 1986) complement one another and contain numerous applications, in particular, to spectral theory. The most advanced theory is presented in (Hörmander 1983–1985); see also (Egorov and Shubin 1988a, 1988b) and references therein. The related FIO method of the canonical operator used to construct approximate eigenfunctions (quasimodes) is presented in (Fedoryuk and Maslov 1976; Maslov 1976, 1987, 1988). Other asymptotic methods of constructing quasimodes are discussed in (Colin de Verdière 1977; Lazutkin 1981, 1987; Babich 1987). Information on the abstract theory of perturbations of discrete spectra is contained in (Friedrichs 1965; Kato 1966). Applications to asymptotic problems for differential operators are given in (Lyusternik and Vishik 1960; Reed and Simon 1972, 1975, 1978, 1979; Nazarov 1987). The questions of averaging theory in spectral problems are considered in (Sánchez-Palencia 1980; Oleinik 1987).

Several books have been devoted to the analysis of special classes of operators. Along with regular boundary value problems, the Schrödinger operator considered in (Berezin and Shubin 1983; Cycon et al. 1987) and, in particular, (Reed and Simon 1972, 1975, 1978, 1979) represents the most interesting example. The books (Joergens and Weidmann 1973; Faddeev and Merkur'ev 1985 (scattering theory); Murtazin and Sadovnichij 1988) and the article (Yafaev 1972) are specially devoted to the multiparticle Schrödinger operator. On the many-dimensional Dirac operator see (Arai 1983). The books and surveys (Eastham 1973; Shubin 1978c, 1979; Johnson 1983; Skriganov 1985; Chulajevskij 1989) are devoted to operators with periodic and almost periodic coefficients. One can learn about the present-day state of the theory of random operators from (Pastur 1973, 1987a, 1987b; Gredeskul et al. 1982; Hörmander 1983–1985; Carmona 1986; Spencer 1986; Carmona and Lacroix 1990; Figotin and Pastur 1992).

The literature on spectral asymptotics is very extensive. The survey (Clark 1967) describes the state of the subject in 1967, and (Birman and Solomyak 1977a) – in 1975. The most detailed exposition of the variational method is contained in (Birman and Solomyak 1974), of the resolvent method for various problems in (Kostyuchenko and Sargsyan 1979; Agmon 1965), and of the hyperbolic equation method in (Kostyuchenko 1968; Greiner 1971; Molchanov 1975; Smith 1981; Taylor 1981). On the ζ-function method see (Seeley 1967; Shubin 1978a). For reviews of direct and inverse results in spectral geometry see (Berger et al. 1971; Singer 1974; Bérard 1986). The approximate spectral projection method is presented in (Shubin 1978a; Levendorskij 1988b, 1990), and applied to non-Weyl asymptotics in (Levendorskij 1988a). The most advanced results obtained with the aid of the hyperbolic equation method can be found in (Duistermaat and Guillemin 1975; Bardos et al. 1982), in Ivrii's book (Ivrii 1984) and his later publications (Ivrii 1985, 1986a, 1986b, 1987), as well as in (Hörmander 1983–1985; Vasil'ev 1986; Safarov 1988a, 1988b).[13]

The problem of whether or not 'one can hear the shape of a drum' was posed by Kac (1966) and, for compact Riemannian surfaces, by Gel'fand (1963). On the most important results on the uniqueness or non-uniqueness of the reconstruction of a differential operator from its discrete spectrum see (McKean 1972; Guillemin and Kazhdan 1980; Urakawa 1982; Gordon and Wilson 1984; Bérard 1986; Buser 1986). The theory of discrete isometry groups is presented in (Shvartsman and Vinberg 1988). On the theory of operators that are close to self-adjoint ones see (Gokhberg and Krein 1965; Agranovich 1977, 1980; Markus and Matsaev 1982).

[13](Added in the English edition.) See also the forthcoming book by Safarov and Vasil'ev.

References*

Agmon, S.

[1965] On kernels, eigenvalues and eigenfunctions of operators related to elliptic problems. Commun. Pure. Appl. Math. *18*, No. 4, 627–663. Zbl.151,202

[1968] Asymptotic formulas with remainder estimates for eigenvalues of elliptic operators. Arch. Ration. Mech. Anal. *28*, No. 3, 165–183. Zbl.159,159

Agranovich, M.S.

[1977] Spectral properties of diffraction problems. In: Vojtovich, N.N., Katsenelenbaum V.Z., Sivov A.N., The Generalized Method of Characteristic Oscillations in Diffraction Theory. Nauka, Moscow, 289–416 (Russian). Zbl.384.35002

[1980] On the convergence of series in terms of the root vectors of operators that are close to self-adjoint ones. Tr. Mosk. Mat. O.-va *41*, 163–186. English transl.: Trans. Mosc. Math. Soc. 1982, No. 1, 167–182 (1982). Zbl.479.47012

Agranovich, M.S., Vishik, M.I.

[1964] Elliptic problems with a parameter and parabolic problems of general form. Usp. Mat. Nauk *19*, No. 3, 53–161. English transl.: Russ. Math. Surv. *19*, No. 3, 53–157 (1964). Zbl.137,296.

Aizenman, M.

[1993] Localization at weak disorder: some elementary bounds. Preprint.

Aizenman, M., Molchanov, S.

[1993] Localization at large disorder and at extreme energies: an elementary derivation. Commun. Math. Phys. *157*, No. 2, 245–278.

Akhiezer, N.I., Glazman, I.M.

[1966] Theory of Linear Operators in Hilbert Space. 2nd ed., Nauka, Moscow (Russian). Zbl.143.365

Anderson, P.W.

[1958] Absence of diffusion in certain random lattices. Phys. Rev. *109*, 1492–1505.

Anosov, D.V., Aranson, S.Kh., Bronshtejn, I.U., Grines, V.Z.

[1985] Smooth dynamical systems. Itogi Nauki Tekh., Ser. Sovrem. Probl. Mat., Fundam. Napravleniya *1*, 151–242. English transl.: Encycl. Math. Sci. *1*. Springer-Verlag, Berlin-Heidelberg-New York, 149–233 (1988). Zbl.658.58002.

Arai, M.

[1983] On essential self-adjointness, distinguished self-adjoint extension and essential spectrum of Dirac operators with matrix valued potentials. Publ. Res. Inst. Math. Sci. *19*, No. 1, 33–57. Zbl.518.35071

Aramaki, J.

[1983] Complex powers of a class of pseudodifferential operators and their applications. Hokkaido Math. J. *12*, No. 2, 199–225. Zbl.548.35111

[1987] On the asymptotic behavior of the spectra of quasi-elliptic pseudodifferential operators in $\mathbb{R}^n$. Tokyo J. Math. *10*, No. 2, 481–505. Zbl.655.35063

---

* For the convenience of the reader, references to reviews in Zentralblatt für Mathematik (Zbl.), compiled using the MATH database, and Jahrbuch über die Fortschritte der Mathematik (FdM.) have, as far as possible, been included in this bibliography.

Arnol'd, V.I.

[1972] Modes and quasimodes. Funkts. Anal. Prilozh. *6*, No. 2, 12–20. English transl.: Funct. Anal. Appl. *6*, 94–101 (1972). Zbl.251.70012

Arnol'd, V.I., Kozlov, V.V., Neishtadt, A.I.

[1985] Mathematical aspects of classical and celestial mechanics. Itogi Nauki Tekh., Ser. Sovrem. Probl. Mat., Fundam. Napravleniya *3*, 5–300. English transl.: Encycl. Math. Sci. *3*. Springer-Verlag, Berlin-Heidelberg-New York, 1–291 (1987). Zbl.612.70002

Aslanyan, A.G., Lidskij, V.B.

[1974] Distribution of the Eigenfrequencies of Thin Elastic Shells. Nauka, Moscow (Russian).

Avron, J.S., Simon, B.

[1978] Analytic properties of band functions. Ann. Phys. *110*, No. 1, 85–101.

[1981] Almost periodic Hill's equation and the rings of Saturn. Phys. Rev. Lett. *46*, No. 17, 1166–1168.

[1982] Almost periodic Schrödinger operators. I. Limit periodic potentials. Commun. Math. Phys. *82*, 101–120. Zbl.484.35069

[1983] Almost periodic Schrödinger operators. II. The integrated density of states. Duke Math. J. *50*, No. 1, 369–391. Zbl.544.35030

Baaj, Saad

[1988] Calcul pseudo-différentiel et produits croisés de C^*-algèbres. C. R. Acad. Sci., Paris, Sér. I *307*, No. 11, 581–586, Zbl.655.46055, No. 12, 663–666. Zbl.770.46027

Babich, V.M.

[1979] The focusing problem and asymptotics of the spectral function of the Laplace-Beltrami operator II. Zap. Nauchn. Semin. Leningr. Otd. Mat. Inst. Steklova *89*, 15–53. English transl.: J. Sov. Math. *19*, 1288–1322 (1982). Zbl.427.35052

[1987] The many-dimensional WKB method or ray method. Its analogues and generalizations. Itogi Nauki Tekh., Ser. Sovrem. Probl. Mat., Fundam. Napravleniya *34*, 93–134, Zbl.657.35119. English transl.: Encycl. Math. Sci. *34*. Springer-Verlag, Berlin-Heidelberg-New York (in preparation).

Bando, S., Urakawa, H.

[1983] Generic properties of the eigenvalue of the Laplacian for compact Riemannian manifolds. Tôhoku Math. J., II. Ser. 35, 155–172. Zbl.534.58038

Bardos, C., Guillot, J., Ralston, J.

[1982] La relation de Poisson pour l'équation des ondes dans un ouvert non borné. Application à la théorie de la diffusion. Commun. Partial Differ. Equations 7, No. 8, 905–958. Zbl.496.35067

Beals, R., Fefferman, C.

[1974] Spatially inhomogeneous pseudodifferential operators. Commun. Pure Appl. Math. *27*, No. 1, 1–24. Zbl.279.35071

Bellissard, J.

[1989] Almost periodicity in solid state physics and C^*-algebras. In: C. Berg and B. Fuglede (eds.), The Harald Bohr Centenary. Mat.-Fys. Medd., K. Dan. Vidensk. Selsk. *42*, No. 3, 35–75. Zbl.678.42007

[1992] Gap labelling theorems for Schrödinger operators. In: M. Waldschmidt et. al (eds.), From Number Theory to Physics. Springer-Verlag, Berlin-Heidelberg-New York, 538–630.

Bellissard, J., Lima, R., Scoppola, E.
[1983] Localization in ν-dimensional incommensurate structures. Commun. Math. Phys. *88*, No. 4, 465–477. Zbl.543.35018

Bellissard, J., Lima, R., Testard, D.
[1985] Almost periodic Schrödinger operators. In: L. Streit (ed.) Mathematics & Physics. Lectures on Recent Results. Vol. 1. World Scientific Publications, Singapore-Philadelphia, 1–64. Zbl.675.34022

Belokolos, E.D.
[1975] A quantum particle in a one-dimensional deformed lattice. Estimates of lacunae dimensions in the spectrum. Teor. Mat. Fiz. *25*, No. 3, 344–357 (Russian).
[1976] A quantum particle in a one-dimensional deformed lattice. Dependence of the energy on the quasimomentum. Teor. Mat. Fiz. *26*, No. 1, 35–41 (Russian).

Bérard, P.H.
[1977] On the wave equation on a compact Riemannian manifold without conjugate points. Math. Z. *155*, No. 2, 249–273. Zbl.341.35052
[1980a] Spectres et groupes cristallographiques I. Domaines euclidiens. Invent. Math. *58*, No. 2, 179–199. Zbl.434.35068
[1980b] Spectres et groupes cristallographiques II. Domaines sphériques. Ann. Inst. Fourier *30*, No. 3, 237–248. Zbl.437.35053
[1986] Spectral geometry: direct and inverse problems. Lect. Notes Math. *1207*, 1–272. Zbl.608.58001
[1989] Variétés riemannienes isospectrales non isométriques. Astérisque *177–178*, 127–154.

Berezanskij, Yu.M.
[1966] Expansions of Self-Adjoint Operators in Terms of Eigenfunctions. Naukova Dumka, Kiev. English transl.: Providence, R.I. 1968. Zbl.167,429

Berezin, F.A., Shubin, M.A.
[1983] The Schrödinger Equation. Moscow State Univ. Press, Moscow. English transl.: Kluwer, Dordrecht 1991. Zbl.546.35002

Berger, M., Gauduchon, P., Mazet, E.
[1971] Le spectre d'une variété Riemannienne. Lect. Notes Math. *194*, 1–251. Zbl.223.53034

Berry, J.-P.
[1981] Tores isospectraux en dimension 3. C. R. Acad. Sci., Paris, Sér. I *292*, No. 2, 163–165. Zbl.464.58018

Besse, A.
[1978] Manifolds All of Whose Geodesics are Closed. Springer-Verlag, Berlin-Heidelberg-New York. Zbl.387.53010

Bezyaev, V.I.
[1978] Asymptotics of the density of states of hypoelliptic almost periodic operators. Mat. Sb., Nov. Ser. *105*, No. 4, 485–511. English transl.: Math. USSR, Sb. *34*, 425–447 (1978). Zbl.382.47026
[1982] Asymptotics of the eigenvalues of hypoelliptic operators on a closed manifold. Mat. Sb., Nov. Ser. *117*, No. 2, 161–180. English transl.: Math. USSR, Sb. *45*, 169–189 (1983). Zbl.485.35073

Biktashev, R.A., Mishchenko, A.S.
[1980] The spectra of unbounded elliptic pseudodifferential operators over C^*-algebras. Vestn. Mosk. Univ., Ser. I, No. 3, 56–58. English transl.: Mosc. Univ. Math. Bull. *35*, No. 3, 59–62 (1980). Zbl.439.47041

Birman, M.Sh.

[1959] On the spectrum of the Schrödinger and Dirac operators. Dokl. Akad. Nauk SSSR *129*, No. 2, 239–241 (Russian). Zbl.104,77

[1961] On the spectrum of singular boundary value problems. Mat. Sb., Nov. Ser. *55*, No. 2, 125–174. English transl.: Am. Math. Soc., Transl., II. Ser. *53*, 23–80 (1966). Zbl.104,326

[1962] Perturbation of the continuous spectrum of an elliptic operator under variation of the boundary and boundary conditions. Vestn. Leningr. Univ. *17*, Ser. Mat. Mekh. Astron., No. 1, 22–55 (Russian). Zbl.149,319

Birman, M.Sh., Solomyak, M.Z.

[1970] On the leading term of the spectral asymptotics for non-smooth elliptic problems. Funkts. Anal. Prilozh. *4*, No. 4, 1–13. English transl.: Funct. Anal. Appl. *4*, 265–275 (1971). Zbl.225.35077

[1972] Spectral asymptotics of non-smooth elliptic operators. I. Tr. Mosk. Mat. O.-va *27*, 3–52. English transl.: Trans. Mosc. Math. Soc. *27*, 1–52 (1975). Zbl.251.35075

[1973] Spectral asymptotics of non-smooth elliptic operators. II. Tr. Mosk. Mat. O.-va *28*, 3–34. English transl.: Trans. Mosc. Math. Soc. *28*, 1–32 (1975). Zbl.281.35079

[1974] Quantitative analysis in Sobolev's imbedding theorems and applications to spectral theory. Tenth Mathematical School, Inst. Mat. Akad. Nauk Ukr. SSR, Kiev. English transl.: Am. Math. Soc., Transl., Vol. 114, Providence (1980). Zbl.426.46019

[1977a] Asymptotics of the spectrum of differential equations. Itogi Nauki Tekh., Ser. Mat. Anal. *14*, 5–58. English transl.: J. Sov. Math. *12*, 247–283 (1979). Zbl.417.35061

[1977b] Asymptotics of the spectrum of pseudodifferential operators with anisotropically homogeneous symbols. Vestn. Leningr. Univ. 1977, No. 13 (Mat. Mekh. Astron., No. 3), 13–21. English transl.: Vestn. Leningr. Univ. Math. *10*, 237–247 (1982). Zbl.377.47033

[1979] Asymptotics of the spectrum of variational problems on solutions of elliptic equations. Sib. Math. Zh. *20*, No. 1, 3–22. English transl.: Sib. Math. J. *20*, 1–15 (1979). Zbl.414.35059

[1980] Spectral Theory of Self-Adjoint Operators in Hilbert Space. Leningr. Gosp. Univ., Leningrad. English transl. Kluwer, Dordrecht etc. 1987. Zbl.744.47017

[1982a] Asymptotics of the spectrum of variational problems on solutions of elliptic systems. Zap. Nauchn. Semin. Leningr. Otd. Mat. Inst. Steklova *115*, 23–35. English transl.: J. Sov. Math. *28*, 633–644 (1985). Zbl.501.35065

[1982b] Asymptotics of the spectrum of pseudodifferential variational problems with constraints. Probl. Mat. Fiz. *10*, 20–36. English transl.: Sel. Math. Sov. *5*, 245–256 (1986). Zbl.602.35086

Bogorodskaya, T.E., Shubin, M.A.

[1986] Variational principle and asymptotic behaviour of the density of states for random pseudodifferential operators. Tr. Semin. Im. I. G. Petrovskogo *11*, 98–117. English transl.: J. Sov. Math. *45*, No. 5, 1440–1454 (1989). Zbl.598.60069

Bojmatov, K.Kh.

[1974] Asymptotics of the spectrum of the Schrödinger operator. Differ. Uravn. *10*, No. 11, 1939–1945. English transl.: Differ. Equations *10*, 1492–1497 (1976). Zbl.292.35063

[1984] Separation theorems, weighted spaces, and their applications. Tr. Mat. Inst. Steklova *170*, 37–76. English transl.: Proc. Steklov Inst. Math. *170*, 39–81 (1987). Zbl.578.46026

Boutet de Monvel, L., Guillemin, V.
[1981] The Spectral Theory of Toeplitz Operators. Princeton Univ. Press, Princeton, N.J. Zbl.469.47021

Bovier, A., Campanino, M., Klein, A., Perez, J.F.
[1988] Smoothness of the density of states in the Anderson model at high disorder. Commun. Math. Phys. *114*, 439–461. Zbl.644.60057

Buser, P.
[1980] Ph.D. Dissertation, Gettinen.
[1986] Isospectral Riemannian surfaces. Ann. Inst. Fourier *36*, No. 2, 167–192. Zbl.579.53036

Buslaev, V.S.
[1975] On the asymptotic behaviour of the spectral characteristics of exterior problems for the Schrödinger operator. Izv. Akad. Nauk SSSR, Ser. Mat. *39*, No. 1, 149–235. English transl.: Math. USSR, Izv. *9*, 139–223 (1976). Zbl.311.35010

Carleman, T.
[1936] Über die asymptotische Verteilung der Eigenwerte partieller Differentialgleichungen. Ber. Sächs. Akad. Wiss. Leipzig *88*, 119–132. Zbl.17,114

Carmona, R.
[1982] Exponential localization in one dimensional disordered systems. Duke Math. J. *49*, No. 1, 191–213. Zbl.491.60058
[1985] One-dimensional Schrödinger operators with random potentials: a survey. Acta Appl. Math. *4*, No. 1, 65–91. Zbl.547.60069
[1986] Random Schrödinger operators. Lect. Notes Math. *1180*, 1–124. Zbl.607.35089

Carmona, R., Lacroix, J.
[1990] Spectral Theory of Random Schrödinger Operators. Birkhäuser, Boston. Zbl.717.60074

Chapman, S.
[1992] Drums that sound the same. Preprint, Stanford Univ.

Chazarain, J.
[1973] Construction de la paramétrix du problème mixte hyperbolique pour l'équation des ondes. C. R. Acad. Sci., Paris, Sér. A *276*, No. 9, 1213–1215. Zbl.253.35058
[1974] Formule de Poisson pour les variétés Riemanniennes. Invent. Math. *24*, No. 1, 65–82. Zbl.281.35028
[1980] Spectre d'un hamiltonien quantique et méchanique classique. Commun. Partial Differ. Equations *5*, No. 6, 595–644. Zbl.437.70014

Chulaevskij, V.A.
[1981] On perturbations of the Schrödinger operator with periodic potential. Usp. Mat. Nauk *36*, No. 5, 203–204. English transl.: Russ. Math. Surv. *36*, No. 5, 143–144 (1981). Zbl.487.34021
[1989] Almost Periodic Operators and Related Non-Linear Integrable Systems. Manchester Univ. Press, Manchester. Zbl.699.47031

Chulaevskij, V.A., Molchanov, S.A.
[1984] The structure of the spectrum of a lacunary-limit-periodic Schrödinger operator. Funkts. Anal. Prilozh. *18*, No. 4, 90–91. English transl.: Funct. Anal. Appl. *18*, 343–344 (1984). Zbl.569.34024

Clark, C.
[1967] The asymptotic distribution of eigenvalues and eigenfunctions for elliptic boundary value problems. SIAM Rev. *9*, No. 4, 627–646. Zbl.159,197

Coburn, L.A., Moyer, R.D., Singer, I.M.
[1973] C^*-algebras of almost periodic pseudodifferential operators. Acta. Math. *130*, No. 3–4, 279–307. Zbl.263.47042

Coddington, E.A., Levinson, N.
[1955] Theory of Ordinary Differential Equations. McGraw-Hill, New York. Zbl.64,330

Colin de Verdière, Y.
[1973] Spectre du laplacien et longueurs des géodésiques périodiques. I, II. Compos. Math. *27*, No. 1, 83–106, No. 2, 159–184. Zbl.272.53034, Zbl.281.53036
[1977] Quasi-modes sur les variétés riemanniennes compactes. Invent. Math. *43*, No. 1, 15–52. Zbl.449.53040
[1979] Sur le spectre des opérateurs elliptiques à bicaractéristiques toutes périodiques. Comment. Math. Helv. *54*, No. 3, 508–522. Zbl.459.58014
[1982] Pseudo-laplaciens. I. Ann. Inst. Fourier *32*, No. 3, 275–286. Zbl.489.58034
[1983] Pseudo-laplaciens. II. Ann. Inst. Fourier *33*, No. 2, 87–113. Zbl.496.58016

Combes, J.N., Briet, P., Duclos, P.
[1987] Spectral properties of Schrödinger operators with trapping potential in the semiclassical limit. Lect. Notes Math. *1285*, 55–72. Zbl.654.47032

Cornfel'd, I.P., Fomin, S.V., Sinai, Ya.G.
[1980] Ergodic Theory. Nauka, Moscow. English transl.: Springer-Verlag, Berlin-Heidelberg-New York, 1982. Zbl.493.28007

Courant, R., Hilbert, D.
[1953] Methods of Mathematical Physics. I. Interscience, New York. Zbl.53,28

Craig, W.
[1983] Pure point spectrum for discrete almost periodic Schrödinger operators. Commun. Math. Phys. *88*, No. 1, 113–131. Zbl.525.35076

Craig, W., Simon, B.
[1983a] Log Hölder continuity of the integrated density of states for stochastic Jacobi matrices. Commun. Math. Phys., *90*, No. 2, 207–218. Zbl.532.60057
[1983b] Subharmonicity of the Lyapunov index. Duke Math. J. *50*, No. 4, 551–560. Zbl.518.35027

Cramer, H., Leadbetter, M.R.
[1967] Stationary and Related Stochastic Processes. Sample Function Properties and Their Applications. John Wiley, New York-Sydney. Zbl.162,211

Cycon, H.L., Froese, R.G., Kirsch, W., Simon, B.
[1987] Schrödinger Operators with Application to Quantum Mechanics and Global Geometry. Springer-Verlag, Berlin-Heidelberg-New York. Zbl.619.47005

Dedik, P.E., Shubin, M.A.
[1980] Random pseudodifferential operators and stabilization of solutions of parabolic equations with random coefficients. Mat. Sb., Nov. Ser. *113*, No. 1, 41–64. English transl.: Math. USSR, Sb. *41*, 33–52 (1982). Zbl.446.35092

Deift, P., Simon, B.
[1983] Almost periodic Schrödinger operators. III. The absolutely continuous spectrum in one dimension. Commun. Math. Phys. *90*, 389–411. Zbl.562.35026

Delyon, F., Levy, Y., Souillard, B.
[1985a] Anderson localization for multidimensional systems at large disorder or low energy. Commun. Math. Phys. *100*, No. 4, 463–470. Zbl.576.60053
[1985b] Anderson localization for one- and quasi one-dimensional systems. J. Stat. Phys. *41*, No. 3–4, 375–388.
[1985c] Approach à la Borland to multidimensional localization. Phys. Rev. Lett. *55*, No. 6, 618–621.

Dinaburg, E.I., Sinai, Ya.G.
[1975] The one-dimensional Schrödinger equation with a quasiperiodic potential, Funkts. Anal. Prilozh. *9*, No. 4, 8–21. English transl.: Funct. Anal. Appl. *9*, 279–289 (1976). Zbl.333.34014

Dixmier, J.
[1969] Les algèbres d'opérateurs dans l'espace hilbertien (algèbres de von Neumann). Gauthier-Villars, Paris. Zbl.175,438

Donsker, M.N., Varadhan, S.R.S.
[1975] Asymptotics for the Wiener sausage. Commun. Pure Appl. Math. *28*, No. 4, 525–565. Zbl.333.60077

Driscoll, B.H.
[1987] Eigenvalues on a domain with discrete rotational symmetry. SIAM J. Math. Anal. *18*, No. 4, 941–953. Zbl.629.58024

Dubrovin, B.A., Krichever, I.M., Novikov, S.P.
[1985] Integrable systems. I. Itogi Nauki Tekh., Ser. Sovrem. Probl. Mat., Fundam. Napravleniya *4*, 179–284. Zbl.591.58013. English transl.: Encycl. Math. Sci. *4*. Springer-Verlag, Berlin-Heidelberg-New York, 173–280 (1990).

Duistermaat, J.J., Guillemin, V.
[1975] The spectrum of positive elliptic operators and periodic bicharacteristics. Invent. Math. *29*, No. 1, 37–79. Zbl.307.35071

Dunford, N., Schwartz, J.T.
[1963] Linear Operators. II. Spectral Theory. Interscience, New York-London. Zbl.128,348

Dynkin, E.B.
[1963] Markov Processes. Gosudarstv. Izdat. Fiz.-Mat. Lit., Moscow. English transl.: Springer-Verlag, Berlin-Heidelberg-New York, 1965. Zbl.132,377

Eastham, M.S.P.
[1973] The Spectral Theory of Periodic Differential Equations. Scott. Acad. Press, Edinburgh-London. Zbl.287.34016

Egorov, Yu.V., Kondrat'ev, V.A.
[1987] An estimate of the number of points of the negative spectrum of the Schrödinger operator. Mat. Sb., Nov. Ser. *134*, No. 4, 556–570. English transl.: Math. USSR, Sb. *62*, 551–566 (1989). Zbl.669.35086

Egorov, Yu.V., Shubin, M.A.
[1988a] Linear partial differential equations. Foundations of the classical theory. Itogi Nauki Tekh., Ser. Sovrem. Probl. Mat., Fundam. Napravleniya *30*, 1–262. English transl.: Encycl. Math. Sci. *30*. Springer-Verlag, Berlin-Heidelberg-New York, 259pp. (1992). Zbl.657.35002
[1988b] Linear partial differential equations. Elements of the modern theory. Itogi Nauki Tekh., Ser. Sovrem. Probl. Mat., Fundam. Napravleniya *31*, 1–125. Zbl.686.35005. English transl.: Encycl. Math. Sci. *31*. Springer-Verlag, Berlin-Heidelberg-New York (in preparation).

Ehjdel'man, S.D., Ivasishen, S.D.
[1970] Investigation of the Green matrix for a homogeneous parabolic boundary value problem. Tr. Mosk. Mat. O.-va *23*, 179–234. English transl.: Trans. Mosc. Math. Soc. *23*, 179–242 (1972). Zbl.224.35046

Elliott, G.A.
[1982] Gaps in the spectrum of an almost periodic Schrödinger operator. C. R. Math. Acad. Sci., Soc. R. Can. *4*, 255–259. Zbl.516.46048
[1986] Gaps in the spectrum of an almost periodic Schrödinger operator. II. In: H. Araki and E. G. Effros (eds.), Geometric Methods in Operator Algebras. Pitman Research Notes in Math. Vol. 123. Longman, London, 181–191. Zbl.639.96059

Eskin, G., Ralston, J., Trubowitz, N.
[1986] The multidimensional inverse spectral problem with a periodic potential. Contemp. Math. *27*, 45–56. Zbl.581.58034

Faddeev, L.D., Merkur'ev, S.P.
[1985] Quantum Scattering Theory for Systems of Several Particles. Nauka, Moscow (Russian). Zbl.585.35078

Fedoryuk, M.V., Maslov, V.P.
[1976] Semiclassical Approximation for the Equations of Quantum Mechanics. Nauka, Moscow. English transl.: Reidel, Dordrecht, 1981. Zbl.449.58002

Fedosov, B.V., Shubin, M.A.
[1978a] The index of random elliptic operators. I. Mat. Sb., Nov. Ser. *106*, No. 1, 108–140. English transl.: Math. USSR, Sb. *34*. 671–699 (1978). Zbl.409.47030
[1978b] The index of random elliptic operators. II. Mat. Sb., Nov. Ser. *106*, No. 3, 455–483. English transl.: Math. USSR, Sb. *35*, 131–156 (1979). Zbl.392.58010

Fefferman, C.L.
[1983] The uncertainty principle. Bull. Am. Math. Soc., New Ser. *9*, No. 2, 129–206. Zbl.526.35080

Fefferman, C.L., Phong, D.
[1980] On the asymptotic eigenvalue distribution of a pseudodifferential operator. Proc. Natl. Acad. Sci. USA 77, No. 10, 5622–5625. Zbl.443.35082

Fejgin, V.I.
[1976] Asymptotic distribution of eigenvalues for hypoelliptic systems in $\mathbb{R}^n$. Mat. Sb., Nov. Ser. *99*, No. 4, 594–614. English transl.: Math. USSR, Sb. *28*, 533-552 (1978). Zbl.336.35079
[1978] New classes of pseudodifferential operators in $\mathbb{R}^n$ and some applications. Tr. Mosk. Mat. O.-va *36*, 155–194. English transl.: Trans. Mosc. Math. Soc. *36*, 153–195 (1979). Zbl.421.35083

[1979] Asymptotic distribution of eigenvalues and the Bohr-Sommerfeld formula. Mat. Sb., Nov. Ser. *110*, No. 1, 66–87. English transl.: Math. USSR, Sb. *38*, 61–81 (1981). Zbl.447.35063

Figotin, A.L.
[1983] Essential self-adjointness and ergodic properties of the Schrödinger operator with a random potential. Dokl. Akad. Nauk Ukr. SSR, Ser. A 1983, No. 8, 18–20 (Russian). Zbl.516.35026

Figotin, A.L., Pastur, L.A.
[1984] An exactly solvable model of a multidimensional incommensurate structure. Commun. Math. Phys. *95*, No. 2, 401–425. Zbl.582.35101
[1992] Spectra of Random and Almost Periodic Operators, Springer-Verlag, Berlin. Zbl.752.47002

Firsova, N.E.
[1986] The direct and inverse scattering problem for the perturbed one-dimensional Hill operator. Mat. Sb., Nov. Ser. *130*, No. 3, 349–385. English transl.: Math. USSR, Sb. *58*, 351–388 (1987). Zbl.616.34017

Freidlin, M.I., Venttsel', A.D.
[1979] Fluctuations in Dynamical Systems Subject to Small Random Perturbations. Nauka, Moscow. Zbl.499.60053. English transl.: Random Perturbations of Dynamical Systems, Springer-Verlag, Berlin-Heidelberg-New York 1984. Zbl.522.60055

Friedrichs, K.O.
[1965] Perturbation of Spectra in Hilbert Space. Am. Math. Soc., Providence, R.I. Zbl.142,110

Froese, R., Herbst, I.
[1982] A new proof of the Mourre estimate. Duke Math. J. *49*, No. 4, 1075–1085. Zbl.514.35025

Fröhlich, J., Martinelli, F., Scoppola, E., Spencer, T.
[1985] Constructive proof of localization in the Anderson tight binding model. Commun. Math. Phys. *101*, No. 1, 21–46. Zbl.573.60096

Fröhlich, J., Spencer, T.
[1983] Absence of diffusion in the Anderson tight binding model for large disorder or low energy. Commun. Math. Phys. *88*, No. 2, 151–184. Zbl.519.60066
[1984] A rigorous approach to Anderson localization. Phys. Rep. *103*, No. 1–4, 9–25.

Garnett, J., Trubowitz, E.
[1984] Gaps and bands of one-dimensional periodic Schrödinger operators. Comment. Math. Helv. *59*, 258–312.
[1987] Gaps and bands of one-dimensional periodic Schrödinger operators. II. Comment. Math. Helv. *62*, 18–37. Zbl.649.34034

Gel'fand, I.M.
[1950] Expansion in terms of the eigenfunctions of an equation with periodic coefficients. Dokl. Akad. Nauk SSSR *73*, No. 6, 1117–1120. Zbl.37,345
[1963] Automorphic functions and the theory of representations. In: Proc. Int. Congr. Math., Stockholm 1962, Inst. Mittag-Leffler, Djursholm, 74–85. Zbl.138,71

Geymonat, G., Grubb, G.
[1977] The essential spectrum of elliptic systems of mixed order. Math. Ann. *227*, 247–276. Zbl.361.35050

Gilkey, P.
[1975] Spectral geometry of a Riemannian manifold. J. Differ. Geom. *10*, No. 4, 601–618. Zbl.316.53035

Gilkey, P., Smith, L.
[1983] The eta-invariant for a class of elliptic boundary value problems. Commun. Pure Appl. Math. *36*, No. 1, 85–131. Zbl.512.58035

Glazman, I.M.
[1963] Direct Methods of Qualitative Spectral Analysis of Singular Differential Operators. Gosudarstv. Izdat. Fiz.-Mat. Lit., Moscow. English transl.: Oldbourne Press, London 1965. Zbl.143,365

Gokhberg, I.Ts., Krein, M.G.
[1965] Introduction to the Theory of Non–Self-Adjoint Linear Operators. Nauka, Moscow. English transl.: Providence, R.I. (1969). Zbl.139,78

Gol'dshejd, I.Ya., Molchanov, S.A., Pastur, L.A.
[1977] The random Schrödinger operator has pure point spectrum. Funkts. Anal. Prilozh. *11*, No. 1, 1–10. English transl.: Funct. Anal. Appl. *11*, 1–8 (1977). Zbl.349.34021

Gordon, A.Ya.
[1976] On the point spectrum of the one-dimensional Schrödinger operator. Usp. Mat. Nauk *31*, No. 4, 257–258 (Russian). Zbl.342.34012
[1979] On the continuous spectrum of the one-dimensional Schrödinger operator. Funkts. Anal. Prilozh. *13*, No. 3, 77–79. English transl.: Funct. Anal. Appl. *13*, 218–220 (1980). Zbl.459.47041

Gordon, C., Webb, D., Wolpert, S.
[1992] Isospectral plane domains and surfaces via Riemannian orbifolds. Invent. Math. *110*, No. 1, 1–22.

Gordon, C.S., Wilson, E.N.
[1984] Isospectral deformations of compact solvmanifolds. J. Differ. Geom. *19*, No. 1, 241–256. Zbl.532.58032

Gredeskul, S.A., Lifshits, I.M., Pastur, L.A.
[1982] Introduction to the Theory of Disordered Systems. Nauka, Moscow (Russian).

Greiner, P.
[1971] An asymptotic expansion for the heat equation. Arch. Ration. Mech. Anal. *41*, No. 3, 163–218. Zbl.238.35038

Grenkova, L.N.
[1981] On the essential self-adjointness of the Schrödinger operator with random potential. Usp. Mat. Nauk *36*, No. 6, 211–212. English transl.: Russ. Math. Surv. *36*, No. 6, 185–186 (1981). Zbl.485.47032

Grenkova, L.N., Molchanov, S.A.
[1988] Asymptotics of Lifshits 'tails' for one-dimensional random systems. Funkts. Anal. Prilozh. *22*, No. 2, 73–74. English transl.: Funct. Anal. Appl. *22*, No. 2, 145–147 (1988). Zbl.661.60042

Grubb, G.
[1978] Spectral asymptotics for Douglis-Nirenberg elliptic systems and pseudodifferential boundary problems. Commun. Partial Differ. Equations *2*, No. 9, 1071–1150. Zbl.381.35034
[1986] Functional Calculus of Pseudodifferential Boundary Problems. Birkhäuser, Boston-Basel-Stuttgart. Zbl.622.35001

Guillemin, V., Kazhdan, D.

[1980] Some inverse spectral results for negatively curved n-manifolds. Proc. Symp. Pure Math. *36*, 153–180.

Guillemin, V., Melrose, R.

[1979] The Poisson summation formula for manifolds with boundary. Adv. Math. *32*, No. 3, 204–232. Zbl.421.35082

Gureev, T.E., Safarov, Yu.G.

[1988] The sharp asymptotics of the spectrum of the Laplace operator on a manifold with periodic geodesics. Tr. Mat. Inst. Steklova *179*, 36–53. English transl.: Proc. Steklov Inst. Math. *179*, 35–53 (1989). Zbl.673.58047

Gusev, A.I.

[1977] The density of states and other spectral invariants of self-adjoint elliptic operators with random coefficients. Mat. Sb., Nov. Ser. *104*, No. 2, 207–226. English transl.: Math. USSR, Sb. *33*, 185–202 (1977). Zbl.376.35050

Gusev, A.I., Shubin, M.A.

[1977] Elliptic operators with random coefficients. In: Limit Theorems for Random Processes. Inst. Mat. Akad. Nauk Ukr. SSR, 93–107 (Russian). Zbl.398.60065

Helffer, B.

[1984] Théorie spectrale pour des opérateurs globalement elliptiques. Astérisque *112*, 1–210. Zbl.541.35002

Helffer, B., Robert, D.

[1981] Comportement semi-classique du spectre des hamiltoniens quantiques elliptiques. Ann. Inst. Fourier *31*, No. 3, 169–223. Zbl.451.35022

[1982a] Propriétés asymptotiques du spectre d'opérateurs pseudodifférentiels sur $\mathbb{R}^n$. Commun. Partial Differ. Equations 7, No. 7, 795–882. Zbl.501.35081

[1982b] Comportement semi-classique du spectre des hamiltoniens quantiques hypoelliptiques. Ann. Sc. Norm. Super. Pisa, Cl. Sci., IV. Ser. *9*, No. 3, 405–431. Zbl.507.35066

Helffer, B., Sjöstrand, J.

[1986] Résonances en limite semi-classique. Mem. Soc. Math. Fr., Nuov. Ser. *24/25*. Zbl.631.35075

[1987] Équation de Schrödinger avec champ magnétique et équation de Harper. Journ. "Equations Dériv. Partielles", St. Jean-De-Monts, *6*, 1–9. Zbl.641.35015

[1988] Structure cantorienne du spectre de l'opérateur de Harper. Sémin., Équations Dériv. Partielles 1987/1988, Exp. No. 12, 1–6. Zbl.666.34012

Hörmander, L.

[1968] The spectral function of an elliptic operator. Acta Math. *121*, 193–218. Zbl.164.132

[1979a] The Weyl calculus of pseudodifferential operators. Commun. Pure Appl. Math. *32*, No. 3, 297–443. Zbl.388.47032

[1979b] On the asymptotic distribution of the eigenvalues of pseudodifferential operators in $\mathbb{R}^n$. Ark. Mat. *17*, 297–313. Zbl.436.35064

[1979c] On the asymptotic distribution of eigenvalues. Proc. Åke Plejel Conference (Uppsala, September 6–8, 1979), 146–154.

[1983–1985] The Analysis of Linear Partial Differential Operators. Vol. 1–4. Springer-Verlag, Berlin-Heidelberg-New York. Zbl.521.35001, Zbl.521.35002, Zbl.601.35001, Zbl.612.35001

Ikeda, A.

[1980a] On the spectrum of a Riemannian manifold of positive constant curvature. Osaka J. Math. *17*, No. 1, 75–95. Zbl.436.58025

[1980b] On lens spaces which are isospectral but not isometric. Ann. Sci. Éc. Norm. Super., IV. Sér. *13*, No. 3, 303–315. Zbl.451.58037

[1983] On spherical space forms which are isospectral but not isometric. J. Math. Soc. Japan *35*, No. 3, 437–444. Zbl.507.58050

Ishii, K.

[1973] Localization of eigenstates and transport phenomena in the one-dimensional disordered system. Supp. Theor. Phys. *53*, 77–138.

Itô, K., McKean, H.P.

[1965] Diffusion Processes and Their Sample Paths. Springer-Verlag, Berlin-Heidelberg-New York. Zbl.127,95

Ivrii, V.Ya.

[1980] On the second term of the spectral asymptotics for the Laplace-Beltrami operator on a manifold with boundary. Funkts. Anal. Prilozh. *14*, No. 2, 25–34. English transl.: Funct. Anal. Appl. *14*, 98–106 (1980). Zbl.435.35059

[1982] On the exact spectral asymptotics for elliptic operators acting in fibre bundles. Funkts. Anal. Prilozh. *16*, No. 2, 30–38. English transl.: Funct. Anal. Appl. *16*, 101–108 (1982). Zbl.505.35066

[1984] Precise spectral asymptotics for elliptic operators acting on fiberings over manifolds with boundary. Lect. Notes Math. *1100*, Springer-Verlag, Berlin-Heidelberg-New York. Zbl.565.35002

[1985] Asymptotics of the discrete spectrum for some operators in $\mathbb{R}^n$. Funkts. Anal. Prilozh. *19*, No. 1, 73–74. English transl.: Funct. Anal. Appl. *19*, 61–62 (1985). Zbl.578.35077

[1986a] On the asymptotic Weyl formula for the Laplace-Beltrami operator on a Riemannian manifold and on a domain with conic singularities at the boundary. Dokl. Akad. Nauk SSSR *288*, No. 1, 35–38. English transl.: Sov. Math., Dokl. *33*, 596–599 (1986). Zbl.652.35091

[1986b] Estimations pour le nombre de valeurs propres négatives de l'opérateur de Schrödinger avec potentiels singuliers. C. R. Acad. Sci., Paris, Sér. I *302*, No. 13, 467–470.

[1986c] Estimations pour le nombre de valeurs propres négatives de l'opérateur de Schrödinger avec potentiels singuliers et application au comportement asymptotique des grandes valeurs propres. C. R. Acad. Sci., Paris, Sér. I *302*, No. 14, 491–494. Zbl.625.35019

[1986d] Estimations pour le nombre de valeurs propres négatives de l'opérateur de Schrödinger avec potentiels singuliers et application au comportement asymptotique des valeur propres d'accumulant vers −0, aux asymptotiques à deux paramètres et à la densité des états. C. R. Acad. Sci., Paris, Sér. I *302*, No. 15, 535–538. Zbl.625.35020

[1987] Estimates of the number of negative eigenvalues of the Schrödinger operator with strong magnetic field. Dokl. Akad. Nauk SSSR *297*, No. 5, 1043–1046. English transl.: Sov. Math., Dokl. *36*, No. 3, 561–564 (1988). Zbl.661.35066

Jingbo Xia

[1993] The spectra of random pseudo-differential operators. Preprint.

Johnson, R.

[1983] A review of recent work on almost periodic differential and difference operators. Acta Appl. Math. *1*, No. 3, 241–261. Zbl.542.34007

Johnson, R., Moser, J.
[1982] The rotation number for almost periodic potentials. Commun. Math. Phys. *84*, No. 3, 403–438. Erratum *90*, No. 2 (1983) 317–318. Zbl.497.35026

Joergens, K., Rellich, F.
[1976] Eigenwerttheorie gewöhnlicher Differentialgleichungen. Springer-Verlag, Berlin-Heidelberg-New York. Zbl.347.34013

Joergens, K., Weidmann, J.
[1973] Spectral properties of Hamiltonian operators. Lect. Notes Math. *319*, 1–148. Zbl.248.35002

Kac, M.
[1959] Probability and Related Topics in Physical Sciences. Interscience, New York. Zbl.87,330
[1966] Can one hear the shape of a drum? Am. Math. Mon. *73*, No. 4, 1–23. Zbl.139,56

Karol', A.I.
[1981] On the ζ-function of a degenerate elliptic Dirichlet boundary value problem. Dokl. Akad. Nauk SSSR *260*, No. 1, 20–22. English transl.: Sov. Math., Dokl. *24*, 209–211 (1981). Zbl.506.35042

Kato, T.
[1966] Perturbation Theory for Linear Operators, Springer-Verlag, Berlin-Heidelberg-New York. Zbl.148,126

Kirsch, W.
[1985] Random Schrödinger operators and the density of states. Lect. Notes Math. *1109*, 68–102. Zbl.567.35021

Kirsch, W., Martinelli, F.
[1982] On the spectrum of Schrödinger operators with random potential. Commun. Math. Phys. *85*, 329–350. Zbl.506.60058
[1983] Large deviations and Lifshitz singularity of the integrated density of states of random Hamiltonians. Commun. Math. Phys. *89*, 27–40. Zbl.517.60071

Kiselev, V.Yu.
[1978] Almost periodic Fourier integral operators and some of their applications. Tr. Semin. Im. I. G. Petrovskogo *3*, 81–97. English transl.: Transl., II. Ser., Am. Math. Soc. *118*, 223–240 (1982). Zbl.457.35069

Kneser, M.
[1967] Lineare Relationen zwischen Darstellungsanzahlen quadratischer Formen. Math. Ann. *168*, 31–39. Zbl.146,59

Kondrat'ev, V.A., Oleinik, O.A.
[1983] Boundary value problems for partial differential equations in non-smooth domains. Usp. Mat. Nauk *38*, No. 2, 3–76. English transl.: Russ. Math. Surv. *38*, No. 2, 1–86 (1983). Zbl.523.35010

Kostyuchenko, A.G.
[1966] Distribution of eigenvalues for singular differential operators. Dokl. Akad. Nauk SSSR *168*, No. 1, 21–24. English transl.: Sov. Math., Dokl. *7*, 584–587 (1966). Zbl.163,134
[1968] Asymptotic behaviour of the spectral function of a self-adjoint elliptic operator. In: Fourth Math. Summer School, Naukova Dumka, Kiev, 42–117 (Russian).

Kostyuchenko, A.G., Levitan, B.M.

[1967] On the asymptotic behaviour of the eigenvalues of the Sturm-Liouville operator problem. Funkts. Anal. Prilozh. *1*, No. 1, 86–96. English transl.: Funct. Anal. Appl. *1*, 75–83 (1967). Zbl.168,123

Kostyuchenko, A.G., Sargsyan, I.S.

[1979] Distribution of Eigenvalues. Nauka, Moscow (Russian). Zbl.478.34022

Kotani, S.

[1982] Lyapunov indices determine absolutely continuous spectra of stationary one-dimensional Schrödinger operators. In: Proc. Taniguchi Symp. Stochastic Analysis, North-Holland, Libr. 32, 225–247. Zbl.549.60058

[1986] Lyapunov exponents and spectra for one-dimensional random Schrödinger operators. Contemp. Math. *50*, 277–286. Zbl.587.60054

Kozhevnikov, A.N.

[1973] Spectral problems for pseudodifferential systems that are elliptic in the sense of Douglis-Nirenberg, and their applications. Mat. Sb., Nov. Ser. *92*, No. 1, 60–88. English transl.: Math. USSR, Sb. *21* 63–90 (1974). Zbl.278.35078

Kozlov, S.M.

[1982] Distribution of the eigenvalues of differential operators in large domains. Usp. Mat. Nauk *37*, No. 5, 185–186. English transl.: Russ. Math. Surv. *37*, No. 5, 180–181 (1982). Zbl.515.35065

[1983] Reducibility of quasiperiodic differential operators and averaging. Tr. Mosk. Mat. O.-va. *46*, 99–123. English transl.: Trans. Mosc. Math. Soc. 1984, No. 2, 101–126 (1984). Zbl.566.35036

Kozlov, S.M., Shubin, M.A.

[1984] On the coincidence of the spectra of random elliptic operators. Mat. Sb., Nov. Ser. *123*, No. 4, 460–476. English transl.: Math. USSR, Sb. *51*, 455–471 (1985). Zbl.568.35091

Kozlov, V.A.

[1983] Estimates of the remainder in formulae for the asymptotic behaviour of the spectrum for linear differential operators. Funkts. Anal. Prilozh. *17*, No. 2, 80–81. English transl.: Funct. Anal. Appl. *17*, 147–149 (1983). Zbl.527.47011

Kuchment, P.A.

[1982] Floquet theory for partial differential equations. Usp. Mat. Nauk *37*, No. 4, 3–52. English transl.: Russ. Math. Surv. *37*, No. 4, 1–60 (1982). Zbl.519.35003

Kunz, H., Souillard, B.

[1980] Sur le spectre des opérateurs aux différences finies aléatories. Commun. Math. Phys. *78*, No. 2, 201–246. Zbl.449.60048

Kuwabara, R.

[1980] On isospectral deformations of Riemannian metrics. Compos. Math. *40*, No. 3, 319–324. Zbl.422.53021

Laptev, A.A.

[1981] Spectral asymptotics of a class of Fourier integral operators. Tr. Mosk. Mat. O.-va *43*, 92–115. English transl.: Trans. Mosc. Math. Soc. 1983, No. 1, 101–127 (1983). Zbl.503.47044

Lazutkin, V.F.
[1981] Convex Billiard and the Eigenfunctions of the Laplace Operator. Leningr. State Univ. Press, Leningrad (Russian). Zbl.532.58031
[1987] Semiclassical asymptotic behaviour of eigenfunctions. Itogi Nauki Tekh., Ser. Sovrem. Probl. Mat., Fundam. Napravleniya *34*, 135–174. Zbl.657.35098. English transl.: Encycl. Math. Sci. *34*. Springer-Verlag, Berlin-Heidelberg-New York (in preparation).

Levendorskij, S.Z.
[1982] Asymptotic distribution of eigenvalues. Izv. Akad. Nauk SSSR, Ser. Mat. *46*, No. 4, 810–852. English transl.: Math. USSR, Izv. *21*, 119–160 (1983). Zbl.529.35060
[1984] Asymptotic behaviour of the spectrum of linear operator pencils. Mat. Sb., Nov. Ser. *124*, No. 2, 251–271. English transl.: Math. USSR, Sb. *52*, 245–266 (1985). Zbl.553.35067
[1985] The approximate spectral projection method. Izv. Akad. Nauk SSSR, Ser. Mat. *49*, No. 6, 1177–1229. English transl.: Math. USSR, Izv. *27*, 451–502 (1986). Zbl.614.35021
[1986a] Asymptotic behaviour of the spectrum of problems with constraints. Mat. Sb., Nov. Ser. *129*, No. 1, 73–84. English transl.: Math. USSR, Sb. *57*, 77–95 (1987). Zbl.615.35063
[1986b] The approximate spectral projector method. Acta. Appl. Math. *7*, 137–197. Zbl.631.35072
[1988] Non-classical spectral asymptotics. Usp. Mat. Nauk *43*, No. 1, 123–157. English transl.: Russ. Math. Surv. *43*, No. 1, 149–192 (1988). Zbl.671.35064
[1990] Asymptotic Distribution of Eigenvalues of Differential Operators. Kluwer Academic Publishers, Dordrecht. Zbl.721.35049

Levitan, B.M.
[1952] On the asymptotic behaviour of the spectral function of a self-adjoint differential equation of order two. Izv. Akad. Nauk SSSR, Ser. Mat. *16*, No. 1, 325–352 (Russian). Zbl.48,324
[1971] Asymptotic behaviour of the spectral function of an elliptic equation. Usp. Mat. Nauk *26*, No. 6, 151–212. English transl.: Russ. Math. Surv. *26*, No. 6, 165–232 (1972). Zbl.236.35035

Levitan, B.M., Otelbaev, M.O.
[1977] On the self-adjointness conditions for the Schrödinger and Dirac operators. Dokl. Akad. Nauk SSSR *235*, No. 4, 768–771. English transl.: Sov. Math., Dokl. *18*, 1044–1048 (1978). Zbl.432.47028

Levitan, B.M., Sabin, A.V.
[1984] Examples of Schrödinger operators with almost periodic potentials and nowhere dense absolutely continuous spectrum. Dokl. Akad Nauk SSSR *276*, No. 3, 539–542. English transl.: Sov. Math., Dokl. *29*, 541–544 (1984). Zbl.586.34019

Levitan, B.M., Sargsyan, I.S.
[1970] Introduction to Spectral Theory. Nauka, Moscow. English transl.: Transl. Math. Monogr. Vol. 39, Providence (1975). Zbl.225.47019

Lions, J.-L., Magenes, E.
[1968] Problèmes aux limites non homogènes. Vol. 1. Dunod, Paris. Zbl.165,108

Louchard, G.
[1969] Mouvement Brownien et valeurs propres du Laplacien. Ann. Inst. Henri Poincaré, Sect. B *4*, No. 4, 331–343. Zbl.181,208

Lyusternik, L.A., Vishik, M.I.
[1960] Solution of some perturbation problems for matrices and self-adjoint differential equations. Usp. Mat. Nauk *15*, No. 3, 3–80. English transl.: Russ. Math. Surv. *15*, 1–73 (1960). Zbl.96,87

Majda, A., Ralston, J.
[1978a] An analogue of Weyl's theorem for unbounded domains. Duke Math. J. *45*, No. 1, 183–196. Zbl.408.35069
[1978b] An analogue of Weyl's theorem for unbounded domains. II. Duke Math. J. *45*, No. 3, 513–536. Zbl.416.35058
[1979] An analogue of Weyl's theorem for unbounded domains. III. An epilogue. Duke Math. J. *46*, No. 4, 725–731. Zbl.433.35055

Malozemov, L.A.
[1988] On the eigenvalues of a perturbed almost periodic operator that are immersed in the continuous spectrum. Usp. Mat. Nauk *43*, No. 4, 211–212. English transl.: Russ. Math. Surv. *43*, No. 4, 219–220 (1988). Zbl.689.47002

Manakov, S.V., Novikov, S.P., Pitaevskij, L.P., Zakharov, V.E.
[1980] Theory of Solitons: the Inverse Problem Method. Nauka, Moscow. English transl.: Consultants Bureau, New York-London (1984). Zbl.598.35001

Marchenko, A.V.
[1979] On the approximation of the spectrum of an elliptic operator with almost periodic coefficients. Mat. Zametki *25*, No. 5, 713–724. English transl.: Math. Notes. *25*, 367–373 (1979). Zbl.436.47038

Marchenko, V.A.
[1974] Periodic Korteweg-de Vries problem. Mat. Sb., Nov. Ser. *95*, No. 3, 331–356. English transl.: Math. USSR, Sb. *24*, 319–344 (1976). Zbl.317.34030
[1977] Sturm-Liouville Operators and Their Applications. Naukova Dumka, Kiev. English transl.: Birkhäuser, Basel etc. (1986). Zbl.399.34022

Marchenko, V.A., Ostrovskij, I.V.
[1975] A characterization of the spectrum of the Hill operator. Mat. Sb., Nov. Ser. *97*, No. 4, 540–606. English transl.: Math. USSR, Sb. *26*, 493–554 (1977). Zbl.327.34021

Markus, A.S., Matsaev, V.I.
[1981] Operators generated by sesquilinear forms and their spectral asymptotics. Mat. Issled. *61*, 86–103 (Russian). Zbl.476.47009
[1982] Comparison theorems for the spectra of linear operators and spectral asymptotics. Tr. Mosk. Mat. O.-va *45*, 133–181. English transl.: Trans. Mosc. Math. Soc. 1984, No. 1, 139–187 (1984). Zbl.532.47012

Martinelli, F., Scoppola, E.
[1986] Introduction to the mathematical theory of Anderson localization. Preprint.

Maslov, V.P.
[1976] The Complex WKB Method and Nonlinear Equations. Nauka, Moscow (Russian). Zbl.449.58001
[1987] Asymptotic Methods for Solving Pseudodifferential Equations. Nauka, Moscow. English transl.: Consultants Bureau, New York-London (1988). Zbl.625.35001
[1988] Asymptotic Methods and Perturbation Theory. Nauka, Moscow (Russian). Zbl.653.35002

Maurin, K.
[1959] Hilbert Space Methods. Państwowe Wydawnictwo Naukowe, Warszawa. English transl.: Pol. Sci. Publ., Warszawa (1967). Zbl.86.309

Maz'ya, V.G.
[1985] Sobolev Spaces. Leningr. State Univ. Press, Leningrad. English transl.: Springer-Verlag, Berlin-Heidelberg-New York, 1985. Zbl.692.46023

Maz'ya, V.G., Nazarov, S.A., Plamenevskij, B.A.
[1984] Asymptotic expansions of the eigenvalues of boundary value problems for the Laplace operator in domains with small holes. Izv. Akad. Nauk SSSR, Ser. Mat. *48*, No. 2, 347–371. English transl.: Math. USSR, Izv. *24*, 321–345 (1985). Zbl.566.35031

McKean, H.P.
[1972] Selberg's trace formula as applied to a compact Riemannian surface. Commun. Pure Appl. Math. *25*, No. 3, 225–246. Zbl.225.30021

McKean, H.P., Singer, I.M.
[1967] Curvature and the eigenvalues of the Laplacian. J. Differ. Geom. *1*, No. 1, 43–69. Zbl.198.443

Melrose, R.
[1982] Scattering theory and the trace of the wave group. J. Funct. Anal. *45*, No. 1, 29–40. Zbl.525.47007

Menikoff, A., Sjöstrand, J.
[1978] On the eigenvalues of a class of hypoelliptic operators. Math. Ann. *235*, 55–85. Zbl.375.35014
[1979] On the eigenvalues of a class of hypoelliptic operators. II. Lect. Notes Math. *755*, 201–247. Zbl.444.35019

Métivier, G.
[1976] Fonction spectrale et valeurs propres d'une classe d'opérateurs non-elliptiques. Commun. Partial Differ. Equations *1*, No. 4, 467–519. Zbl.376.35012
[1977] Valeurs propres de problèmes aux limites elliptiques irrégulières. Bull. Soc. Math. Fr. *51-52*, 125–219. Zbl.401.35088
[1978] Valeurs propres d'opérateurs définis par la restriction de systèmes variationelles à des sous-espaces. J. Math. Pures Appl., IX. Sér. *57*, No. 2, 133–156. Zbl.328.35029
[1981] Spectral asymptotics for the ∂-Neumann problem. Duke Math. J. *48*, No. 4, 779–806. Zbl.489.35064
[1982] Estimation du reste en théorie spectrale. Journ. "Equations Dériv. Partielles", St. Jean-De-Monts 1982, No. 1, 1–22. Zbl.498.35066

Micheletti, A.-M.
[1972] Perturbazione dello spettro dell'operatore di Laplace, in relazione ad una variazione del campo. Ann. Sc. Norm. Super. Pisa, III. Sér. *26*, 151–169. Zbl.234.35073

Milnor, J.
[1964] Eigenvalues of the Laplace operator on certain manifolds. Proc. Nat. Acad. Sci. USA *51*, No. 4, 542. Zbl.124,312

Minakshisundaram, S., Plejel, A.
[1949] Some properties of the eigenfunctions of the Laplace operator on Riemannian manifolds. Canad. J. Math. *1*, No. 2, 242–256. Zbl.41,427

Mohamed, A.
[1982] Étude spectrale d'opérateurs hypoelliptiques à caractéristiques multiple. I. Ann. Inst. Fourier *32*, No. 3, 39–90. Zbl.488.35079
[1983] Étude spectrale d'opérateurs hypoelliptiques à caractéristiques multiple. II. Commun. Partial Differ. Equations *8*, No. 3, 247–316. Zbl.522.35069

Molchanov, S.A.

[1975] Diffusion processes and Riemannian geometry. Usp. Mat. Nauk *30*, No. 1, 3–59. English transl.: Russ. Math. Surv. *30*, No. 1, 1–63 (1975). Zbl.305.53041

[1978] Construction of the eigenfunctions of one-dimensional disordered structures. Izv. Akad. Nauk SSSR, Ser. Mat. *42*, No. 11, 70–103. English transl.: Math. USSR, Izv. *12*, 69–101 (1978). Zbl.386.34029

Moser, J.

[1981] An example of a Schrödinger equation with almost periodic potential and nowhere dense spectrum. Comment. Math. Helv. *56*, No. 2, 198–224. Zbl.477.34018

Moser, J., Pöschel, J.

[1984] An extension of a result by Dinaburg and Sinai on quasiperiodic potentials. Comment. Math. Helv. *59*, No. 1, 39–85. Zbl.533.34023

Mott, N.F., Twose, Q.D.

[1961] The theory of impurity conduction. Adv. Phys. *10*, No. 1, 107–163.

Murray, F.J., Neumann, J. von

[1936] On rings of operators. Ann. Math., II. Ser. *37*, 116–229. Zbl.14,161

Murtazin, Kh.Kh., Sadovnichij, V.A.

[1988] Spectral Analysis of the Multiparticle Schrödinger Operator. Moscow State Univ. Press, Moscow (Russian).

Mynbaev, K.T., Otelbaev, M.O.

[1988] Weighted Functional Spaces and the Spectra of Differential Operators. Nauka, Moscow (Russian). Zbl.651.46037

Naimark, M.A.

[1969] Linear Differential Operators. 2nd ed., Nauka, Moscow. English transl.: F. Ungar Publ., New York (1967/1968). Zbl.193,41, Zbl.57,71

Nazarov, S.A.

[1987] Asymptotic Expansions of Eigenvalues. Leningr. State Univ. Press, Leningrad (Russian).

Nikol'skij, S.M.

[1977] Approximation of Functions of Several Variables and Embedding Theorems. 2nd ed., Nauka, Moscow. English transl.: Springer-Verlag, Berlin-Heidelberg-New York, 1975. Zbl.496.46020, Zbl.185,379

Novikov, S.P.

[1983] Two-dimensional Schrödinger operators in periodic fields. Itogi Nauki Tekh., Ser. Sovrem. Probl. Mat. *23*, 3–32. English transl.: J. Sov. Math. *28*, 1–20 (1985). Zbl.564.35083

Oleinik, O.A.

[1987] Homogenization problems in elasticity. Spectra of singularly perturbed operators. Lond. Math. Soc. Lect. Note Ser. *122*, 53–95. Zbl.667.73009

Oseledets, V.I.

[1968] Multiplicative ergodic theorem. Characteristic Lyapunov exponents of dynamical systems. Tr. Mosk. Mat. O.-va *19*, 179–210. English transl.: Trans. Mosc. Math. Soc. *19*, 197–231 (1969). Zbl.236.93034

Ozawa, S.

[1983] An asymptotic formula for the eigenvalues of the Laplacian in a three-dimensional domain with a small hole. J. Fac. Sci., Univ. Tokyo, Sect. I A *30*, No. 2, 243–257. Zbl.541.35060

Pastur, L.A.
[1971] Self-averagibility of the number of states of the Schrödinger operator with a random potential. In: Mathematical Physics and Functional Analysis. Physical Engineering Institute of Low Temperatures, Acad. Sci. Ukr. SSR, Kharkov, 111–116 (Russian).
[1973] Spectra of random self-adjoint operators. Usp. Mat. Nauk *28*, No. 1, 3–64. English transl.: Russ. Math. Surv. *28*, No. 1, 1–67 (1973). Zbl.268.60034
[1974] On the spectrum of random Jacobian matrices and the Schrödinger operator on the whole axis with a random potential. Preprint, Physical Engineering Institute of Low Temperatures, Acad. Sci. Ukr. SSR, Kharkov. Appeared in: Limit Theor. Random Process, Akad. Nauk Ukr. SSR, Inst. Mat., 152–166 (1977) (Russian). Zbl.399.60056
[1977] The behaviour of certain Wiener integrals as $t \to \infty$ and the density of states of the Schrödinger equation with a random potential. Teor. Mat. Fiz. *32*, No. 1, 88–95. English transl.: Theor. Math. Phys. *32*, 615–620 (1978). Zbl.353.60053
[1980] Spectral properties of disordered systems in the one-body approximation. Commun. Math. Phys. *75*, No. 1, 179–196. Zbl.429.60099
[1985] Spectral properties of random and almost periodic differential and finite difference operators. In: Statistical Physics and Dynamical Systems (Köszeg, 1984), Birkhäuser, Boston, Mass., 49–67. Zbl.575.35085
[1987a] Spectral theory of random self-adjoint operators. Itogi Nauki Tekh., Ser. Teor. Veroyatn., Mat. Stat., Teor. Kibern. *25*, 3–67. English transl.: J. Sov. Math. *46*, No. 4, 1979–2021 (1989). Zbl.661.60079
[1987b] Spectral properties of random and almost periodic operators. Sov. Sci. Rev., Sect. C, Math. Phys. Rev. *6*, 1–112. Zbl.627.60059

Patodi, V.K.
[1971] Curvature and the eigenforms of the Laplace operator. J. Differ. Geom. *5*, No. 3, 233–249. Zbl.211,539

Pavlov, B.S., Shushkov, A.A.
[1988] Extension theory and potentials of radius 'zero' with internal structure. Mat. Sb., Nov. Ser. *137*, No. 2, 147–183. English transl.: Math. USSR, Sb. *65*, No. 1, 147–183 (1990). Zbl.709.47046

Perry, P.A.
[1984] Exponential bounds and semifiniteness of point spectrum for n-body Schrödinger operators. Commun. Math. Phys. *92*, No. 2, 481–483. Zbl.546.35051

Petkov, V., Stojanov, L.
[1988] On the number of periodic reflecting rays in generic domains. Ergodic Theory Dyn. Syst. *8*, No. 1, 81–91. Zbl.668.58005

Pöschel, J.
[1983] Examples of discrete Schrödinger operators with pure point spectrum. Commun. Math. Phys. *88*, No. 4, 447–463. Zbl.532.35070

Rauch, J.
[1980] Perturbation theory for eigenvalues and resonances of Schrödinger Hamiltonians. J. Funct. Anal. *35*, No. 3, 304–315. Zbl.432.35013

Reed, M., Simon, B.
[1972] Methods of Modern Mathematical Physics. Vol. 1. Functional Analysis. Academic Press, New York-London. Zbl.242.46001
[1975] Methods of Modern Mathematical Physics. Vol. 2. Fourier Analysis, Self-Adjointness. Academic Press, New York-London. Zbl.308.47002
[1978] Methods of Modern Mathematical Physics. Vol. 4. Analysis of Operators. Academic Press, New York-London. Zbl.401.47001

[1979] Methods of Modern Mathematical Physics. Vol. 3. Scattering Theory. Academic Press, New York-London. Zbl.405.47007

Robert, D.

[1982] Comportement asymptotique des valeurs propres d'opérateurs du type Schrödinger à potentiel "dégénéré." J. Math. Pures Appl., IX. Sér. *61*, 275–300. Zbl.511.35069

Rofe-Beketov, F.S.

[1984] Kneser constants and effective masses for band potentials. Dokl. Akad. Nauk SSSR *276*, No. 2, 356–359. English transl.: Sov. Phys., Dokl. *29*, 391–392 (1984). Zbl.597.34016

Rojtburd, V.L.

[1976] On the semiclassical asymptotics of the spectrum of a pseudodifferential operator. Usp. Mat. Nauk *31*, No. 4, 275–276. Zbl.343.35060

Royer, G.

[1980] Croissance exponentielle de produits markoviens de matrices aléatoires. Ann. Inst. Henri Poincaré, Nouv. Ser., Sect. B *16*, 49–62. Zbl.433.60073

Rozenblum, G.V.

[1972a] The distribution of the discrete spectrum for singular differential operators. Dokl. Akad. Nauk SSSR *202*, No. 5, 1012–1015. English transl.: Sov. Math., Dokl. *13*, 245–249 (1972). Zbl.249.35069

[1972b] On the eigenvalues of the first boundary value problem in unbounded domains. Mat. Sb., Nov. Ser. *89*, No. 2, 234–247. English transl.: Math. USSR, Sb. *18*, 235–248 (1973). Zbl.267.35063

[1974] Asymptotics of the eigenvalues of the Schrödinger operator. Mat. Sb., Nov. Ser. *93*, No. 3, 346–367. English transl.: Math. USSR, Sb. *22*, 349–371 (1975). Zbl.304.35070

[1975] Remarks on estimates of the spectrum of differential operators. In: Proceedings of the Sixth Winter School on Mathematical Programming and Related Subjects (Drogobych, 1973). Functional Analysis and Its Applications, Tsentral. Ekenom.-Mat. Inst., Akad. Nauk SSSR, Moscow, 236–244 (Russian).

[1977] Asymptotics of the negative discrete spectrum of the Schrödinger operator. Mat. Zametki *21*, No. 5, 399–407. English transl.: Math. Notes *21*, 222–227 (1977). Zbl.347.35064

Rüssmann, H.

[1980] On the one-dimensional Schrödinger equation with a quasiperiodic potential. Ann. N.Y. Acad. Sci. *357*, 90–107. Zbl.477.34007

Safarov, Yu.G.

[1986] Asymptotics of the spectrum of a pseudodifferential operator with periodic bicharacteristics. Zap. Nauchn. Semin. Leningr. Otd. Mat. Inst. Steklova *152*, 94–104. English transl.: J. Sov. Math. *40*, No. 5, 645–652 (1988). Zbl.621.35071

[1987] On the spectral asymptotics of the transmission problem. Acta Appl. Math. *10*, No. 1, 101–130. Zbl.643.58032

[1988a] Asymptotics of the spectral function of a positive elliptic operator without the non-trapping condition. Funkts. Anal. Prilozh. *22*, No. 3, 53–65. English transl.: Funct. Anal. Appl. *22*, No. 3, 213–223 (1988). Zbl.679.35074

[1988b] Sharp spectral asymptotics of a boundary value problem and periodic billiards. Izv. Akad. Nauk SSSR, Ser. Mat. *52*, No. 6, 1230–1251. English transl.: Math. USSR, Izv. *33*, 553–573 (1989). Zbl.682.35082

Safarov, Yu.G, Vasil'ev, D.G.
[1988] Branching Hamiltonian billiards. Dokl. Akad. Nauk SSSR *301*, No. 2, 271–275. English transl.: Sov. Math., Dokl. *38*, No. 1, 64–68 (1989). Zbl.671.58012

Sánchez-Palencia, E.
[1980] Non-homogeneous Media and Vibration Theory. Lect. Notes Phys. *127*, Springer-Verlag, Berlin-Heidelberg-New York. Zbl.432.70002

Savin, A.V.
[1988] Asymptotic expansion of the density of states for one-dimensional Schrödinger and Dirac operators with almost periodic and random potentials. Sb. Nauchn. Tr., IFTP, Moscow (Russian).

Seeley, R.T.
[1967] Complex powers of an elliptic operator. Proc. Symp. Pure Math. *10*, 288–307. Zbl.159,155
[1969] Analytic extension of the trace associated with elliptic boundary value problems. Am. J. Math. *91*, No. 6, 963–983. Zbl.191,119
[1978] A sharp asymptotic remainder estimate for the eigenvalues of the Laplacian in a domain of $\mathbb{R}^3$. Adv. Math. *29*, No. 3, 244–269. Zbl.382.35043
[1980] An estimate near the boundary for the spectral function of the Laplace operator. Am. J. Math. *102*, No. 3, 869–902. Zbl.447.35029

Serre, J-P.
[1970] Cours d'arithmétique. Presses Universitaires de France, Paris. Zbl.225.12002

Shenk, D., Shubin, M.A.
[1985] Asymptotic expansion of the density of states and the spectral function of the Hill operator. Mat. Sb., Nov. Ser. *128*, No. 4, 474–491. English transl.: Math. USSR, Sb. *56*, 473–490 (1987). Zbl.624.34018

Shilov, G.E.
[1961] Mathematical Analysis. Special Course. 2nd ed., Gosudarstv. Izdat. Fiz.-Mat. Lit., Moscow. English transl.: Pergamon Press, London etc. 1965. Zbl.97,36

Shubin, M.A.
[1974] Differential and pseudodifferential operators in spaces of almost periodic functions. Mat. Sb., Nov. Ser. *95*, No. 4, 560–587. English transl.: Math. USSR, Sb. *24*, 547–573 (1976). Zbl.311.47022
[1975] On the essential self-adjointness of uniformly hypoelliptic operators. Vestn. Mosk. Univ., Ser. I *30*, No. 2, 91–94. English transl.: Mosc. Univ. Math. Bull. *30*, No. 112, 147–150 (1975). Zbl.295.47054
[1976a] Theorems on the coincidence of the spectra of an almost periodic pseudodifferential operator in $L_2(\mathbb{R}^n)$ and $B_2(\mathbb{R}^n)$. Sib. Mat. Zh. *17*, No. 1, 200–215. English transl.: Sib. Math. J. *17*, 158–170 (1976). Zbl.324.47030
[1976b] Almost periodic pseudodifferential operators and von Neumann algebras. Tr. Mosk. Mat. O.-va *35*, 103–164. English transl.: Trans. Mosc. Math. Soc. *35*, 103–166 (1979). Zbl.405.47035
[1978a] Pseudodifferential Operators and Spectral Theory. Nauka, Moscow. English transl.: Springer-Verlag, Berlin-Heidelberg-New York, 1987. Zbl.451.47064
[1978b] The density of states for self-adjoint elliptic operators with almost periodic coefficients. Tr. Semin. Im. I. G. Petrovskogo *3*, 243–275. English transl.: Transl., II. Ser., Am. Math. Soc. *118*, 307–339 (1982). Zbl.493.35041
[1978c] Almost periodic functions and partial differential operators. Usp. Mat. Nauk *33*, No. 2, 3–47. English transl.: Russ. Math. Surv. *33*, No. 2, 1–52 (1978). Zbl.383.47025

[1979] Spectral theory and the index of an elliptic operator with almost periodic coefficients. Usp. Mat. Nauk *34*, No. 2, 95–135. English transl.: Russ. Math. Surv. *34*, No. 2, 109–157 (1979). Zbl.431.47027

[1993] Discrete magnetic Laplacian. Preprint (to appear in Commun. Math. Phys.).

Shubin, M.A., Tulovskij, V.N.

[1973] On the asymptotic distribution of the eigenvalues of pseudodifferential operators in $\mathbb{R}^n$. Mat. Sb., Nov. Ser. *92*, No. 4, 571–588. English transl.: Math. USSR, Sb. *21*, 565–583 (1975). Zbl.286.35059

Shvartsman, O.V., Vinberg, Eh.B.

[1988] Discrete groups of motions of spaces of constant curvature. Itogi Nauki Tekh., Ser. Sovrem. Probl. Mat., Fundam. Napravleniya *29*, 147–259 (Russian). Zbl.699.22017. English transl.: Encycl. Math. Sci. *29*, Springer-Verlag, Berlin-Heidelberg-New York, 139–248, 1993.

Simon, B.

[1982] Almost periodic Schrödinger operators: a review. Adv. Appl. Math. *3*, No. 3, 463–490. Zbl.545.34023

[1983a] Nonclassical eigenvalue asymptotics. J. Funct. Anal. *53*, No. 1, 84–98. Zbl.529.35064

[1983b] Semiclassical analysis of low lying eigenvalues. I. Non-degenerate minima: asymptotic expansions. Ann. Inst. Henri Poincaré, Sect. A *38*, No. 2, 295–307. Errata: Ann. Inst. Henri Poincaré, Phys. Theor. *40*, 224 (1984). Zbl.526.35027, Zbl.537.35023

[1984a] Semiclassical analysis of low lying eigenvalues. II. Tunnelling. Ann. Math., II. Ser. *120*, 89–118. Zbl.626.35070

[1984b] Semiclassical analysis of low lying eigenvalues. III. Width of the ground state band in strongly coupled solids. Ann. Phys. *158*, No. 2, 415–420. Zbl.596.35028

[1985] Almost periodic Schrödinger operators. IV. The Maryland model. Ann. Phys. *159*, No. 1, 157–183. Zbl.595.35032

[1987] Internal Lifschitz tails. J. Stat. Phys. *46*, No. 5–6, 911–918. Zbl.683.60079

Simon, B., Taylor, M.

[1985] Harmonic analysis on SL(2, $\mathbb{R}$) and smoothness of the density of states in the one-dimensional Anderson model. Commun. Math. Phys. *101*, No. 1, 1–19. Zbl.577.60067

Simon, B., Taylor, M., Wolff, T.

[1985] Some rigorous results for the Anderson model. Phys. Rev. Lett. *54*, No. 14, 1589–1592.

Simon, B., Wolff, T.

[1986] Singular continuous spectrum under rank one perturbation and localization for random Hamiltonians. Commun. Pure Appl. Math. *39*, No. 1, 75–90. Zbl.609.47001

Sinai, Ya.G.

[1985] On the structure of the spectrum of the difference Schrödinger operator with an almost periodic potential near the left boundary. Funkts. Anal. Prilozh. *19*, No. 1, 42–48. English transl.: Funct. Anal. Appl. *19*, 34–39 (1985). Zbl.574.39004

[1987] Anderson localization for one-dimensional difference Schrödinger operator with quasiperiodic potential. J. Stat. Phys. *46*, No. 5–6, 861–909. Zbl.682.34023

Singer, I.M.
[1974] Eigenvalues of the Laplacian and invariants of manifolds. In: Proc. Int. Congr. Math., Vancouver 1974, Vol. 1, 187–200. Zbl.345.58014

Sjöstrand, J.
[1980] On the eigenvalues of a class of hypoelliptic operators. IV. Ann. Inst. Fourier *30*, No. 2, 109–169. Zbl.417.47024

Skriganov, M.M.
[1985] Geometric and arithmetic methods in the spectral theory of many-dimensional periodic operators. Tr. Mat. Inst. Steklova *171*, 122pp. English transl.: Proc. Steklov Inst. Math. *171*, 121pp. (1987). Zbl.567.47004

Slivnyak, I.M.
[1966] On the spectrum of the Schrödinger operator with random potential. Zh. Vychisl. Mat. Mat. Fiz. *6*, No. 6, 1104–1108. English transl.: USSR Comput. Math. Math. Phys. *6*, 221–228.

Smagin, S.A.
[1976] Fractional powers of hypoelliptic systems in $\mathbb{R}^n$. Mat. Sb., Nov. Ser. *99*, No. 3, 331–341. English transl.: Math. USSR, Sb. *28*, 291–300 (1978). Zbl.346.35029

Smith, L.
[1981] The asymptotics of the heat equation for a boundary value problem. Invent. Math. *63*, No. 3, 467–493. Zbl.487.35012

Solomyak, M.Z.
[1985] Asymptotics of the spectrum of the Schrödinger operator with non-regular homogeneous potential. Mat. Sb., Nov. Ser. *127*, No. 1, 21–39. English transl.: Math. USSR, Sb. *55*, 19–37 (1986). Zbl.583.35083

Solomyak, M.Z., Vulis, I.L.
[1974] Spectral asymptotics of degenerate second-order elliptic operators. Izv. Akad. Nauk SSSR, Ser. Mat. *38*, No. 6, 1362–1392. English transl.: Math. USSR, Izv. *8*, 1343–1371 (1976). Zbl.346.35078

Souillard, B.
[1986] Spectral properties of discrete and continuous random Schrödinger operators: a review. Preprint. Appeared in: Random media, IMA. Vol. Math. Appl. *7*, 279–291 (1987). Zbl.659.60089

Spencer, T.C.
[1986] The Schrödinger equation with a random potential – a mathematical review. In: K. Osterwalder and R. Stora (eds.), Phénomènes critiques, systèmes aléatoires, théories de gauge, Summer Sch. Theor. Phys., Sess. 43, Les Houches, 1984, North-Holland Publishing Co., Amsterdam-New York, 895–942. Zbl.655.60050

Stanton, N., Tartakoff, D.
[1984] The heat equation for the $\bar{\partial}_s$-Laplacian. Commun. Partial Differ. Equations *9*, No. 7, 597–686. Zbl.577.35065

Sunada, T.
[1985] Riemannian coverings and isospectral manifolds. Ann. Math., II. Ser. *121*, 248–277.

Tamura, H.
[1981] Asymptotic formulas with sharp remainder estimate for bound states of Schrödinger operators. I. J. Anal. Math. *40*, 166–182. Zbl.501.35063
[1982a] Asymptotic formulas with sharp remainder estimates for bound states of Schrödinger operators. II. J. Anal. Math. *41*, 85–108. Zbl.552.35058
[1982b] Asymptotic formulae with remainder estimates for eigenvalues of Schrödinger operators. Commun. Partial Differ. Equations *7*, No. 1, 1–53. Zbl.525.35064
[1984] The asymptotic formula for the number of bound states in the strong coupling limit. J. Math. Soc. Japan *36*, No. 3, 355–374. Zbl.612.35101

Tanikawa, M.
[1979] The spectrum of the Laplacian and smooth deformation of Riemannian metric. Proc. Japan Acad., Ser. A *55*, No. 4, 125–127. Zbl.465.58028

Tanno, S.
[1980] A characterization of canonical spheres by the spectrum. Math. Z. *175*, No. 3, 267–274. Zbl.431.53037

Taylor, M.
[1981] Pseudodifferential Operators. Princeton Univ. Press, Princeton, N.J. Zbl.453.47026

Thomas, L.E.
[1973] Time dependent approach to scattering from impurities in a crystal. Commun. Math. Phys. *33*, No. 4, 335–343.

Thouless, D.
[1972] A relation between the density of states and range of localization for one-dimensional random systems. J. Phys. C *5*, 77–81.

Titchmarsh, E.
[1946] Eigenfunction Expansion Associated with Second-Order Differential Equations. Clarendon Press, Oxford. Zbl.61,135

Trèves, F.
[1982] Introduction to Pseudodifferential and Fourier Integral Operators. Vol. 1, 2. Plenum Press, New York-London. Zbl.453.47027

Tulovskij, V.N.
[1971] Distribution of eigenvalues for differential operators with constant coefficients. Funkts. Anal. Prilozh. *5*, No. 3, 85–90. English transl.: Funct. Anal. Appl. *5*, 241–245 (1972). Zbl.244.35068

Urakawa, H.
[1982] Bounded domains which are isospectral but not congruent. Ann. Sci. Éc. Norm. Supér., IV. Sér. *15*, No. 3, 441–456. Zbl.505.58036

Vainberg, B.R.
[1987] On the asymptotic expansion of the spectral function of an elliptic operator in $\mathbb{R}^n$. Tr. Semin. Im. I. G. Petrovskogo *12*, 75–87. English transl.: J. Sov. Math. *47*, 2537–2546 (1989). Zbl.682.47027
[1988] Asymptotic behaviour as $t \to \infty$ of the solutions of exterior mixed problems for hyperbolic equations and semiclassics. Itogi Nauki Tekh., Ser. Sovrem. Probl. Mat., Fundam. Napravleniya *34*, 57–92. Zbl.657.35080. English transl.: Encycl. Math. Sci. *34*. Springer-Verlag, Berlin-Heidelberg-New York (in preparation).

Vasil'ev, D.G.
[1984] Two-term asymptotics of the spectrum of a boundary value problem under internal reflection of general form. Funkts. Anal. Prilozh. *18*, No. 4, 1–13. English transl.: Funct. Anal. Appl. *18*, 267–277 (1984). Zbl.574.35032
[1986] Asymptotics of the spectrum of a boundary value problem. Tr. Mosk. Mat. O.-va *49*, 167–237. English transl.: Trans. Mosc. Math. Soc. 1987, 173–245 (1987). Zbl.623.58024

Veliev, O.A.
[1987] Asymptotic formulae for the eigenvalues of a periodic Schrödinger operator and the Bethe-Sommerfeld conjecture. Funkts. Anal. Prilozh. *21*, No. 2, 1–15. English transl.: Funct. Anal. Appl. *21*, 87–100 (1987). Zbl.638.47049

Venkov, A.B.
[1979] Spectral theory of automorphic functions, Selberg's zeta-function, and certain problems of analytic number theory and mathematical physics. Usp. Mat. Nauk *34*, No. 3, 69–135. English transl.: Russ. Math. Surv. *34*, No. 3, 79–153 (1979). Zbl.414.10025

Venttsel', A.D.
[1975] A Course in the Theory of Stochastic Processes. Nauka, Moscow. English transl.: McGraw-Hill, New York 1981. Zbl.502.60001

Vignéras, M.-F.
[1980] Variétés riemanniennes isospectrales et non isométriques. Ann. Math., II. Ser. *112*, No. 1, 21–32. Zbl.445.53026

Volovoj, A.V.
[1987] Improved two-term asymptotics of the distribution function of the eigenvalues of an elliptic operator on a compact manifold. Dokl. Akad. Nauk SSSR *294*, No. 5, 1037–1041. English transl.: Sov. Math., Dokl. *35*, No. 3, 604–607 (1987). Zbl.701.35122

Vulis, I.L.
[1976] Spectral asymptotics of elliptic operators of any order with strong degeneration. Zap. Nauchn. Semin. Leningr. Otd. Mat. Inst. Steklova *59*, 25–30. English transl.: J. Sov. Math. *8*, 212–216 (1978). Zbl.346.35079

Weyl, H.
[1911] Das asymptotische Verteilungsgesetz der Eigenwerte linearer partieller Differentialgleichungen (mit einer Anwendung auf die Theorie der Hohlraumstrahlung). Math. Ann. *71*, 441–479. FdM.43,436
[1912] Über die Abhängigkeit der Eigenschwingungen einer Membran von deren Begrenzung. J. Reine Angew. Math. *141*, 1–11. FdM.43,948

Witt, E.
[1941] Eine Identität zwischen Modulformen zweiten Grades. Abh. Math. Semin. Hansische Univ. *14*, 323–337. Zbl.25,17

Wolpert, S.
[1978] The eigenvalue spectrum as moduli for flat tori. Trans. Am. Math. Soc. *244*, 313–321. Zbl.405.58021
[1979] The length spectra as moduli for compact Riemannian surfaces. Ann. Math., II. Ser. *109*, No. 2, 323–351. Zbl.441.30055

Yafaev, D.R.
[1972] On the discrete spectrum of the three-particle Schrödinger operator. Dokl. Akad. Nauk SSSR *206*, No. 1, 68–70 (Russian).
[1976] On the point spectrum in the quantum-mechanical many-body problem. Izv. Akad. Nauk SSSR, Ser. Mat. *40*, No. 4, 908–948. English transl.: Math. USSR, Izv. *10*, 861–896 (1977). Zbl.344.35024

Zheludev, V.A.
[1970] On the perturbation of the spectrum of the one-dimensional self-adjoint Schrödinger operator with a periodic potential. Probl. Mat. Fiz. 4, 61–82. Zbl.271.34031

Ziman, J.M.
[1972] Principles of the Theory of Solids. Cambridge Univ. Press, Cambridge. Zbl.408.70007
[1979] Models of Disorder. The Theoretical Physics of Homogeneously Disordered Systems. Cambridge Univ. Press, Cambridge.

Author Index

Subject Index

Printing: Saladruck, Berlin
Binding: Buchbinderei Lüderitz & Bauer, Berlin

GPSR Compliance

The European Union's (EU) General Product Safety Regulation (GPSR) is a set of rules that requires consumer products to be safe and our obligations to ensure this.

If you have any concerns about our products, you can contact us on ProductSafety@springernature.com

In case Publisher is established outside the EU, the EU authorized representative is:

Springer Nature Customer Service Center GmbH
Europaplatz 3
69115 Heidelberg, Germany

Batch number: 10682213

Printed by Printforce, the Netherlands